Advanced Microsystems
for Automotive Applications 2009

Gereon Meyer · Jürgen Valldorf · Wolfgang Gessner

Advanced Microsystems for Automotive Applications 2009

Smart Systems for Safety, Sustainability, and Comfort

 Springer

Dr. Gereon Meyer
VDI/VDE Innovation + Technik GmbH
Steinplatz 1
10623 Berlin
Germany
gmeyer@vdivde-it.de

Wolfgang Gessner
VDI/VDE Innovation + Technik GmbH
Steinplatz 1
10623 Berlin
Germany
gessner@vdivde-it.de

Dr. Jürgen Valldorf
VDI/VDE Innovation + Technik GmbH
Steinplatz 1
10623 Berlin
Germany
valldorf@vdivde-it.de

ISBN 978-3-642-00744-6   e-ISBN 978-3-642-00745-3
DOI 10.1007/978-3-642-00745-3
Springer Dordrecht Heidelberg London New York

Coverdesign: deblik, Berlin

Printed on acid-free paper

Springer is part of Springer Science+Business Media (www.springer.com)

Preface

The current economic crisis is cutting the automotive sector to the quick. Public authorities worldwide are now faced with requests for providing loans and accepting guarantees and even for putting large automotive companies under state control. Assessing the long-term benefits of such help and weighing the needs of different sectors against each other poses a major challenge for the national policies. Given the upcoming change of customer preferences and state regulations towards safety, sustainability and comfort of a car, the automotive industry is particularly called to prove its ability to make necessary innovations available in order to accelerate its pace to come out of the crisis. Consequently the Green Car is assuming a prominent role in the current debate.

Various power train concepts are currently under discussion for the Green Car including extremely optimised internal combustion engines, hybrid drives and battery-electric traction. Electrical cars are the most appealing option because they are free of local emissions and provide the opportunity to use primary energy from sources other than crude oil for transport. Well to wheel analysis show that their green-house gas emissions can be rated negligibly small if electricity from renewable sources like wind and solar is used. The mass introduction of electrical cars, however, is still a few years down the road, given the necessity to completely rethink the vehicle's concept: Novel solutions are needed for energy storage, traction, range extension, energy efficiency, power control and the overall system integration. All these are topics of advanced industrial research. Fatally, it is just the industrial research departments doing such work that in times of crisis have to fear to sustain budget cuts.

The promotion of Green Cars, in particular the electrical vehicle, has to rely on joint commitments by the industry and the public authorities. In terms of the electrical vehicle, first steps towards such public private partnership were recently taken by EPoSS, the European Technology Platform on Smart Systems Integration. Starting with an expert workshop on smart systems for the electrical vehicle carried out together with the European Commission in June 2008, EPoSS systematically built-up a basis and provided essential support to the Green Cars Initiative as part of the European Economic Recovery Plan.

The 13th International Forum on Advanced Microsystems for Automotive Applications (AMAA) taking place in Berlin on May 5-6, 2009 is presenting even more of the advanced research work for the way out of the crisis. With EPoSS being one of the organisers of the conference this year the focus is put on smart miniaturised systems and ICT solutions for future mobility. For the first time in the history of the AMAA a specific topic was chosen: "Smart Systems for Safety, Sustainability and Comfort". With the Enterprise Europe Network (EEN) being the main supporter of this initiative, the role of innovative SMEs in the value chain of future mobility is highlighted.

The papers published in this book were selected from more than 40 submissions. Highlights include presentations on the scope of two European projects related to the Green Car issue: EEVERT aiming at the development of energy efficient vehicles for road transport, and E3Car doing research on nanoelectronics as the key enabler for electrical vehicles. Furthermore, several papers are describing first results of the Intersafe II project that is dealing with cooperative intersection safety. Likewise prominent topics are a detector for left behind occupants, novel driver assistance systems and several other highly integrated components for various applications in the automobile.

I like to thank all authors for submitting very interesting papers and for assisting the editors during the publication process. Furthermore, I like to thank the AMAA Steering Committee for its continuous support and advice. Particular thanks are addressed to EPoSS and to the EEN as well as to the industrial sponsors who helped us to built up the financial backbone for the AMAA 2009.

I also like to express my gratitude and acknowledgements to the team of VDI/VDE Innovation + Technik GmbH, particularly Laure Quintin who is running the AMAA office, Michael Strietzel who is responsible for the technical preparation of this publication, and especially the chairmen of the 2009 event, Dr. Gereon Meyer and Dr. Jürgen Valldorf.

Berlin, May 2009

Wolfgang Gessner

Funding Authorities

European Commission

Berlin Senate Office for Economics, Technology and Women's Issues

Brandenburg Ministry for Economics

Supporting Organisations

European Council for Automotive R&D (EUCAR)

European Association of Automotive Suppliers (CLEPA)

Advanced Driver Assistance Systems in Europe (ADASE)

Zentralverband Elektrotechnik- und Elektronikindustrie e.V. (ZVEI)

Hanser Automotive

mst|*news*

enabling MNT

Organisers

European Technology Platform on Smart Systems Integration (EPoSS)

Enterprise Europe Network Berlin-Brandenburg

VDI|VDE Innovation + Technik GmbH

Honorary Committee

Eugenio Razelli	President and CEO Magneti Marelli S.P.A., Italy
Rémi Kaiser	Director Technology and Quality Delphi Automotive Systems Europe, France
Nevio di Giusto	President and CEO Fiat Research Center, Italy
Karl-Thomas Neumann	CEO, Continental AG, Germany

Steering Committee

Mike Babala	TRW Automotive, Livonia MI, USA
Serge Boverie	Continental AG, Toulouse, France
Prof. Dr. Geoff Callow	Technical & Engineering Consulting, London, UK
Bernhard Fuchsbauer	Audi AG, Ingolstadt, Germany
Kay Fürstenberg	IBEO GmbH, Hamburg, Germany
Wolfgang Gessner	VDI/VDE-IT, Berlin, Germany
Roger Grace	Roger Grace Associates, Naples FL, USA
Dr. Klaus Gresser	BMW Forschung und Technik GmbH, Munich, Germany
Henrik Jakobsen	SensoNor A.S., Horten, Norway
Horst Kornemann	Continental AG Frankfurt am Main, Germany
Hannu Laatikainen	VTI Technologies Oy, Vantaa, Finland
Dr. Roland Müller-Fiedler	Robert Bosch GmbH, Stuttgart, Germany
Paul Mulvanny	QinetiQ Ltd., Farnborough, UK
Dr. Andy Noble	Ricardo Consulting Engineers Ltd., Shoreham-by-Sea, UK
Dr. Ulf Palmquist	EUCAR, Brussels, Belgium
Dr. Pietro Perlo	Fiat Research Center, Orbassano, Italy
Dr. Detlef E. Ricken	Delphi Delco Electronics Europe GmbH, Rüsselsheim, Germany
Christian Rousseau	Renault SA, Guyancourt, France
Patric Salomon	4M2C, Berlin, Germany
Ernst Schmidt	BMW AG, Munich, Germany
Dr. Florian Solzbacher	University of Utah, Salt Lake City UT, USA
Egon Vetter	Ceramet Technologies Ltd., Melbourne, Australia
Hans-Christian von der Wense	Freescale GmbH, Munich, Germany

Conference Chairs:

Dr. Gereon Meyer	VDI/VDE-IT, Berlin, Germany
Dr. Jürgen Valldorf	VDI/VDE-IT, Berlin, Germany

Table of Contents

Green Cars

Safety

Driver Assistance

Components and Generic Sensor Technologies

Appendices

Green Cars

Energy Efficient Vehicles for Road Transport – EE-VERT

D. Ward, MIRA Ltd.
M. Granström, Volvo Technology AB
A. Ferré, Lear Corporation

Abstract

EE-VERT is a project funded under the Seventh Framework Programme of the European Commission concerned with improving the energy efficiency of road vehicles. In particular EE-VERT targets a 10% reduction in fuel consumption and CO_2 generation as measured on the standard NEDC. The central concept to EE-VERT is the deployment of an overall energy management strategy that co-ordinates energy demand and consumption by a network of smart generators and smart actuators. This paper describes the EE-VERT concept and the activities that will be undertaken to realise the project's goals.

1   Introduction

Road transport is the second largest producer of "greenhouse gases" within the European Union (EU) [1]. The European Commission has a declared strategy to reduce the overall production of greenhouse gases and in particular carbon dioxide (CO_2). Although overall greenhouse gas emissions in the EU fell by 5% over the period 1990–2004, the emissions of CO_2 from road transport rose by 26% in the same period. While improvements in vehicle technology have led to reductions in CO_2 emissions, for example the average CO_2 emissions in new vehicles was 186 g km^{-1} in 1995 but 163 g km^{-1} in 2004; increases in traffic volumes and congestion have been responsible for the net growth in CO_2 emissions in the sector. Reducing the CO_2 generated by road transport is therefore a key aspect of the overall strategy to reduce the production of greenhouse gases. To this end binding targets have been set with the objective of reducing the average emissions of CO_2 from new passenger cars in the EU to 130 g km^{-1} in 2012.

Technologies such as those to be developed in the EE-VERT project are a key part of achieving these required improvements. In particular, the significant improvements to be delivered by the EE-VERT project are needed in order to make road transport and the vehicles used in road transport a net contributor to decreasing overall levels of CO_2 generation in Europe. As the volumes of road vehicles in service and the CO_2 they produce are significantly greater than other surface transport means, EE-VERT presents an opportunity to make a substantial impact at the European level.

1.1 Existing Approaches

The majority of research efforts for CO_2 reduction up to now have been focussed on novel powertrain developments (particularly hybrid concepts). A hybrid vehicle has a relatively high potential to reduce CO_2 emissions but it presently requires cost-intensive and drastic technical modifications. Besides the higher costs these technical modifications also lead to the mass of the vehicle being considerably increased. Due to higher vehicle prices and lower economic efficiency hybrid vehicles have penetrated the automotive market relatively slowly in Europe, where up to now Diesel-engined vehicles have been the choice of the consumer for more efficient vehicles. Hence, hybrid vehicles can provide only a slowly increasing effect for the reduction of CO_2 emissions in the medium term.

A further aspect concerns the concepts for energy generation and consumption in the vehicle, particularly for auxiliary systems. In general, the energy required for auxiliary systems (e.g. power steering, water pump) is generated and consumed continuously, regardless of demand. Similarly, the energy generation for the vehicle's electrical system operates continuously. Some technologies have been deployed to the market that partially address these inefficiencies, for example electric power-assisted steering (EPAS/EHPAS) and regenerative alternators [2]. Nevertheless, these technologies may be viewed as "islands" of improving energy efficiency, but an overall energy management strategy has yet to be deployed.

From the above discussion, it can be seen that there are still significant potentials for energy savings in vehicles with conventional powertrains. Addressing the unoptimised electrical generators and consumers of today's vehicles can have a high impact on CO_2 reduction within the next ten years, affecting improvements for the whole range (types and models) of road transport vehicles. These improvements will also benefit hybrid vehicles.

1.2 The EE-VERT Approach

The EE-VERT concept is to combine several different approaches to energy saving within an overall energy management concept. The approaches include:

- ▶ Greater efficiency in energy generation with a new concept for the electrical machine and an optimised operation strategy;
- ▶ Improved efficiency in energy consumption, through electrification of auxiliaries and the use of more efficient electrical machines such as brushless DC motors;
- ▶ Energy recovery from wasted energy such as the use of thermoelectric devices to recover waste energy from the exhaust system;
- ▶ Energy scavenging from unused sources of energy, for example the use of solar cells;
- ▶ The use of short-term and long-term energy storage for recovered energy. Examples of the former include ultracapacitors; and of the latter more efficient battery technologies.

Note that EE-VERT will not carry out research into areas such as ultracapacitors, battery technologies and thermoelectric generators that are already being developed in other research activities. However, EE-VERT will develop an overall energy management strategy that permits the efficient integration of these technologies into a vehicle equipped with smart energy generation and smart actuators.

Fig. 1 shows an estimate of the benefit of EE-VERT technologies in a passenger car, where total savings of around 10% in fuel consumption and CO_2 emissions can be realised in the New European Driving Cycle (NEDC) through a combination of higher efficiency and minimising the energy demand for vehicle operation (4%), and overall energy management making use of energy recovery from sources on-board the vehicle (4–6%). In addition further measures that are not measurable on the NEDC but which can give a benefit of up to 4% in real-life usage are predictive algorithms for energy optimised operation (2%) and the use of solar panels to generate energy (2%).

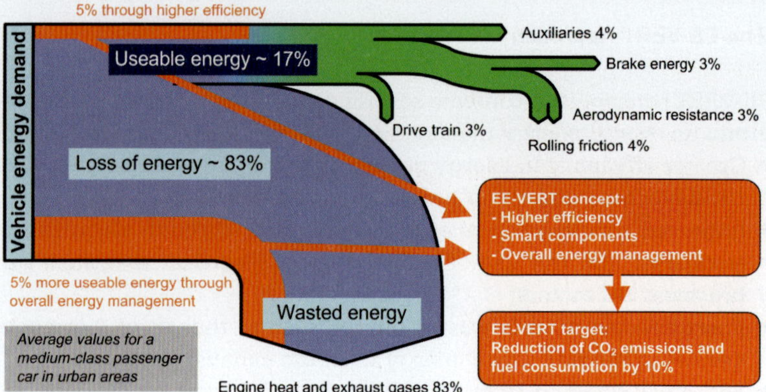

Fig. 1.   Estimate of energy savings in a passenger car

2   EE-VERT Concepts

In this section further aspects of the concepts that will be developed in EE-VERT are described.

2.1   Electrical Architecture

The electrical system in conventional vehicles consists of a generator (the alternator) mechanically linked to the engine, an energy storage system (12V battery) and several consumers (loads) of which the power consumption can be actively controlled.

As discussed in the introduction, to increase the energy efficiency of the system, it is necessary to introduce new components able to recovery energy from other sources (e.g. kinetic energy recovery, exhaust gas thermo-electric recovery or solar cells) and place this recovered energy in short-term or long-term storage depending on the present and predicted demand from energy consumers.

It is possible to incorporate those energy recovery systems into the present-day electrical architecture. However, the simplicity of the electrical system severely limits the efficiency of energy recovery and energy management strategies that may be implemented. Firstly, there is a limitation on the maximum voltage allowed in the electrical distribution system. Since new power generation components with higher efficiency usually work at higher voltages, the efficiency of energy recovery systems is low. Secondly, energy management strategies are mainly limited to the control of power consumption of the loads.

Therefore, to permit integration of multiple generation, actuation and storage devices with different optimal operating voltages and usage profiles, a radical change in the electrical architecture is required. Consequently, it is necessary to divide the electrical distribution system into two parts: one electrical distribution system (14 V) for the existing auxiliary systems (such as lighting and radio) and a main electrical distribution system with a broader voltage range (14 V – 35 V) for efficient energy recovery. Both systems are connected by a DC/DC converter and incorporate different energy storage elements (ultracapacitors, high recharge batteries) according to the system and vehicle requirements. Fig. 2 shows a conceptual view of the proposed electrical architecture, with the two electrical distribution systems and the different elements in them: power generators, storage and consumers and the energy management and power control units (PCU) for control. This electrical architecture is scalable and permits the easy integration of additional elements such as solar cells to supply quiescent current demand as well as collecting energy during vehicle use.

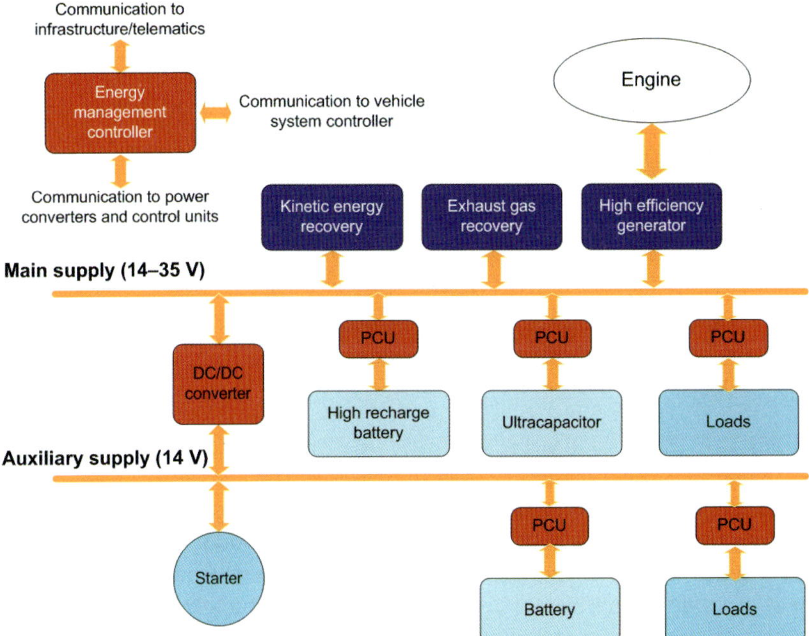

Fig. 2.   EE-VERT electrical architecture concept

2.2 Power Generation

In the EE-VERT system electrical power generation is scheduled according to the demand from consumers. The improved EE-VERT generator will use a new type of electrical machine with higher efficiency at each operating point and optimised management of the generator operation. The operation management will lead to generator operation in regimes with a higher efficiency level.

The maximum efficiency of a state-of-the-art generator is 60 or 65%. However, the average efficiency is only about 40 or 45% in standard operation without operation management. Fig. 3 shows the operational area of a standard generator in comparison to the operation area of the proposed EE-VERT generator concept during the NEDC. Due to the low electrical power demand and the efficiency characteristics the average efficiency of a standard generator is only about 45% during the NEDC.

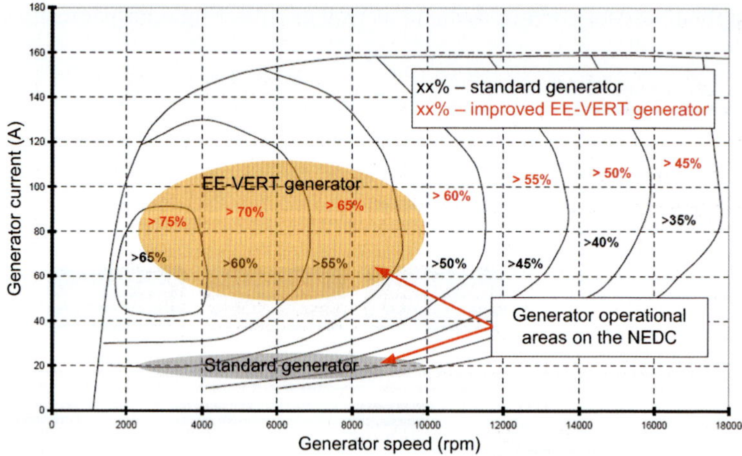

Fig. 3.   Efficiency diagram for a standard generator

The efficiency for converting mechanical energy to electrical energy will be improved from on average 45% to on average 65% through novel design of the electrical machine and the operational strategy. Furthermore the electrification of auxiliaries will have a positive influence on the efficiency as well, due to a higher electrical power demand during the NEDC. The increased electrical power demand is compensated with considerable less mechanical power demand due to a demand-oriented operation of electrified auxiliaries.

The generator will also realise additional functionalities such as brake energy recuperation by a temporarily increased generator output power on a floating voltage level. Together with new energy storage devices this will lead to a further benefit for the reduction of fuel consumption.

Since a generator is necessary in every conventional vehicle, the generator therefore plays a key role within the EE-VERT system approach.

2.3 Overall Energy Management

In present-day vehicles, even in those regarded as the state-of-the-art, electrical power is generated with little knowledge of the actual loads, and some auxiliary systems consume power continually regardless of demand. In addition, increased electrification of auxiliary systems promises mass reduction and efficiency gains, but this can only be accomplished if the energy generation and distribution is continuously optimised and adapted to the current driving conditions and power demands.

In the EE-VERT system it is therefore envisaged that the electrical power generation will be scheduled according to the demand from consumers. Furthermore, energy can be recovered from other sources (e.g. regenerative braking, waste heat, solar cells) and placed in short-term or long-term storage depending on the present and predicted demand. This concept requires a radical change in the electrical architecture (as described in Section 2.1) to permit integration of multiple generation, actuation and storage devices with different optimal operating voltages and usage profiles. The EE-VERT system architecture will combine the contributions to energy saving possible from individual components and permit them to run optimally.

The architecture will deploy a distributed network of smart components, whose characteristics are co-ordinated to optimise their interaction and their efficiency. In addition to this, it is crucial to manage the use of different types of energy, such as electrical, mechanical or thermal energy, which means that an overall vehicle optimisation and management concept needs to be developed. Also, influence on energy consumption in the vehicle by the driver behaviour will be addressed, as will be the possibilities to extend the number of energy generating and harvesting devices.

A further aspect is the need of safety-related systems. During the FP6 EASIS project [3], a number of hardware and software architectures for integrated safety systems were proposed. In particular for the more safety-relevant systems it was determined that a degree of redundancy in the power supply is needed. Whilst not a primary research focus of EE-VERT, the architecture to be proposed will incorporate support for such requirements.

Finally it is envisaged that a distributed network of smart components will require some form of standardised communications to permit the exchange of energy-related data. It is anticipated that EE-VERT will create the possibility of self-organising networks of smart components, and will therefore develop requirements for such communications that can be offered for international standardisation through a body such as ISO.

3   Conclusions

This paper has presented the approach to energy efficiency and CO_2 reduction that will be developed during the EC-funded EE-VERT project. Central to the EE-VERT approach is the deployment of an overall energy management strategy in road vehicles that will integrate novel concepts for power generation and smart actuators, as well as permitting the integration of concepts for energy harvesting being developed elsewhere. While the focus of EE-VERT is vehicles with a conventional powertrain, EE-VERT will also realise improvements in hybrid vehicles that will assist in the deployment of cost-effective alternative-powertrain vehicles.

4   Acknowledgements

EE-VERT is a project funded by the European Commission under the Seventh Framework Programme, grant reference SCS7-GA-2008-218598. The partners are MIRA Ltd. (UK, Co-ordinator), Volvo Technology AB (SE), Robert Bosch GmbH (DE), Lear Corporation Holding Spain SL (ES), Engineering Center Steyr GmbH & Co KG (AT), FH Joanneum Gesellschaft mbH (AT), Univeristatae Politehnica din Timisora (RO) and S.C. Beespeed Automatizari S.R.L. (RO)

References

[1] Commission Communication COM/2007/0019 "Results of the review of the Community Strategy to reduce CO_2 emissions from passenger cars and light-commercial vehicles". Available at http://eur-lex.europa.eu/LexUriServ/LexUriServ.do?uri=COM:2007:0019:FIN:EN:HTML

[2] Voelcker J, "2008 BMW 5-Series: Regenerative braking in a nonhybrid", part of "Top 10 Tech Cars" feature in *IEEE Spectrum*, April 2007.

[3] EASIS – Electronic Architecture and System Engineering for Integrated Safety Systems, available at http://www.easis-online.org

David Ward
MIRA Limited
Watling Street
Nuneaton
CV10 0TU
UK
david.ward@mira.co.uk

Magnus Granström
Volvo Technology Corporation
Mechatronics and Software
SE-412 88 Göteborg
Sweden
magnus.granstrom@volvo.com

Antoni Ferré
Applied Research and Tech. Development
Lear Corporation
Electrical and Electronics Division
C/. Fusters , 54
43800 Valls (Tarragona)
Spain
aferre@lear.com

Keywords: road vehicles, energy efficiency, energy management, CO_2 reduction

Nanoelectronics: Key Enabler for Energy Efficient Electrical Vehicles

R. John, Infineon Technologies
O. Vermesan, SINTEF
M. Ottella, P. Perlo, Fiat Research Centre

Abstract

Future Electric (EVs) and Hybrid Electric Vehicles (HEVs) will provide more flexibility when choosing between primary energy sources, including those which are renewable. In general conventional ICEs vehicles transform between only 17 and 22% (depending on power train) of the fuel chemical energy with a typical primary energy consumption of 550-600 Wh/km (0.06 l/km). Efficient electrically powered trains can achieve conversion efficiencies greater than 75% from the batteries to the wheels, which corresponds to consumption in primary energy of about 390 Wh/km in the case where electricity is produced by conventional carbon based power plants, or only 180 Wh/km where the electricity is produced solely by renewable energy. The partial recovery of kinetic energy during braking gives rise to further improvement in the overall efficiency. The development of advanced smart electronic systems in power trains is therefore essential for delivering a considerable energy saving in terms of the most critical sources (oil and natural gas - NG). This paper presents the advances made in the overall power electronic modules for electric and hybrid vehicles, and which are addressed in the E^3Car project.

1 E^3Car Concepts and Main Ideas

Sufficient reduction in fossil fuels consumption and dependence, and their impact on climate is achieved only through technological breakthroughs.

It is now widely accepted that for the first time in the history of mobility Energy/Power storage, power electronics and electrical motors have globally reached a level of development allowing the progression to affordable and performing electrified vehicles for a variety of applications. However, in each of these areas there is still a considerable distance to go before a full electrical vehicle deployment.

The E^{3}Car project addresses the overall set of electronic modules and subsystems (Fig. 1) required for achieving an efficient power train and interfaces to the electrical power/energy sources.

Optimisation of battery technology, smart mechatronic actuators and the power train management unit (in particular embedded algorithms and software) are not in the focus of this project but will be part of forthcoming initiatives.

The development of the EV market and subsequent penetration are evolving in terms of availability of infrastructure and population density.

New charging technology is required to facilitate utilities that control the duration and extent of electrical vehicle charges, allowing the capability to reduce or increase the loads on the electrical grid. Such new systems will be installed in the vehicle and will collect information from the driver, ensuring that the vehicle battery is fully charged as required. Balancing the power consumption of the EV with the capacity of the electrical grid requires processing data from thousands of EVs to create a smooth load profile of the grid operator action in order to avoid power spikes or having to engage back up sources of generation.

 Electric and hybrid electric vehicles will create challenges to the semiconductor industry since the electrical vehicles incorporate fundamentally different ways to generate, transform and store energy. Depending on the technical approach of these vehicles, some electronic building blocks will become obsolete (ignition control, direct injection, exhaust emission, gear control, etc.), while other components and functional units will be required (traction control, energy management, lighting control, navigation and terrain based management, distributed sensors and actuators, integrated embedded systems).

This process of continuous electrification will happen in the next few years and it will change the semiconductor industry landscape and the vehicles architectures and the structure of the supply chain within the automotive industry.

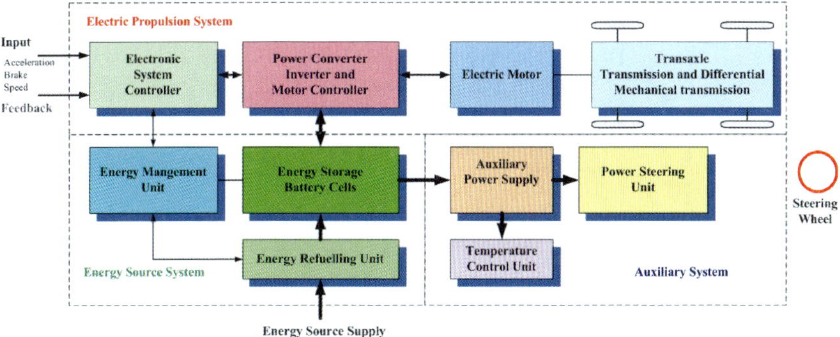

Fig. 1.   Electrical vehicle functional components

At both subsystem and system levels the activities are focused on the development of devices and circuits as follows:

▶ Power Conversion - AC-DC, DC-DC, high power modules
▶ Power and Energy Management - Smart battery management, super capacitors, alternative energy sources and E-grid integration
▶ Power Distribution Network - Power switches, high current sensors and safety fail mode switches
▶ Smart Dynamic Monitoring - Information systems and feedback loops based on Si Sensors

The electrical vehicles require powerful and efficient DC-DC converters to ensure easy integration without handling of large cooling systems to maintain operating temperature. The converters adapt the battery voltage to the level required by different devices such as motors, light, navigation computer, radio, power windows, dashboard displays, infotainment, communication systems etc. The AC-DC and DC-DC power converters contain microcontrollers, driver circuitry and power electronic elements and IGBTs and MOSFETs are the key component for these converters, the needs for these components will increase significantly.

The nanoelectronics technologies, devices, circuits and modules developed will address a broad application spectrum, which is encapsulated in Fig. 2.

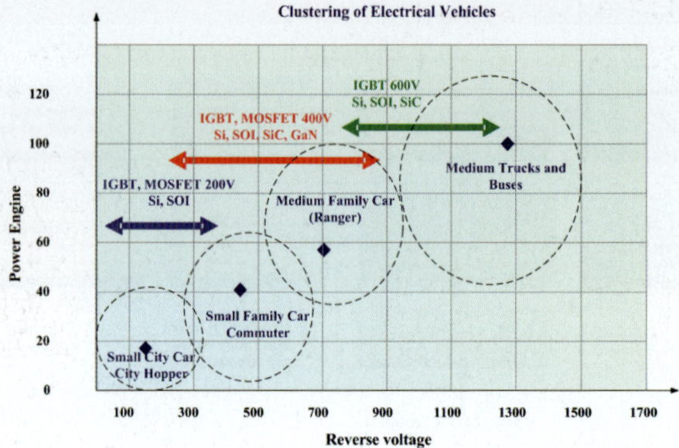

Fig. 2.   Nanoelectronics technologies and clustering of electrical vehicles

Future EVs will fulfil the requirements of urban and suburban traffic, which differs depending on the vehicle type and size. The different market segments comprise small passenger vehicles, second family vehicle, the family vehicle or intermediate vehicle segment, high class segment, commercial delivery vans, trucks, minibuses and urban buses and also electric bicycles and scooters. The nanoelectronics technologies and modules developed will address the needs of theses segment. The types of electrical vehicles addressed by the E^{3}Car project are presented in Tab. 1.

Electrical Vehicle Class	Small City Car/Vehicle (City Hopper)	Small Family Car/Vehicle (Commuter)	Medium Family Car/Vehicle (Ranger)	Medium Tracks and Busses
Body Style	2 Door 2 Passengers	2 Door 2-3 Passengers	4 Door 4 Passengers	2-4 Door 2-10 Passengers
Battery Type	Different types	Li-Ion	Li-Ion	Li-Ion
Peak Power	25 kW	50 kW	75 kW	125 kW
Continuous Power	20 kW	47 kW	65 kW	100 kW
Generator Power	15 kW	45 kW	60 kW	100 kW
Recharge Time@240V	6 hours	6 hours	6 hours	10 hours
0 to 100 km/h time (seconds)	< 10 s	8 s	8 s	20 s
Mass/Weight	400 kg	1000 kg	1500 kg	4000 kg
Full EV Range	100 km	200 km	200 km	100 km
Weight/Peak Power Ratio	1.6	2	2	3.2

Tab. 1.   Types of electrical vehicles

The subsytems developed will be tested on vehicles such as Audi, Think, and Fiat Phylla. (Fig. 3).

Fig. 3.   European Electrical Vehicle Concepts (Think, Fiat)

2   Semiconductor Technologies for Power Conversion and Control

The research activities aim to develop an innovative high temperature suitable interconnection technology, which, in spite of resulting heat dissipation in the range of kWs, enables highest integration density, fulfilling automotive environmental requirements.

The semiconductor content of electric and hybrid electric electrical vehicles is much higher than in internal combustion vehicles and the new power train of the electric vehicles is the electrified power train that requires significant additional power electronic components.

The use of high voltage batteries require modifications within the in vehicle electronic architecture. The battery module contains power management circuitry balancing the energy streams between the battery's cells while monitoring the operating conditions. The circuitry uses microcontrollers as well as discrete transistors and power MOSFETs.

In the electrical vehicles the functional units typically driven by mechanical parts and belts will be driven electrically. The auxiliary units, like the air conditioning compressor that has a power demand of several kWs will be connected to the high voltage circuits, which leads to increased efficiency. The increase in nanoelectronics semiconductor will be implemented within the electrified power train modules such as DC-DC converters and the motor control circuits. Fuel based air condition systems might be used as well for preheating the vehicle in winter, or in cold climate regions, since the transformation of fuel to heat is very efficient. In the electrical vehicles the electrical and kinetic power flow has to be balanced constantly and these type of vehicles support a wide variety of operating modes that requires distributed microprocessors and powerful control algorithms. The controller for the electric engine determines the driving characteristics of the electrical vehicle and executes a strategic monitoring, control and steering task. The efficiency and the robustness of the algorithms and the control and processing units will be the area where the OEMs and tier ones will create the added value in the future.

The analog, mixed signal components used in the DC-DC and AC-DC converters, and the development of hybrid modules for electrical vehicles will increase significantly in the next years.

Referring to Fig. 3, the various subsystems are supposed to work at different current and voltage levels, which in turn require the development of the two following semiconductor technologies targeted within the project:
▶ MOS/DMOS
▶ IGBT

This choice is based on the vehicle types that are expected to come into mass production within the next 3-5 years, i.e. those with 60-80 kW peak power at the wheels as the best compromise between vehicle performances, cost of components and market. Widening the target to include larger vehicles necessitates very high power electrical and electronic components, which will in turn require a revision of electrical power train layout to cope with power dissipation needs. In this case the use of materials (e.g. SiC or GaN) working at higher temperatures, with efficiencies of up to and above 98.5%, is currently under investigation. However, it must be pointed out that the major added value of the vehicles targeted in E^3Car will be associated with power-energy management strategies.

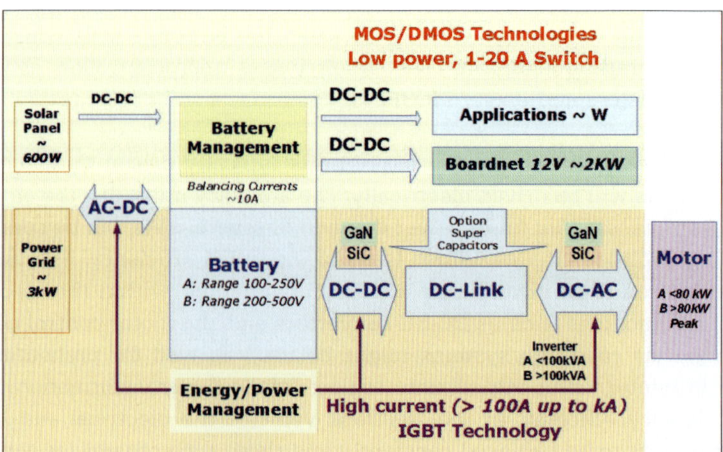

Fig. 4. E^3Car Technology development

MOSFET and IGBT are the predominant power semiconductors in HEV applications. Due to the unipolar characteristic, the switching losses of a MOSFET are significantly lower than those of an IGBT. As a result applications with high switching frequency (>100 kHz) employ the use of MOSFETs while applications with low switching frequencies (<10 kHz) are typically dominated by IGBTs. The unipolar characteristic also leads to a resistive transfer characteristic. In contrast the transfer characteristic of the IGBT shows a threshold voltage of about 0.8V due to the pn junction at the back side of the IGBT. It is only above this voltage that a resistive characteristic can be observed.

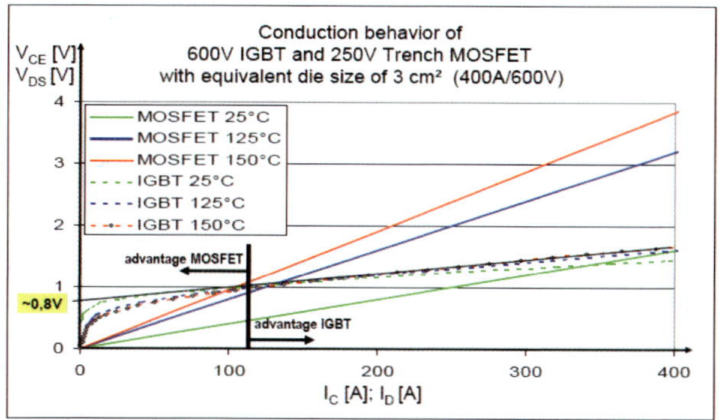

Fig. 5.   Characteristics of IGBT and MOSFET devices

Due to this resistive characteristic, IGBT semiconductors are better suited to applications that utilise higher current densities as IGBT copes better with increased junction temperatures. Another relevant differentiator is the anti parallel diode which is needed for drive application. It is integrated in the structure of the MOSFET, while it needs to be added for the IGBT either externally or on the same chip. To compare the two chip technologies in an application, the power losses of semiconductors of the same area and in a given system are shown below in Tab. 2 and Tab.3.

DC-link voltage	Type of semiconductor	Module (3x) (Typ: 62 nm) (Si area incl. diodes)	Phase current	Total losses in converter @ Tj=125°C in W Converter efficiency			
				1kHz	*5kHz*	*16kHz*	*20kHz*
200 V	*150 V Trench MOS*	*1800 mm² Si area*	*410 A*	*3530* *95.0%*	*3715* *94.7%*	*4226* *93.9%*	*4414* *93.7%*
350 V	*600 V CoolMOS*	*1800 mm² Si area*	*234 A*	*4433* *93.7%*	*4457* *93.6%*	*4523* *93,5%*	*4547* *93,5%*
350 V	*600 V IGBT3*	*FF400R06KE3*	*234 A*	*894* *98.7%*	*1092* *98.4%*	*1626* *97.7%*	*1818* *97.4%*

Tab. 2.   Overview 1

For the calculated full hybrid converter, a system based on a DC-link voltage of 350 V using a 600 V IGBT shows best performance. Even if the DC link voltage is reduced to 200 V the 600 V IGBT has lower losses over the whole switching frequency range than a dedicated 250 V MOSFET.

DC-link voltage	Type of semiconductor	Module (3x) (Typ: Econ-oPACK3)	Phase current	Total losses in converter @ Tj=150°C in W Converter efficiency			
				1kHz	*5kHz*	*16kHz*	*20kHz*
200 V	250 V MOSFET	420 mm² Si area	90 A	977 93.5%	991 93.4%	1027 93.2%	1036 93.1%
200 V	600 V IGBT3	FS75R06KE3	90 A	450 97.0%	504 96.6%	618 95.9%	660 95.6%
350 V	600 V IGBT3	FS75R06KE3	52 A	212 98.6%	256 98.3%	378 97.5%	426 97.2%

Tab. 3.   Overview 2

The results for a component system, with a maximum rating of 15kW equipped with 420 mm² of power semiconductor, are presented in the Tab. 3. For the calculated mild hybrid converter a system based on a DC-link voltage of 350 V using a 600 V IGBT gives the best converter performance. The 600 V IGBTs are double sided processed and extremely thin with a thickness of only 70 µm. The same silicon components are employed for a voltage link below 250 V. Optimising in terms of this direct voltage link will require a further reduction in the silicon components; a thickness of about 40 µm is typically required with double sided processing (Fig. 6).

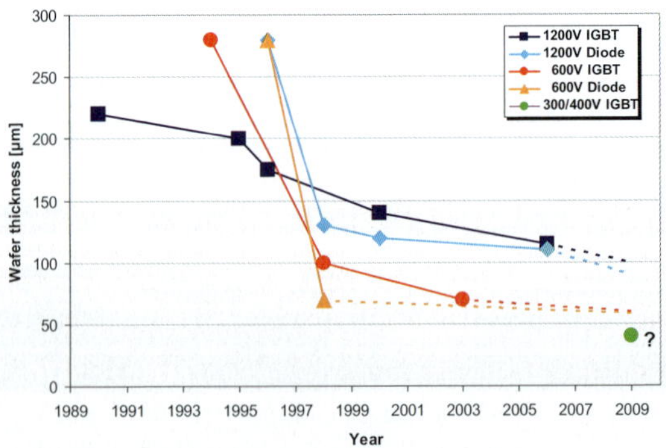

Fig. 6.   Wafer thickness for different high voltage semiconductor technologies

Electronic power supply applications in the automotive sector have the highest requirements with regard to power semiconductors. These have to provide a high level of efficiency by keeping static and dynamic losses at a minimum, while running at high operating temperatures. Furthermore, these components need to be very light and small in volume, while ensuring that the resulting system costs are kept as low as possible.

An example of this new power Si-based semiconductor generation is the 600 V IGBT provided by Infineon Technologies AG, currently withstanding junction temperatures of up to 200°C. However, the class of vehicles investigated in this project are powered by battery voltages of under 250 V, for which a dedicated IGBT 400 V technology is being developed, as well as requiring the production and handling of extremely thin wafer and chips (about 40 µm), which in turn demands new component design and the development of new machines.

The availability of these components permits optimal application of the silicon technology to EVs and enables high converter efficiency (reduction of losses of power converter by 10-20%, which is an improvement of converter efficiency by 9%).

On the CMOS/DMOS technology side, the BCD-on-SOI provides a bundle of promising inherent features:
- ▶ Fully dielectric isolation
- ▶ Very low leakage at high temperatures
- ▶ Latch-up-free operation over complete temperature range
- ▶ Less parasitic for robust design and high EMI performance
- ▶ Capability to handle extreme high junction temperatures and power dissipation
- ▶ Small size (0.35/0.18 µm) - low need of wafer area, enhanced integration density

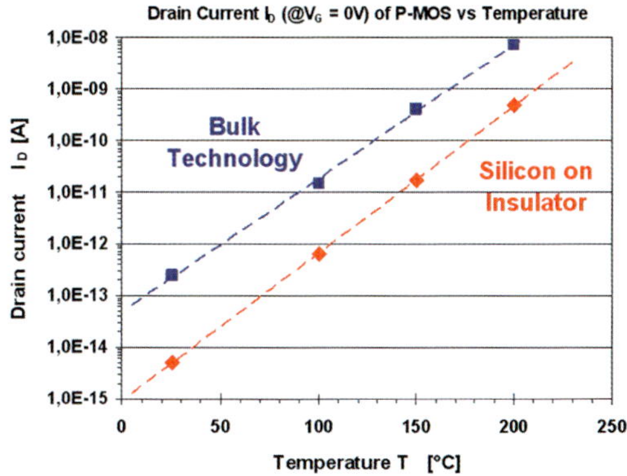

Fig. 7.   Leakage current variations with temperature for Bulk and SOI technology

Fig. 7 shows the advantages of using SOI technology to reduce leakage currents at high temperature operating conditions required by electrical vehicle modules and sub systems.

3   System Reliability and Robustness

Generally, acceptance of power electronics is hugely dependent on reliability. At present, simple rules of thumb are used in the design of power electronic systems for the automotive industry, e.g. heat is one of the most significant causes of degradation in power semiconductors and studies have shown that even a 7-10°C increase in junction temperature can result in halving the MTBF of the device. Therefore, in many applications the switch is chosen such that for the worst case scenario, i.e steady-state operating conditions, the transistor junction temperature does not exceed 125°C. Such approaches do not instil confidence in terms of lifetime reliability and cost effectiveness and new approaches and developments of devices with higher operating temperature ranges are required. In addition, niche EV applications in the current market do not provide extended databases and appropriate experience of failure modes.

A key aim of the E^3Car project will be to adapt and extend the work of the "robustness validation" concepts developed by the automotive semiconductor industry, combining traditional approaches in system and component reliability i.e., testing large sample sizes for long periods assuming a binomial distribution, which is not practical when reliability goals are higher than 90-95%. In particular, the newly released SAE Recommended Practice J1879-Oct2007 Handbook for Robustness Validation of Semiconductor Devices in Automotive Applications, and SAE Recommended Practice J1211-Apr2008 Handbook for Robustness Validation of Automotive Electrical/Electronic Modules, and other Handbooks that are in the pipeline, are combined with the development of enhanced qualification methodologies and accelerated testing techniques for PEBBs and PCBBs. Closely related to this, the development of a more detailed investigation and understanding of sub-system and device mission profiles, derived from vehicle mission profiles, is currently underway. This will allow more a realistic specification of the operational environment for an improved and more effective interaction throughout the supply chain, avoiding inadequate or over-engineered designs.

4 Summary and Discussion

The future needs, in terms of urban and universal mobility, will create a demand for cost efficient and regenerative mobility. The universal vehicle, with operating ranges of more then 600 km with one tank stop, will be transformed by the need of different mobility solutions. Several customer segments (families, groups of owners, environmental friendly focused individuals, etc.) are interested in urban mobility vehicles with low emissions and low energy cost. These customer segments require a second vehicle that is suitable for town and city driving with about 100-200 km distance between charging.

EVs and HEVs will occupy different market segments and will address different society needs. EVs will penetrate the market segments represented by local transport users, in communities in which electrical energy is cheap and easy to access. On the other hand, HEVs will penetrate the market segments comprising drivers who want long–range, less-polluting motor vehicles. The penetration into these markets will be significantly dependent on many factors like technology performance, cost, regulations, etc.

The transition to the Electrical Mobility requires major efforts at a European level to strengthen the whole supply chain from power electronic components to the system integrators (car manufacturers). In this context, nanoelectronics technology, devices, circuits and sub systems are key enabler elements for building the new generation of energy efficient electrical vehicles and green cars. The E^3Car project addresses this necessity by combining, in a bottom-up approach, the contribution of major European silicon foundries, silicon design houses, Tier 2, Tier 1, and OEMs.

In particular Infineon Technologies AG, STMicroelectronics, AMI Semiconductor Belgium BVBA, OKMETIC, austriamicrosystems AG and ATMEL play major roles as the technological providers, in strong cooperation with automotive component and subcomponent providers such as Bosch, Siemens, Philips, Epyon, VTI, and Valeo.

Excellence centres such as SINTEF, Fraunhofer Institute, Centre Nacional de Microelectronica, Fundacion CiDETEC, VTT, Alcatel-Thales III-V Lab, Tyndall, Institut Mikroelectronickych Aplikaci, CISC Semiconductor, Consiglio Nazionale delle Ricerche, FH Joanneum, Technische Universitaet Wien, Brno University of Technology, CEA-LETI, provide their expertise in various frontier research field and product application aspects.

Major European OEMs as Fiat and Audi and electric vehicle manufactures as Think and Elbil supply specification and tes ting platforms.

Reiner John
Infineon Technologies
Im Campeon 1-12
85579 Neubiberg
Germany
reiner.john@infineon.com

Ovidiu Vermesan
SINTEF ICT
Forskningsvn. 1
N-0314 Oslo
Norway
ovidiu.vermesan@sintef.no

Marco Ottella
Centro Ricerche FIAT S.C.p.A.
Strada Torino 50
10043 Orbassano (TO)
ITALY
marco.ottella@crf.it

Pietro Perlo
Centro Ricerche FIAT
Strada Torino 50
10043 Orbassano (TO)
Italy
pietro.perlo@crf.it

Keywords: nanoelectronics, electrical vehicle, power management, power conversion, power distribution, smart dynamic monitoring

Safety

Left Behind Occupant Recognition Based on Human Tremor Detection via Accelerometers Mounted at the Car Body

C. Fischer, Bergische Uni Wuppertal / Delphi Delco Electronics & Safety
T. Fischer, Delphi Delco Electronics & Safety
B. Tibken, Bergische Universität Wuppertal

Abstract

The aim is an additional sensor system that is able to detect left behind occupants in parked cars. This should avoid the decrease of fatalities found in unattended or oversight individuals in vehicles. Based on acceleration measurements directly at the car chassis information about the occupancy is extracted. At the beginning of this paper the theory of the signal source, the used car model and the applied classification algorithms is shortly given. Afterwards several measurement results are presented and become analyzed with regard to a following automatic classification. The next step is the evaluation of different classification algorithms and the explanation of the performance on the acceleration datasets that were collected during this research. Many different classification algorithms are available, at this point the support vector machine (SVM), k-nearest-neighbor (k-NN), probabilistic neural network (PNN), decision tree and clustering were observed.

1   Motivation

Passenger recognition that can detect the occupancy or even the number of occupants of a car has a wide range of passenger safety applications. One of them is the delivery of additional information for post-crash systems like the European eCall or the American OnStar. These systems provide information about the position of the crashed car, the heaviness of the accident and other needed information for the rescue team. In some cases motionless occupants were found too late. An occupant detection system could supply the number of occupants or alert until everybody got recovered. Another potential application is the left behind alert for parking vehicles. The NHTSA (National Highway Traffic Safety Administration) initiated a research [1] which shows that hyperthermia is the second leading cause of death for children in non-traffic fatalities in the U. S. (19.6%); the first one is backover (49.5%). Based on this fact the

NHTSA and KidsAndCars.org induced a campaign against leaving kids unattended behind in cars. If this does not decrease the number of hyperthermia deaths in vehicles, the NHTSA is going to introduce a general law that prohibits the leaving behind of living individuals in automobiles. To provide our costumers the possibility to follow the upcoming law, Delphi started the development of an intelligent sensor system that is able to detect left behind occupants in parked vehicles.

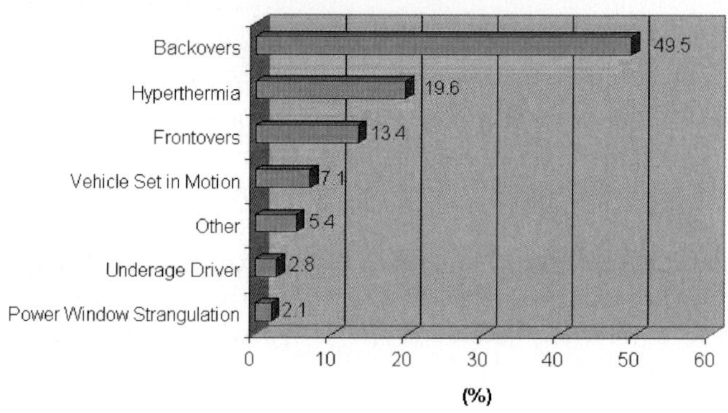

Fig. 1.   U. S. Fatalities for non-traffic accidents by type
(Source: www.kidsandcars.org as of Apr. 2008)

2   Basics

For the recognition of human beings, many solutions are available at the market. They use different signal sources produced by living individuals. In the following table a short overview of possibly usable indications is given.

The previous table shows the pros and cons regarding only the technical interests for the use of these indications. Aspects like low acceptability on inside cameras by the customer or design issues by the vehicle manufacturer do not matter to the technical feasibility of the introduced approach. Due to the expected time for the occupant detection of 15 to 20 seconds systems based on e. g. carbon dioxide sensors are not usable for the requested task. These sensors offer a very high response time compared to the given time for the occupancy recognition. In the presented article accelerometers directly mounted to the car body are used to recognize occupants in a parked car. Accelerometers become more and more popular in the automotive industry, which is founded on the decreasing production costs and on the mechanical robustness of these sensing devices.

Several indications of living individuals		
Signal source	*Pro*	*Contra*
Vibration induced by heart-beat, breathing, muscle tremor	*Not suppressible*	*Low signal magnitude*
Heat (Body temperature)	*Low system requirements*	*Shadowing effects; external heating*
Weight of the occupant	*Easy measu-rable*	*Additional sensor needed to detect only living individuals*
Respiration (CO$_2$)	*Good indica-tion for living creatures*	*High response time, especially for occupants asleep*
Image recognition (face or body contour)	*High informati-on content*	*Shadowing effects; different illumination conditions; hard to detect only living individuals and no dolls*

Measurements at a setup car were made to analyze the vibration signals for the appearance of human heartbeat or respiration. The acceleration data showed an indication for the occupancy that is not related to heartbeat or respiration. Intensive researches pointed out a correlation between the observed frequencies and a medical cause – called human tremor.

2.1 The Human Tremor

The tremor of human beings is an unintentional, rhythmic, oscillating muscle movement which can not be suppressed by the individual itself. This and the fact that it is present by every normal person makes it interesting for the development of a left behind occupant detection. One of the known unhealthy tremors is the Parkinson's disease.

Fig. 2 displays a healthy human tremor spectrum directly measured at the body of a test person [2]. Medical researches are more interested in the classification of the movement disorders caused by unhealthy forms of tremor, so it is hard to obtain qualified information about the healthy tremor which occurs by almost every person. In general this muscle contractions are separated in two cases – rest and action tremor. The action tremor results from static exposure, physical exercise or even physical exhaust. The last one is well known by athletes who go to the physical limits.

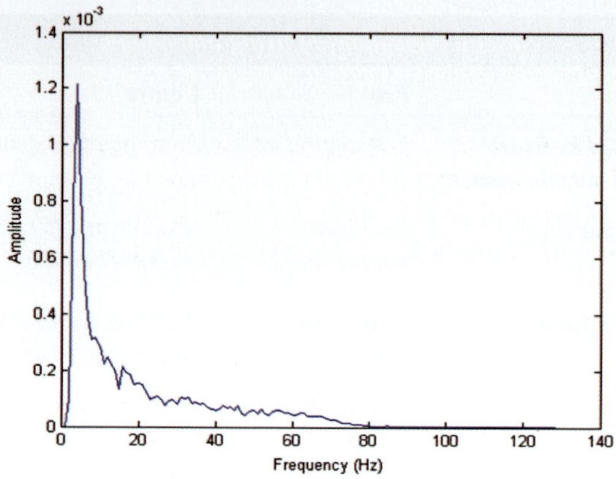

Fig. 2.   Frequency Spectrum of a Human Tremor directly measured at the Body [2]

More important for a left behind detection is the rest tremor. This kind of tremor occurs without any load. A medical hypothesis for the existence of this muscle behavior is the tension of the body against earth's gravity. The frequency and magnitude variation between different persons depending on the conditions like too much caffeine or less sleep are an additional challenge for automatic occupant recognition.

The table below displays the medical classification of human tremor frequencies [3].

Medical classification of human tremor frequencies	
Low-frequency tremor	Below 4 Hz
Mid-frequency tremor	4 up to 7 Hz
High-frequency tremor	7 up to 15 Hz

For healthy adults a mid-frequency tremor is normal, because of the previous named dependencies on the conditions of the test person a range between 4 and 12 Hz has to be observed. The analyzed disease pattern of the unhealthy tremor forms in medical papers and researches provide not enough characteristics about the healthy one, so that there is also a need for extractable characteristics for the automatic classification. Important signal features will be pointed out in the measurement and classification parts of this article.

2.2 The Used Classification Methods

In a further step different classification algorithms will be determined on the collected data. This will expose the most acceptable algorithm for the focused task of occupancy detection in a parked car. Five of the most popular methods will be evaluated, but before a short explanation of theses decision techniques is given in the following section.

First of all a simple algorithm called k-nearest-neighbor (k-NN) is used, which is also known as a lazy learning classifier. This means that there is no previous training needed like it is for the other mentioned classifiers. k is the number of the nearest neighbors that have to be considered for the decision making process (Fig. 3).

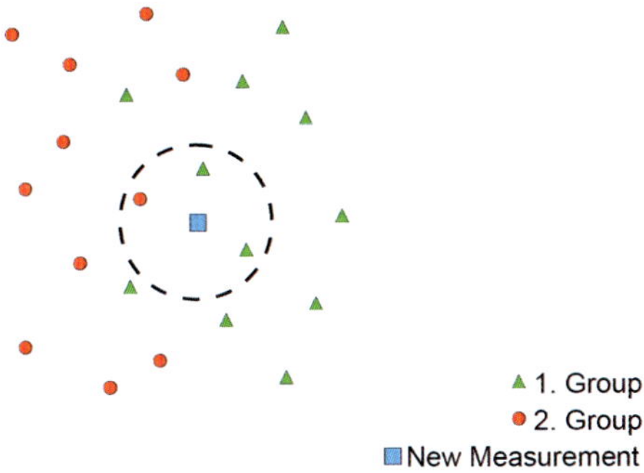

Fig. 3.   Example of the k-NN Method (k = 3)

With the known group membership of the neighbors a probability of membership for the new measured data point is determined. This is applied by a majority decision which can be additionally weighted by a distance metric. The most common one is the Euclidean distance.

Another method for classification can be implemented by data clustering. Cluster algorithms try to separate the groups geometrically (Fig. 4)

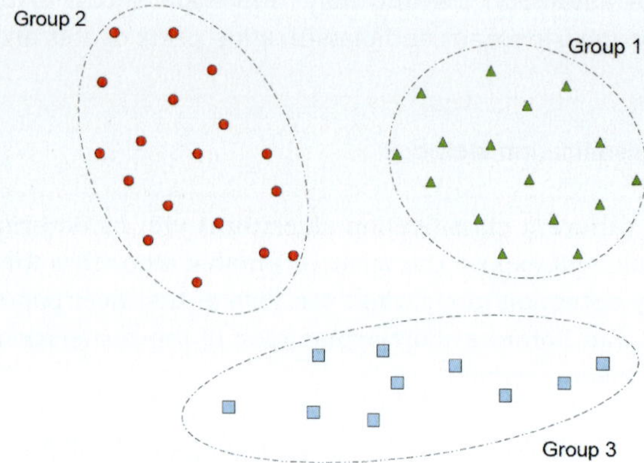

Fig. 4.   Example of a Three Group Clustering

As exemplarily shown in Fig. 4 the training dataset is cut into data clouds. These clouds represent a membership which in case of a good clustering contains only data points with the same real world membership. After this training part a new measurement will be assigned to one of the predefined clusters by probability calculations.

The third used classifier is a decision tree. Decision trees are often used to find the direct way between input and the possibly best output. If it is based on statistical mathematics it can also be used for classification applications. The used one in this paper is called J-48 which is founded on the C4.5 algorithm developed by Ross Quinlan. It is a non-binary method that splits the training data into leaf nodes (Fig. 5).

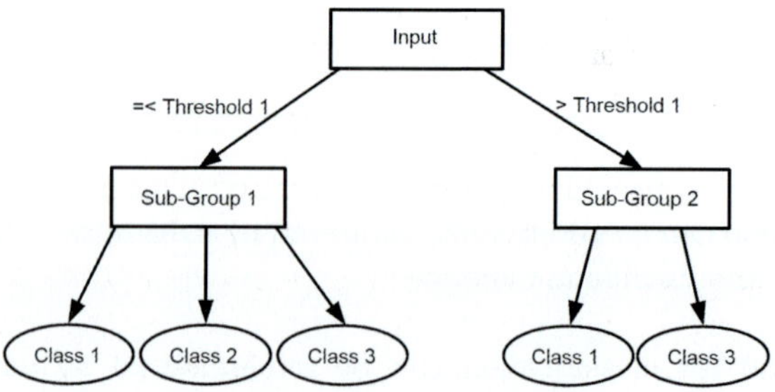

Fig. 5.   Example of a Decision Tree for Three Classes

After a decision tree is built it becomes pruned to reduce the number of nodes and so reduce the complexity of the tree to the minimum needed level. Past this training phase new data is sorted by the created decision rules until it ends in a leaf node that represents the most possible class for the new measurement.

A more biological related approach is the idea of artificial neural networks (ANN). These methods are often used in pattern or image recognition applications, because they can build a decision model for very complex non-linear problems.

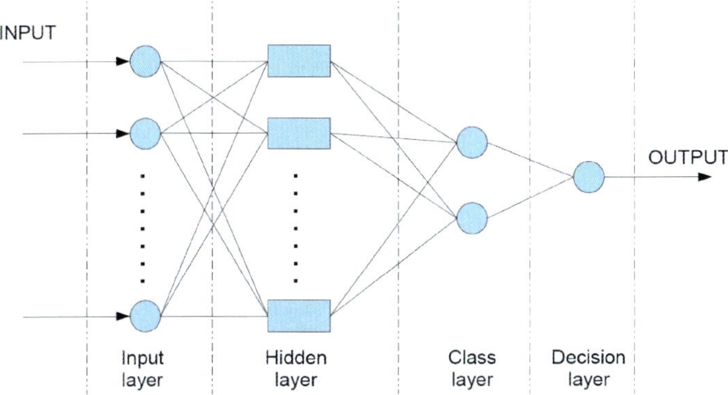

Fig. 6.   Four Layer artificial Neural Network

Fig. 6 shows a schematic of a neural network. Many versions of neural networks are available; the used one in this work is the probabilistic neural network (PNN). The probabilistic neural network estimates the probability density functions for each class based on the training samples. During the learning (training) process of the network the connections between the neurons get weighted, biased or even deleted. At the end a probabilistic density function for each class depending on the input layer is trained. The decision layer uses the largest vote to predict the target class.

The last classifier becomes more and more popular which is also founded in the increasing performance of today's computers. Because of this it is possible to calculate support vector machines with more complex kernel functions than only linear ones. The support vector machine (SVM) is originally applied in the deterministic optimisation [4] and was developed by Vladimir Vapnik and Aleksei Chervonenkis in the 1960's. Before the support vector machine can be applied, it has to be trained. Out of this training phase a support vector for a following decision process is provided. This vector describes the maximum-margin hyperplane between the datasets (Fig. 7).

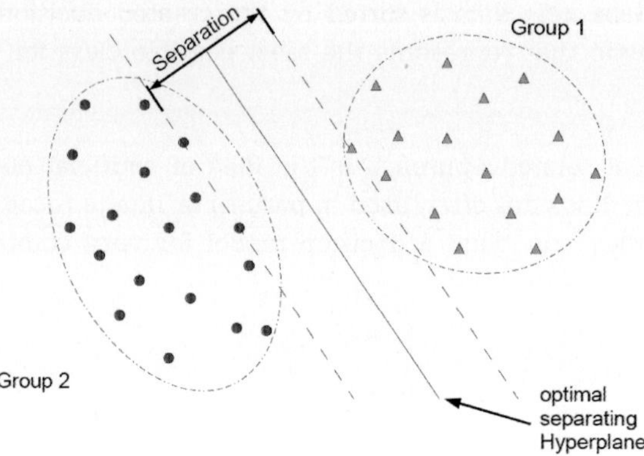

Fig. 7. Support Vector Machine in two-dimensional Space for a Binary Classification

Before the previously mentioned classifiers can be used a feature extraction has to be applied to the measured data. This reduces the input data dimension with less information loss and will be discussed in the classification part of this article.

3 Laboratory Experiments

This section deals with the experiments and the measurements done at the test setup. At the outset the theoretical considerations with reference to the test setup are introduced to convey a first impression and an understanding for the following measurements. Subsequently the measured signals with different combinations of passengers are presented; also the influence of passing traffic will be reflected. In a concluding part the results of the measurements are summarized.

3.1 The Experimental Setup

To get a previous overview of the expected signals and a definition of the needed setup requirements a model was built. Based on this model the transferring system components were identified and system limits were defined. The medical description of the signal source – tremor – limits the observation spectrum up to 15 Hz. Because of this a restriction of the involved mechanical components and their properties is given.

A car on suspensions can be approximated by a composite spring-damper-mass system (Fig. 8).

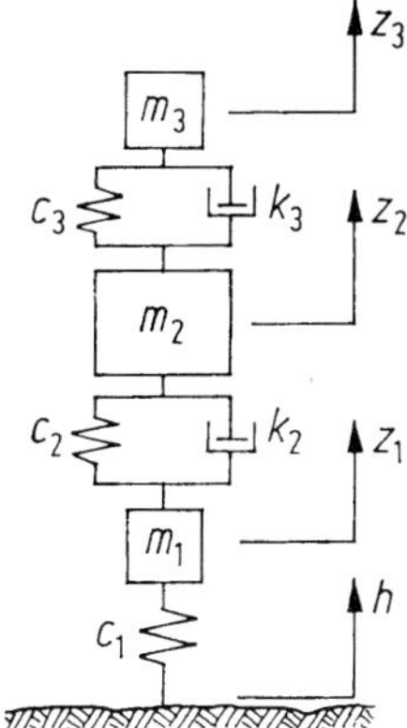

Fig. 8.   Spring-Damper-Mass System [5]

In Fig. 8 the spring-damper-mass system for one tire is schematically illustrated. The passenger with its mass m_3 is positioned on a seat which is represented by the spring coefficient c_3 and the damping characteristic k_3. m_2 describes the proportionate mass of a quarter of the car chassis.

The natural frequencies of the car body are chosen by its manufacturer and the first resonance frequency is set above 30 Hz [5, 6]. So the chassis can be approximated as a rigid compound with a total mass of $4*m_2$. A FE-Simulation (finite elements simulation) or a multi-body system (MBS) is not necessary, because the first bending and torsion vibrations occur not in the observed frequency range [6]. The chassis with seats and passenger(s) is mounted on the chassis suspension which consists of the suspension (c_2, k_2) and the wheel (m_1, c_1). The tire is dynamically modelled with a spring constant c_1 and normally without any damping characteristics, because the attenuation is insignificant for increasing velocity and tire pressure [5]. In the static case – a parking car – the damping can be unattended as long as the tire pressure is correct. Decreasing tire pressure means increasing attenuation due to the tire.

The occurring natural frequencies of the previous described system are summarized in Fig. 9.

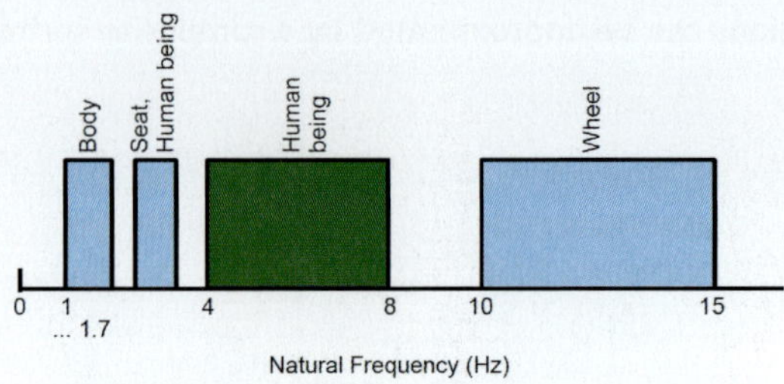

Fig. 9.   Separation of the natural frequencies as in [5]

In Fig. 7 is the natural frequency of human beings also represents the frequency range of the highest sensitivity on external vibrations impressed on a person [5]. The transfer function of the seat has got an attenuating character above 5 Hz.

Out of these preliminary considerations and first measurements at a test setup with accelerometers, a model was built up in Matlab's Simulink environment. The base for this model is a wireframe for the car body including the previously described spring-damper-mass systems (Fig. 10).

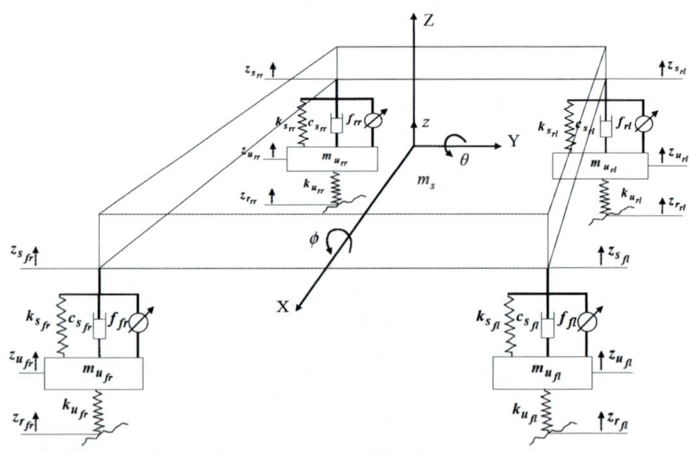

Fig. 10.   Exemplary Car Suspension Model [7]

Model components were placed with real values of their mass, dimension and other characteristics [8]. Due to the gained information out of this theoretical consideration two test cars were assembled; one middle class sedan and one high class sedan. The presented data in this article was measured at the mid class car, but the results of both cars are similar.

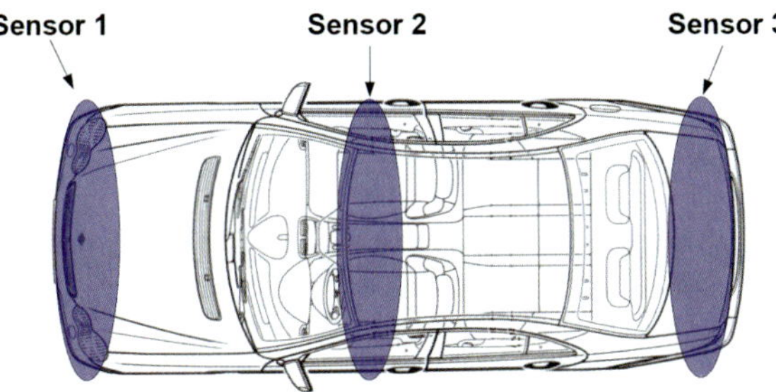

Fig. 11.   Sensor positions at the Test Car

Three high sensitive analog accelerometers (1200 mV/g) were directly mount-ed to the car body. They were placed at the front, mid and rear of both cars (Fig. 11). The sensor output voltages become pre-amplified and band-pass filtered before the three signals are digitized simultaneously by a 16 bit ADC (Analog-Digital-Converter, Fig. 12).

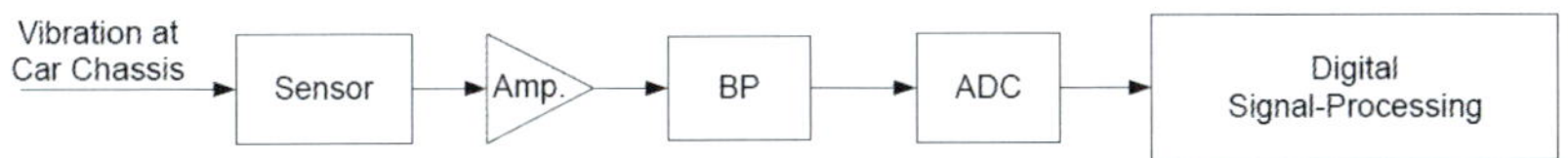

Fig. 12.   Schematic Signal Path

The used filters have got a low cut-off-frequency at 1 Hz and a high cut-off-fre-quency at 50 Hz. This is on the one hand an anti-alias filter above 50 Hz and on the other hand a filter which removes modification in the angle between the sensor and gravity (DC signal) below 1 Hz. The used PCB circuit is shown in Fig. 13.

Fig. 13.   PCB Circuit with the Amplifiers and the Band-Pass Filters

After the preprocessing the digital time-domain signals are available for analy-sis and later classification evaluation.

3.2 Experimental Measurements

With the described setup several measurements in different arrangements of occupancy and vehicle conditions were observed.

Different test scenarios		
External conditions	▶ ▶	*Busy road, loud music* *Quiet side road, garage*
Passenger conditions	▶ ▶ ▶ ▶	*Empty* *Different human beings (Children, adults, etc.)* *Different numbers of passengers* *Animals (Dogs)*
Vehicle conditions	▶ ▶ ▶	*With/without parking brake* *Variation on tire pressure* *Loading status (e. g. trunk)*

Within this work all in the table listed scenarios were tested and assessed. Some of the recorded signal spectra are presented below to provide an impression of the occurring signal frequencies. For the assessment of the technical possibilities of the developed system it was interesting if the number of passengers could be determined. In Fig. 14 the recorded spectra of two adult test persons are compared.

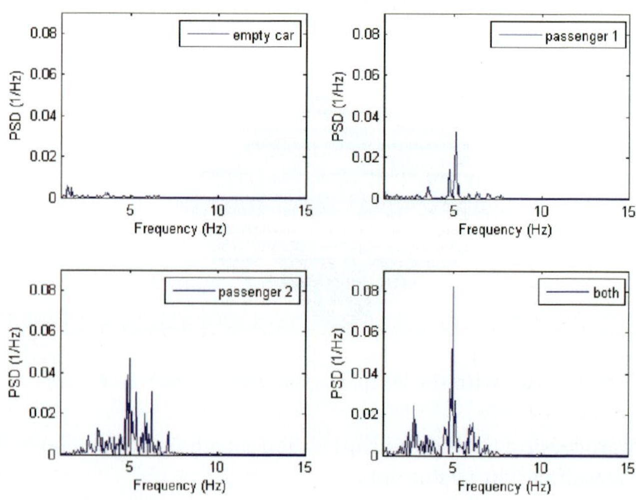

Fig. 14.   Separation between two adult Test Persons

As shown the stimulated spectrum of frequency varies between the first and the second person in width and magnitude. Even if researches into the identification of persons with their tremor are promising [9], the measured signals at the car chassis provide only limited conclusions on the number of passengers.

Fig. 14 displays that the dominating frequencies of both test persons are located in the same range. For the identification of single persons it would be necessary to create a "tremor fingerprint" which unites spectral characteristic features. This is no subject of this work, because the identification of the occupancy status is sufficient for the intended purpose.

A dependency on the tire pressure could be noticed in measurements with varying the pressure [10]. Due to increasing drop in pressure the magnitude of the recorded tremor signals decreases, nevertheless detection with 50% of the correct pressure is possible. As soon as the tires are completely without pressure and the car is standing on the rims, only the tremor signal of an adult is strong enough to be occasional detected in the spectrum. This is also observable in the signal-to-noise ratio which decreases to a quarter of the normal value. The influence of the parking brake and/or the loading conditions of the car was over all measurements insignificant.

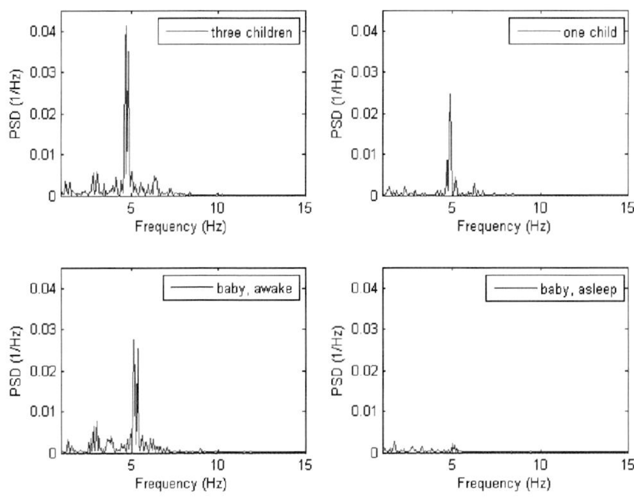

Fig. 15.   Spectra produced by Children

Fig. 15 displays measurements including children in different seating scenarios. They are normally easy to recognize, because children can not really sit motionless in a car as long as they are awake. Thus the action tremor is additionally measured with the rest tremor. In case of sleeping children the rest tremor is the only signal source. This leads nearly to a permanent loss of signal, especially for the two year old baby (Fig. 15, down right). During the measurements with different passengers several scenarios with dogs were also taken. These animals showed a good detectable tremor which is in the same frequency range like humans tremor, so that only a detection of living individuals and no further separation between human beings and dogs is possible.

The last considered testing scenario during this evaluation is represented by one of the most occurring parking conditions; parking nearby a busy road. Several measurements with and without different occupants were taken directly at a road with plenty of heavy loaded traffic. Fig. 16 compares a time-domain signal of an empty vehicle to one of an occupied car at the same position at the passing traffic track.

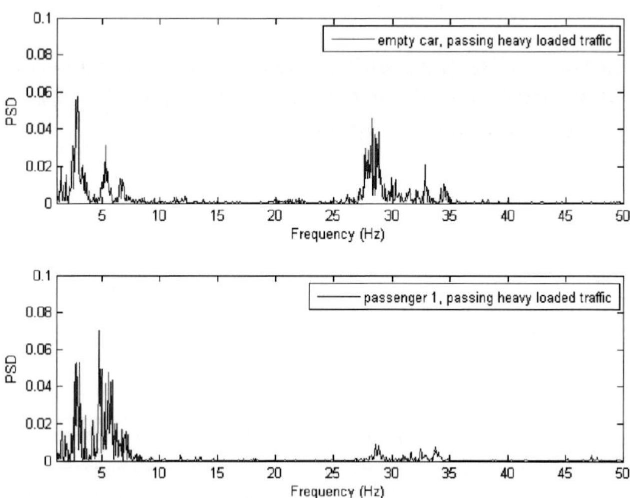

Fig. 16.   Spectra at very busy Road (top: empty Car, bottom: occupied Car)

Beside a distracting signal at 27 Hz also stimulation in the wanted signal (until 15 Hz) can be observed. This circumstance would lead to a false detection if only the pure frequency-domain would be considered. Passing traffic is limited in time, because of this an additional rating in the time-frequency-domain (STFT – Short Term Fourier Transform) is done.

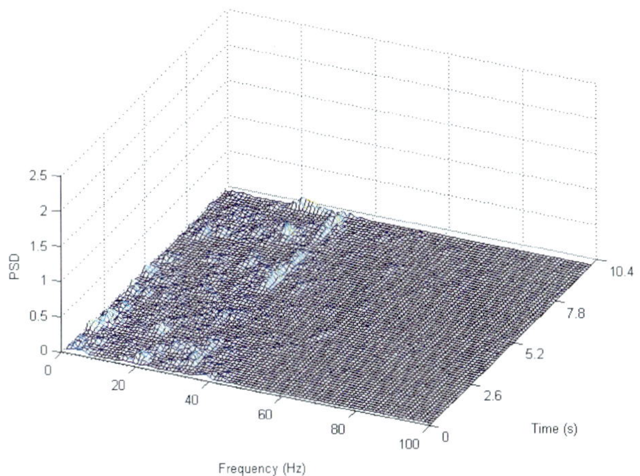

Fig. 17.   Time-Frequency-Domain of an empty Car in calm Environment

Fig. 17 displays the STFT-spectrum of an empty car parked in a calm environment. It shows the expected noise produced by the sensor and also the attenuation by the analog filter at 50 Hz is visible.

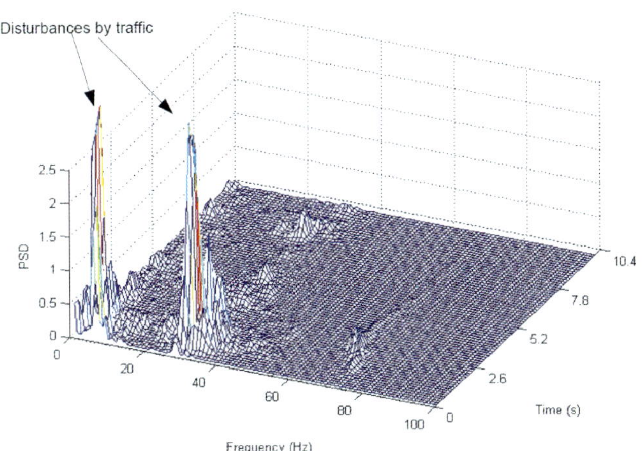

Fig. 18.   Time-Frequency-Domain of an empty Car disturbed by passing
Traffic

In contrast to Fig. 17 the car parked at a busy track (Fig. 18) exhibits short time disturbances by passing traffic. Two short pulses, one at 15 Hz and the other one at 30 Hz are displayed. Due to the time-frequency resolution they can be identified as disturbances and the risk of false detection decreases.

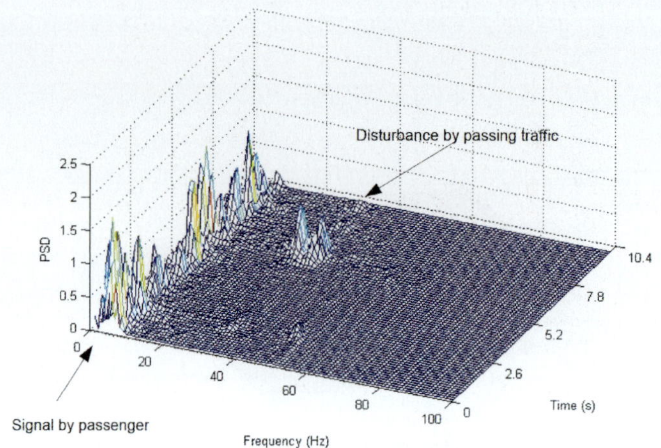

Fig. 19. Time-Frequency-Domain of an occupied Car disturbed by passing
Traffic

In case of an occupied vehicle parked at a busy road the human tremor is
constant over time at 5 Hz (Fig. 19), whereas a short peak at 30 Hz occurs; a
separation between empty and occupied is possible owing to the STFT. The
shown disturbances could only be observed during passing heavy load traffic;
normal car traffic did not produce effectual vibrations to disturb the measured
signals.

At the end a benchmark of the three accelerometers on the taken measurements was done. Equation 1 shows the used signal-to-noise ratio based on the
signal power of the two occupancy states of the vehicle.

$$SNR = 10 * \log_{10}\left(\frac{P_{occupied}}{P_{empty}}\right) \tag{1}$$

The comparison (Fig. 20) of the three mounted sensors points out that the third
sensor (mounted in the rear of the car) provides the best signal-to-noise ratio
at any measurement condition. This seems to be related to the large distance
between the sensor and the center of mass in the oscillating system.

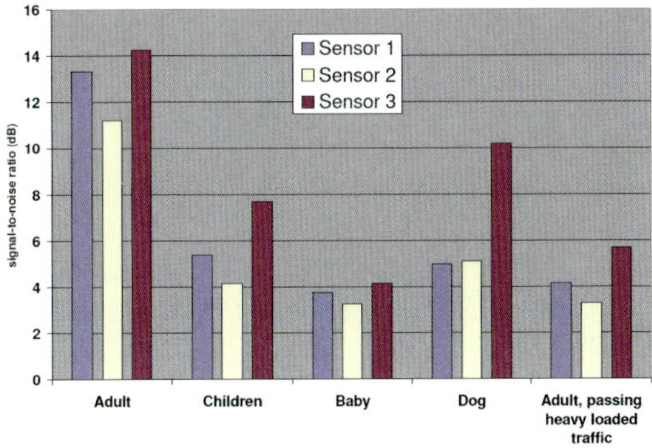

Fig. 20.   Comparison of the three Sensor Positions by Standard Situations

Although the rear sensor provides the main information about the status of the car the other sensors provide redundancy information which ensures a better classification.

4   Classification

Based on the collected data with the described test setup the five mentioned classifier were ranked. In preparation for this ranking a feature extraction on the digital filtered sensor signal took place. Several transformations were applied to the filtered time domain signal (e. g. FFT – Fast Fourier Transform, STFT – Short Term Fourier Transform, WT - Wavelet-Transform) and spectral characteristics out of these transformations were extracted [11]. These values were combined to a feature vector of a size of 105 items per measurement. The information out of over 1500 measurements was preprocessed like this to build the basis for the following classifier evaluation.

In practice classifier algorithms are tested in a k-fold cross-validation [12]; commonly a 10-fold cross-validation is used to determine the performance of classification methods. This process portions the dataset into $k = 10$ subsets. Afterwards this routine trains the classifier with k-1 sets and uses the left set to validate the trained method. This procedure runs $k = 10$ times and provides comparable performance information, composed in a confusion matrix (see table below).

Confusion Matrix for binary classification		
True Pred.	*0*	*1*
0	*True Negatives (TN)*	*False Negatives (FN)*
1	*False Positives (FP)*	*True Positives (TP)*

Due to this table several parameters like the classifier accuracy (Eq. 2) can be determined.

$$accuracy = \frac{TP + TN}{TP + TN + FP + FN}$$

(2)

Fig. 21 shows the determined accuracy of the different decision methods.

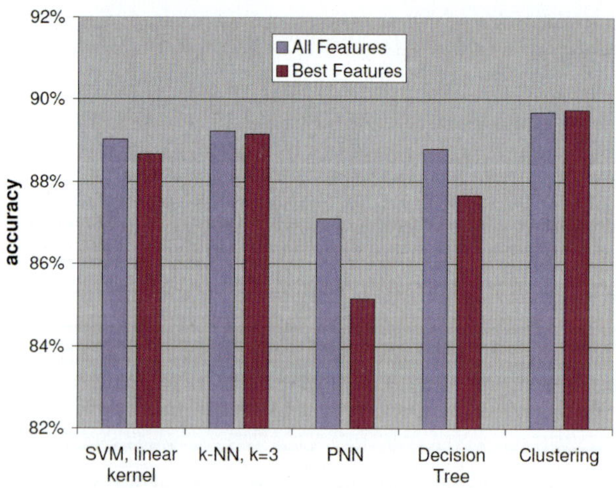

Fig. 21. Ranking of the Classification Algorithms

The algorithms are compared in two cases. Case one uses the complete feature vector data as basis and in case two only the best features were used. To find out which features provide the most information a second ranking took place [13]. It pointed out that the most reliable characteristics for the focused task are out of the frequency-, time-frequency- and wavelet-domain. With this information the dimension of the feature vector can be shrunk from 105 items down to 54 items; 51% of the size of the original feature vector.

The decision method based on data clustering provides both feature vectors the highest accuracy (over 89%), but the other tested classifiers except probabilistic neural network are useful too.

5   Conclusions and Further Work

Even the tremor of motionless passengers normally produces detectable vibrations at the chassis of a parked car. This signal can be converted into voltage information with the help of high sensitive accelerometers. The introduced test setup was used to evaluate a left behind occupant detection based on acceleration measurements. The tremor produced signal varies from person to person in magnitude and center frequency. Because of this several passenger and scenario combinations were measured and analyzed to train classifiers for the automatic detection of the occupancy of a parked vehicle.

Based on the obtained data five popular classification algorithms were evaluated in a 10-fold cross-validation process. The worst classifier for the focused task is the probabilistic neural network with a minimum accuracy of 85%. The applied clustering algorithm showed the best results (over 89%) in the tested group even with a reduced feature vector of 51% of the primary feature vector.

Based on the results presented in this article the best classification methods will be implemented in a test car to verify the performance of each of them for every day use. Especially tests with passengers in the age between 5 and 7 years will be observed to improve the system for the recognition of left behind children. Furthermore several interior sensor systems (e. g. Ultrasonic or infrared devices) will be verified for a system fusion that increases the classifier accuracy.

References

[1]   NHTSA, "Data Collection Study: Deaths and Injuries resulting from certain Non-Traffic and Non-Crash events", 2004.

[2]   Engin, M., et al., "The classification of human tremor signals using artificial neural networks", Expert Systems with Applications, Vol. 33, 2007.

[3]   Lyons, Kelly E., "Handbook of essential tremor and other tremor disorders", Marcel Dekker Inc., 2005.

[4]   Albani, M., "Numerische Optimierung, Aufbau und Test eines Sensorlayouts zur Sitzbelegungserkennung mit Hilfe von unterschiedlichen Algorithmen unter Verwendung von verschiedenen Optimierungskriterien", Faculty of Electrical, Information and Media Engineering, University of Wuppertal, Germany, 2005.

[5]   Mitschke, M., "Dynamik der Kraftfahrzeuge – Band B Schwingungen", 2.Edition, 1984.

[6]   Spickenreuther, M., "Funktionsmodell der Karosserie zur Auslegung des Schwingungskomforts im Gesamtfahrzeug", Dissertation, Lehrstuhl fuer Fahrzeugtechnik der Technischen Universitaet Muenchen, 2006.

[7]   Chamseddine, A., Noura, H., Ouladsine, M., "Sensor location for Actuator Fault Diagnosis in Vehicle Active Suspension", 17th IEEE International Conference on Control Applications, San Antonia, Texas, USA, September 2008.

[8]   Draghici, B., "Development of a model for the identification of the occupants in automobiles relying on low-g acceleration measurements", Faculty of Electrical, Information and Media Engineering, University of Wuppertal, Germany, 2008.

[9]   Strachan, S., Murray-Smith, R., "Muscle Tremor as an Input Mechanism", UIST 2004, Santa Fe, 2004.

[10]   Bahadir, C., "Evaluation of different classification algorithms for occupancy detection systems in vehicles with low-g based acceleration measurements", Faculty of Electrical, Information and Media Engineering, University of Wuppertal, Germany, 2008.

[11]   Jakubowski, J., et al., "Higher Order Statistics and Neural Network for Tremor Recognition", IEEE Transactions on biomedical Engineering, Vol. 49, No. 2, February 2002.

[12]   Ye, N., et al., "The Handbook of Data Mining", Lawrence Erlbaum Associates Inc., 2003.

[13]   Fischer, C., Fischer, T., Tibken, B., "Sensierung von zurückgelassenen Personen in geparkten Fahrzeugen mittels Beschleunigungsmessungen direkt an der Fahrzeugkarosserie", 3.Tagung Sensoren im Automobil, 2009.

Christian Fischer, Thomas Fischer
Delphi Delco Electronics & Safety
Customer Technology Center Wuppertal
Delphiplatz 1
42119 Wuppertal
Germany
christian.fischer2@delphi.com
thomas.fischer@delphi.com

Bernd Tibken
Bergische Universität Wuppertal
Fachbereich E – Elektrotechnik, Informationstechnik, Medientechnik
Lehrstuhl für Automatisierungstechnik/Regelungstechnik
Rainer-Gruenter Str. 21
42097 Wuppertal
Germany
tibken@uni-wuppertal.de

Keywords: occupant detection, classification, support vector machine, k-nearest-neighbor, artificial neural network, clustering, decision tree, human tremor, hyperthermia, accelerometer

Novel Pre-Crash-Actuator-System based on SMA for Enhancing Side Impact Safety

E. Zimmerman, V. Muntean, Faurecia Innenraum Systeme
T. Melz, Bj. Seipel, Th. Koch,
Fraunhofer-Institute for Structural Durability and System Reliability

Abstract

The European IP APROSYS (Advanced PROtection SYStems) is currently being finished. The overall project goal was to decrease the number of fatalities on European roads. Within Sub-Project 6 a novel pre-crash side impact protection system was developed which improves structural behaviour in side crash. The main focus is to reduce intrusions. The given paper presents research activities which were carried out within SP6. The pre-crash-system combines a sensor unit, a data processing or data fusion unit and at least one reversible high-speed actuator. As outcome of this study about the pre-crash-system, a crash load redirection to the unstruck side was found to be most powerful and a suitable actuator was developed which takes away the crash loads directly from the incoming object at the door. This was achieved by creating a rigid connection from the struck door to other stiff car regions. By this, the energy absorbing process starts earlier and involves more structural parts. This system changes the crash deformation modes completely. Both, the B-pillar as well as the door intrusions are being significantly reduced, especially in regions being most critical for the occupant.

1   Introduction

The wish to decrease the number of accidents and their consequences in road transportation has been growing stronger and stronger in recent years. The high number of injured traffic participants and for sure the number of fatalities will not be accepted any longer by the public opinion. Requirements for better crash avoidance and crash protection have been continuously increased in severity and are coming from all sides: legal entities for homologation, insurances and from the customers and their organizations, especially with the new requirements which will be valid from 2012.

It is still a must to improve crashworthiness of cars in general with respect to self and partner protection. The research activities there have not come to a final state, even not for the most relevant crash scenarios mirrored in the defined test protocols used in all countries over the world. Crash situations, especially, require car structures and restraint systems to adapt to the broad variability existing in real crash scenarios (both the occupants of a car: size, age and gender, position and the crash situations: severity, type, impacting object).

Future vehicles will have to be much more equipped with intelligent safety systems including sensors, actuators, advanced technologies, advanced materials which will enable to correctly communicate, understand and seize the passengers (type, position) and traffic partners (collision avoidance and compatibility) and finally to engage the proper safety devices to ensure best protection. Side impact protection, due to its particularities: small survival space, few energy absorbing structures, little time and space to ensure the maximum functionality of restraint systems, seems to be an accident scenario which would immediately benefit from such an adaptive system.

Answering to these demands, the EU launched a project called APROSYS (Advanced PROtection SYStems) within their 6th framework programme. This article presents a part of the work conducted within APROSYS sub project 6, dealing with the development of an active pre-crash system to increase occupant protection in a side impact accident. The car model chosen for the implementation of the pre-crash system was the DC Neon, model year 94-99.

A pre-crash system in general, is composed of a sensor system which is supposed to sense the dangerous traffic partners and define its characteristics, a decision unit and an actuator to perform the needed action to improve the crash performance then to adapt to crash situations. The work to develop and furthermore to integrate such pre-crash actuator system is presented in the next chapters.

2   Side Impact - a severe Scenario

An accident is classified as being a side impact if a vehicle impacts laterally from whatever angle with another road object. In Europe, from accident statistics, the side impact is ranked the second after the frontal impact. This puts up demands to focus research work on development of further measures to increase the protection in side impact.

Compared to frontal impact, in lateral impact the acceleration of the two impacting objects is completely different. The occupant in the target car becomes accelerated in the direction of the hitting vehicle, in the same direction as the structural deformation whilst in frontal crash, the occupant is decelerated to a complete halt in impact direction. The critical phenomenon is the sudden change of direction of the acceleration, which needs to be managed by the safety and restraint devices implemented in the car. This leads to a special pattern for biomechanical injuries of the occupants. The fatalities usually occur when the biomechanical limits for the head deceleration and thorax loads are exceeded. However, severe injuries can occur at the pelvis and lower extremities if the contact with the interior trim is too hard. To reduce the severity of a side impact two strategies have to be considered: reduce the structural deformation of the car body in regions where the occupant life is jeopardized and soften the contact of the occupant with the interior trim. For APROSYS SP6 however, it was decided to work on intrusion reduction only, to focus on devices which try to reduce the deformation of the car structure.

Looking closer at a side impact accident with respect to structural deformation, three very different crash scenarios can be distinguished, which are, car to car, car to road infrastructure and motorcycle to car impact. To ensure protection in all these scenarios a safety system must be developed which is able to cope with the specific protection needs of those very different load cases. The main focus of the research activities was though on the impact with the deformable barrier for which an extensive analysis of the structural behavior at side crash was performed.

As a reference, a full car crash according to EuroNCAP protocol was performed, which leads to a good understanding of the structural crash modes in side crash. Intrusion measurements were done by acceleration sensors mounted at significant locations on the B-pillar, rocker, and door panel on both car sides as well as post crash intrusion measurements. As an outcome of this study relevant structures driving deformation (B-pillar, rocker, roof, floorboard, doors) were identified and analyzed.

From there, the B-pillar – is considered to be the car structure driving the intrusion in side crash; the B-pillar, by its plastic hinges gives the deformation pattern of the complete side structure. It does not follow a continuous bending line, but the B-pillar hinges at 5 plastic hinges (Fig. 1). It can be observed that the hinge at the height of the belt line intrudes the most into the passenger compartment.

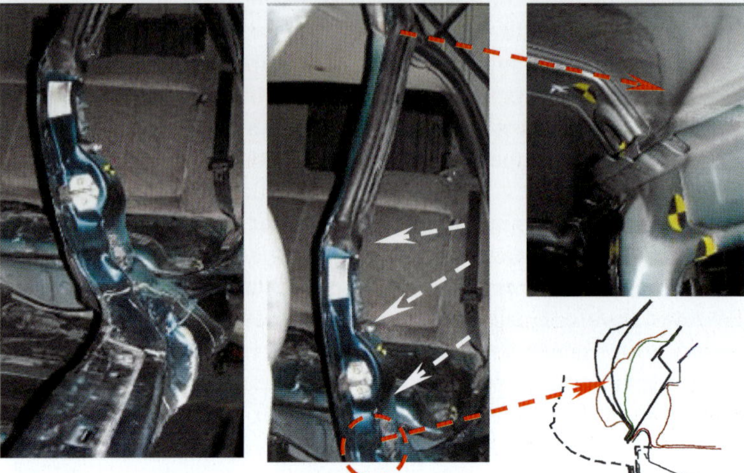

Fig. 1.   B-pillar hinging in side crash with MDB

Different concepts to reduce intrusion were investigated. They consisted of either connecting the doors to the car body in different configurations, or supporting B-pillar or redirecting crash loads towards stiff regions, especially non impacted structures like B-pillar, floorboard and crash box.

Pre-crash systems are able to exactly respond to the side impact challenges: small survival space, few deformable structures available and a very small time span in which the impact event occurs. Pre-crash systems represent a gain of time and space. APROSYS was set to show that a pre-crash system would work in general. The sensor system should show its ability to sense an obstacle, data fusion should provide a trigger signal for the actuator and the actuator should successfully reinforce the car structure and reduce intrusion.

3   Layout of the Aprosys Protection System

From the investigated concepts it was found that, the best way to reduce intrusion was a structural actuator which redirects the loads to the unstruck side, where the rocker meets the crash box and the floor. This was considered to be a very stiff region with a low risk of injuries for the occupants, where the crash loads can be transferred. The actuator presented in this paper is designed in a way that it takes the crash loads directly from outer sheet metal of the struck door at the very first contact phase and directs them towards the previously defined stiff region on the unstruck side. In this way, it actively opens a new load path for the big amount of crash loads and consequently unloads the B-pillar. On top, it reduces the intrusion not only at the B-pillar but in regions where the occupant sits.

Generally, the actuator system comprises of 3 regions as depicted in Fig. 2:
- ▶ Door seat coupling (corresponding to struck side) where the tube actuator is located
- ▶ Seat-center console-seat coupling
- ▶ "tambour spring" – predefined rotation hinge located on the unstruck side at rocker to floorboard connection Broken down to components, the actuator system is composed of a transversal tube (5) (Fig. 2), reinforcements to prevent buckling (6 and 7), additional supports of the beam on the unstruck side (8, 9, 10, 11) and active parts (2 and 4) actuated to close the load mechanism.

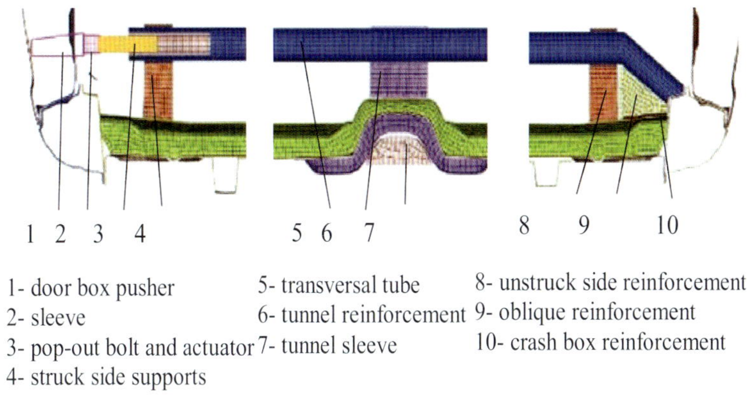

1- door box pusher
2- sleeve
3- pop-out bolt and actuator
4- struck side supports

5- transversal tube
6- tunnel reinforcement
7- tunnel sleeve

8- unstruck side reinforcement
9- oblique reinforcement
10- crash box reinforcement

Fig. 2.   Actuator layout in simulation

The principle of the active mechanism is shown in Fig. 3. The active mechanism (the actuators) is located on the struck side and consists of two parts. The first one is a rotating door beam which in an undeployed state stands upright and does not hinder the normal traveling of the glass. When fired, this beam rotates into a horizontal position filling the complete door box. The second part is a tube with a popping out bolt which is mounted in the seat, transversally just below the knees. In normal driving conditions the space between seat and door is free, when deployed, the bolt creates a tight engagement of the door and the transversal tube mounted in the seat.

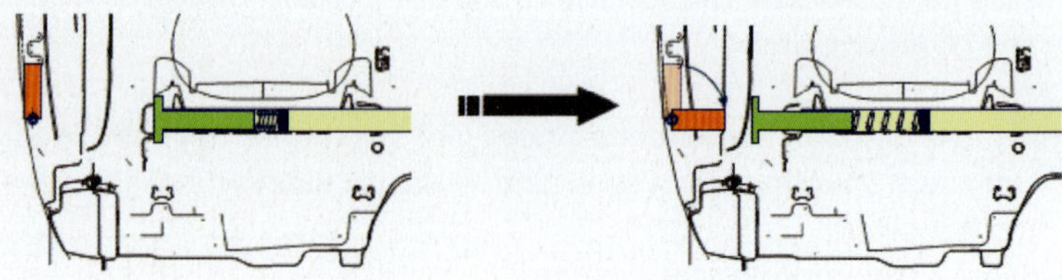

Fig. 3.   Active part function

The reinforcements stabilize the actuator system behavior under crash loads and prevent it from buckling or any other undesired intrusion in the passenger environment. The tambour spring located on the unstruck side is meant to take the crash loads from the barrier and put them to the car structure by a pre-defined deformation pattern. The activated system forms a closed chain of parts; it picks up the loads directly where the occupant sits and directs them to a stiff region on the unstruck side. By this the actuator system is installing a new load path which unloads the B-pillar and gives survival space directly to the occupants.

4   SMA Actuators

4.1   SMA Tube Actuator

The linear movement to close the load path is done by a SMA-Tube actuator. Several requirements [1] on the actuator were derived to fulfill the needed functions. As a summary, the actuator must be capable to pop out a bolt with a travel of 100 mm in a time span of not more than 200 ms. Nevertheless, as set within APROSYS, the actuator must be reversible in case of a false trigger signal. This would be a benefit to pyrotechnical techniques used for other restraint systems. The development of this actuator was done by the Fraunhofer Institute-LBF, Darmstadt.

The actuator comprises: housing, bolt, locking elements, springs, and power electronics. The bolt engages with a sleeve at the end of its travel (Fig. 4).

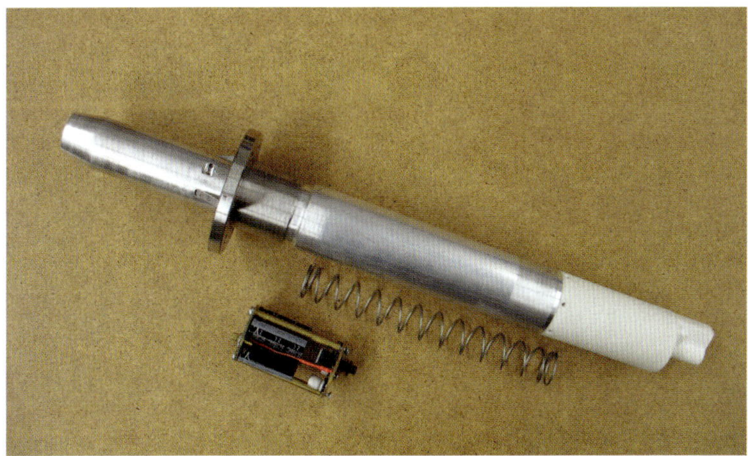

Fig. 4.   Tube actuator with driving spring and integrated electronics.

Fig. 5 shows a schematic drawing of the tube actuator. A bolt (1), which normally rests in a housing, can be driven by a pre-stressed steel-spring (3) from its initial position (inside the seat frame) to a working position. At this working position the top of the bolt is guided by a counterpart at the door trim (4). Small moveable radially mounted locking elements (2) lock the bolt there and prevent an unhindered backward motion. The release mechanism that initiates the movement of the bolt consists of two main parts; a small lever arm that interacts with the rear end of the bolt (5) and a shape memory alloy wire (6) that during activation pulls the lever arm releasing the bolt.

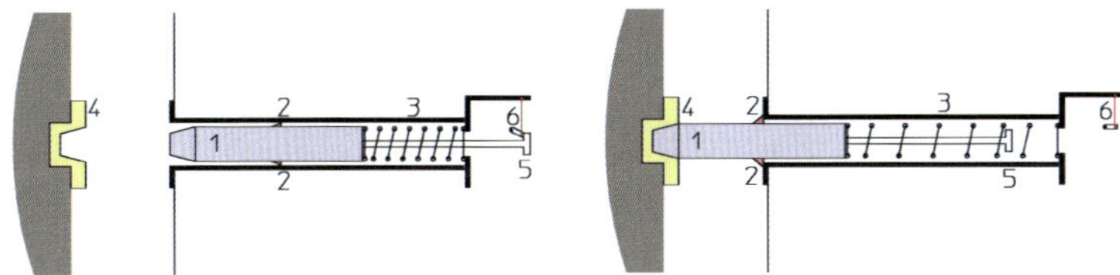

Fig. 5.   Tube actuator mechanisms (left: not deployed; right: deployed)

Like this, no permanent damages neither on the own car (interior trims) nor on the impacting object (e.g. passenger) are implied. To react in crash timing (a matter of milliseconds!) the heating process is supported by a capacitor and is controlled electronically. The electronics are optimized to consume not more energy than provided by the car battery. It ensures that the actuator complies with standard car equipment. The capacitor makes the activation completely independent from the car battery, for example in case if during impact, the car power supply would fail.

4.2 SMA Rotating Door Beam Actuator

The design of the active rotating door beam is in line with the design of a static "rotating" beam, which was used in the Chrysler Neon test cars for crash tests. This design was obtained as a result of FE calculations. It was decided not to integrate this demonstrator for crash tests, but to use the static "rotating" door beam. The activation of the rotating door beam is the same as for the developed tube actuator (See Fig. 6.). The rotating door beam which in an undeployed state stands upright and does not hinder the normal traveling of the glass (7) is integrated inside the door (8). A pre-stressed steel-torsion-spring (4) is mounted inside the beam. The spring is held by a clamp (2), which will be released by a shape memory alloy wire (1), if an activation signal is received. In this case the beam (5) rotates around a joint (3) and the counter part for the tube actuator (9) is brought into right position. In case of a positive false alarm, the rotated beam returns into its initial position driven by a second shape memory alloy wire (6).

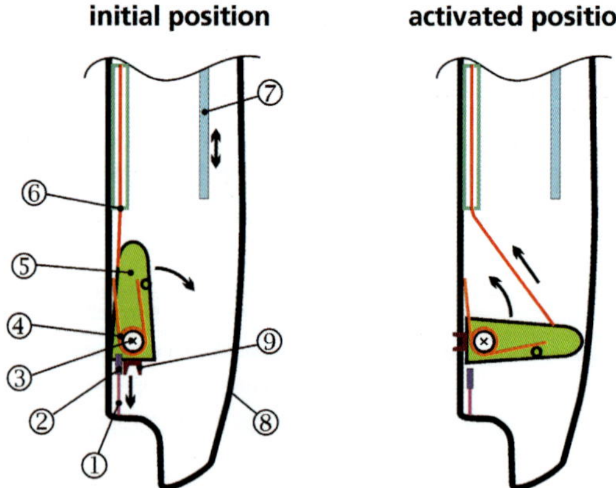

Fig. 6.   Schematic Rotating Beam

5   Performance of Actuator System

The actuator integration and the development process were strongly supported by CAE simulations. The full car model of the DC Neon was validated for side impact for both MDB and Pole impact scenarios. However such a validation is limited. With respect to this, the on going validation of the CAE models is mandatory as a part of the development process. This process was strongly supported by Cidaut, Spain. Therefore, tests to relate the real performance of the actuator system to the CAE environment in which it was developed were organized. The tests can roughly be grouped in two families:

> ▶ Component testing to validate intermediate steps from the development process; especially to asses the mechanical performance and the reaction time of the SMA actuator
> ▶ Full car testing to validate the complete pre-crash system in the DC Neon.

5.1 Component Testing

Two sets of requirements were broken down to the actuator, which concern the mechanical resistance to crash loads and the reaction time. Mechanical resistance: Integrated in the car, the actuator is submitted to a high amount of loads [3]. As any structure, the overall actuator performance is ruled by the weakest feature in the system chain. For the SMA tube actuator, this are the locking elements (see Fig. 4). The representation of the small locking elements in CAE was considered not feasible. Very small finite elements would have to be used to capture generic behavior, which is not possible due to the numerical limitations of finite elements explicit codes. This leads to a special attention in the development process of the locking elements. Dynamic pendulum tests have been conducted to ensure the required mechanical performance. The results are depicted in Fig. 7. In addition, quasi static tests have been performed and the results show the fulfillment of the required mechanical performance.

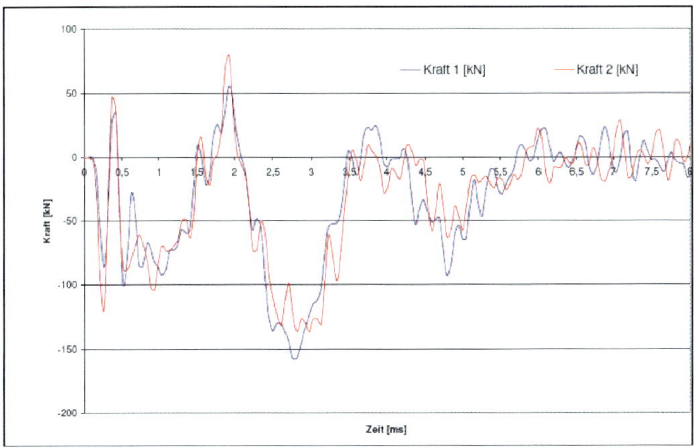

Fig. 7.   Pendulum test force output

Reaction time: The pre-crash system must deliver the trigger for the actuator within 200 ms before crash. The reaction time of the actuator is a go/no go for the entire pre-crash system developed within Aprosys. A delay in the actuator deployment leads to a malfunction of the complete protection system.

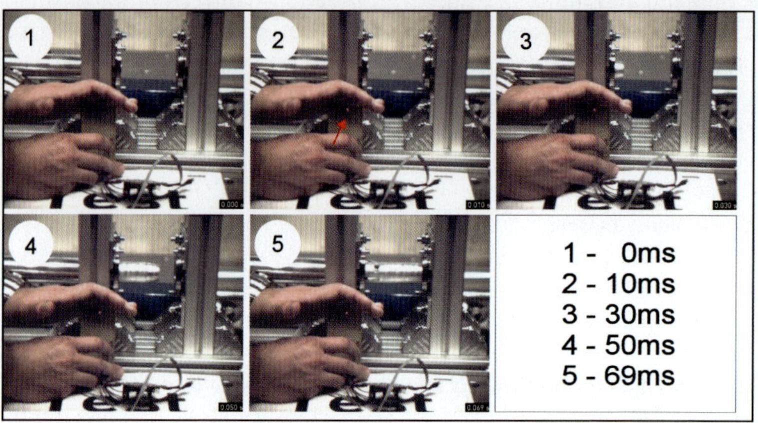

Fig. 8.   Tube actuator performance test, timing measured with high-speed-camera

The test shows a proper reaction time and a clear repeatability (see Fig. 8). It started at time 0ms, after 10 ms the trigger signal was send to the CAN interface of the actuator (illustrated by red LED, second picture). 20 ms later the bolt has significantly started to move and reached after another 39 ms its end-position. So all in all the response time of the tube actuator could be identified to 59 ms. The actuator moves the bolt ($\sim$400 g) for a travel of 115 mm in about 60 ms which is well in the project target 200 ms, a time span given by the sensor system.

5.2   Full Car Testing

Four crashes were performed referred herewith as: "reference", "deployed actuator", "active 01 actuator" and "optimized actuator". All tests were performed according to EuroNCAP side impact protocol with deformable barrier. The subjected cars were DC Neon model year '94-99. To capture intrusion and to deliver qualitative and quantitative information, a special crash measurement system has been installed. It consisted of on board cameras to capture from a close distance the actuator performance and several crash accelerometers to capture dynamically the deformation modes of the car structure. On top, pre and post geometrical measurement of the car body has been done to have the confidence in the results. The reference car contains no actuator at all; in the test car "deployed actuator", the actuator is integrated in a deployed state. The car "active 01" contains an actuator which is activated by an external sensor. This external sensor is a floorboard switch which the barrier wheel contacts within 200 ms before impact. This is the same case for the "optimized actuator". The optimization refers to mechanical changes and not to the mode of deployment.

As a summary, the static measurement results are listed below for all crashes with respect to the B-pillar belt line intrusion:
- ▶ "Reference" vehicle with a maximum intrusion of 488 mm;
- ▶ "Deployed actuator" vehicle with a maximum intrusion of 412 mm;
- ▶ "Dynamic actuator 01" vehicle with a maximum intrusion of 444 mm;

The DC Neon has a rather weak car structure, which explains the big amount of deformation on the three tests. However, a closer look on Fig. 9 points to the fact that the intrusion reduction in the "deployed actuator" amounts to more than 50 mm. This reduction is spread allover the door but especially where the occupant sits. This translates in valuable survival space!

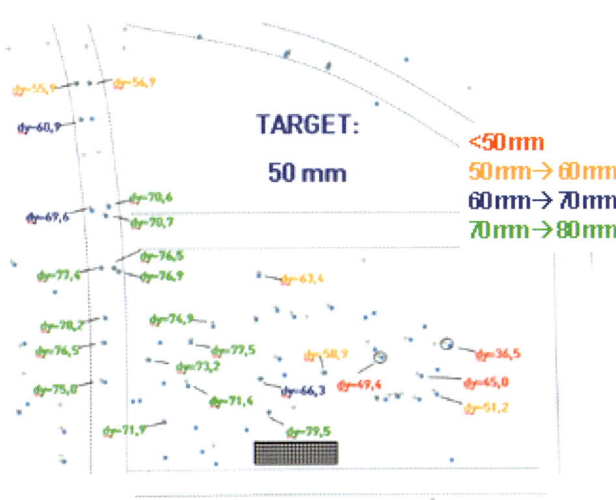

Fig. 9.   Performance of deployed actuator.

A striking improvement can be seen at the steering wheel, which confirms the reduction of intrusion all over the door and not only locally.

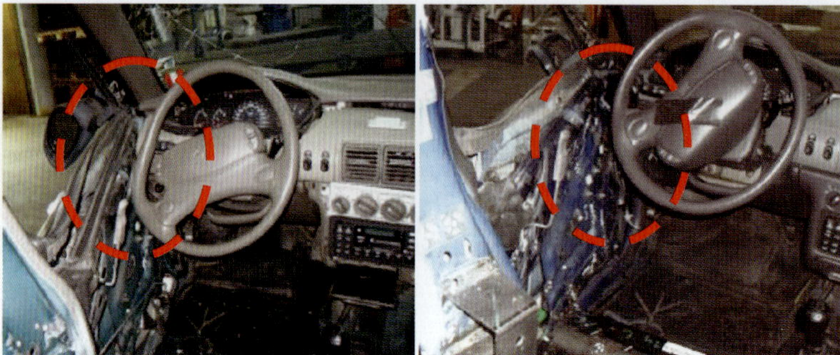

Fig. 10. Steering wheel deformation

In Fig. 10 left picture, the steering wheel of the reference vehicle is severely distorted in comparison with the deployed actuator vehicle (right picture) which is barely touched by the door. The intrusion reduction at the door on a spread region is one major gain of the actuator system integration in the car. Additionally, the actuator system brings overall benefits which enable to increase the side impact structural performance, such as:

▶ Restraint system has more space and time to deploy more efficiently;
▶ Contact time is delayed and with this contact forces are reduced as well as the other biomechanical values;
▶ Door is kept away from overriding the rocker; in this way, almost no contact with the seat in the lower region exists;
▶ Crash modes of the side structure are changed, the crash ends sooner.

Optimization Process:

As a consequence of the test session, a better understanding of the actuator mechanisms has been achieved. This leads further on to an optimization loop; several changes have been implemented in the actuator system and are summarized below:

▶ The bolt made out of steel to increase mechanical resistance; this leads to an increase of reaction time, however, after a mass optimization, the reaction time was brought to 90 ms which is in the target;
▶ The door box was made deformable, to be able to absorb crash energy and to lower the impact forces by this allow deformation in the system;
▶ The transversal tube has been reduced in diameter from 50 mm to 30 mm; however, the actuator housing was kept to initial dimensions;
▶ All reinforcements were reduced in thickness and stiffness;
▶ The seat was lifted into the mid height position.

At the B-pillar belt line, the optimized actuator has a maximum intrusion of 418 mm. This shows that by reducing the overall actuator stiffness, the same effect of intrusion reduction is obtained.

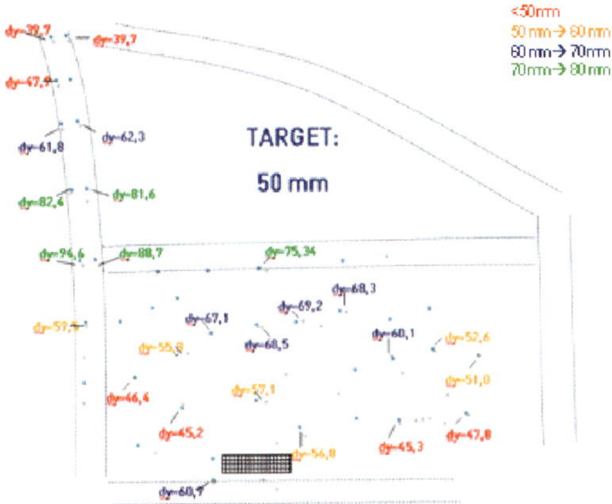

Fig. 11. Performance of the optimized actuator.

In Fig. 11, the map of intrusion reduction is depicted. As for the deployed actuator vehicle, the intrusion is reduced on a well spread region, giving survival space for the occupant and for the restraint systems.

The table below summarizes the intrusion values from the crash tests:

Test vehicle	Maximum intrusion
"Reference"	488 mm
"Deployed Actuator"	412 mm
"Active actuator 01"	444 mm
"Optimized actuator"	418 mm

6   Protection with Active Parts

To increase the crashworthiness of a car the two, structural integrity and occupant handling need to be taken into account. It is to be expected that only adaptive safety systems will be able to cope with demands arising from the large variety of crash scenarios (not covered by protocols). Adaptivity has two aspects, the one is to deploy parts or to bring them in position then to withstand crash loads. Or load paths are rearranged by connecting parts in a different way, again to offer a better protection level for a specific load case. As all those adaptive changes need some time to happen, trigger information is needed a bit before these adaptive measures are loaded by the crash. This is the reason why adaptive measures and pre-crash systems go together. Another key idea which was followed up in the concept presented here is the conversion of space. The safety devices when deployed, occupy space which in normal driving situations is not available for protection measures. This is especially beneficial in side impact where space for energy absorption is extremely small. Vehicle crashworthiness can be increased by integrating active safety systems targeting to act on the occupant directly.

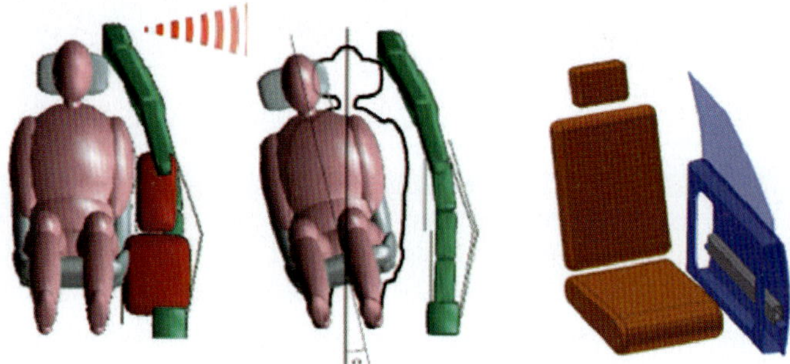

Fig. 12.   Possibilities for active interior.

Door trim, centre console and cockpits, all these interior structures today contribute passively to the occupant protection; but could contain active devices in the future. The door trim offers the possibility to implement active pads to control the contact stiffness (Fig. 12). The seat itself offers a very special active protection potential; it could rotate about X car axis or translate along Y car axis to move the passenger out of critical areas and gain survival space (Fig 12).

7   Summary

The work presented herewith was carried out within APROSYS SP6 which deals with the development of a pre-crash system with SMA actuator to reduce structural intrusion in side impact. The research activities were focused on the DC Neon, model year 1994-1999. The pre-crash system developed with APROSYS SP6 comprises: a pre-crash sensor (fusion of RADAR and video information) and an actuator system (a new mechanical system integrated in the seats and in door to create an additional load path for side impact; the active part consist of a SMA actuator and has the function to close the new load chain). The fusion of the pre-crash sensor delivers a pre-crash trigger signal 200 ms before the crash. The SMA actuator mainly consists of a bolt-spring mechanism which uses a SMA wire as indirect drive for release. The reaction time is 60 ms for the Aluminum bolt and 90 ms for the steel one.

A set of crash tests was organized in order to asses the real actuator performance and to give consistency to the complete development work. It consists of one reference crash test and two crash tests with actuator system integrated. An additional optimization loop was conducted which included measures to generally reduce the actuator system dimensions, therefore to ease the actuator integration. The actuator system delivers an improvement in structural intrusion of more than 50 mm up to 70 mm on a well spread region, but especially in the region where the occupant sits. Enlarging survival space directly where the occupant sits makes any car more crashworthy for all types of side impact scenarios which might not be covered by current crash protocols. The introduced pre-crash actuator system occupies space which is not available in normal driving situation i.e. the clearance in the door box to allow the glass to travel up and down. Making this additional space available for protection measures plus the opening of a new transversal load path from the earliest beginning on makes this actuator system so well performing.

References

[1] Käsgen, J., Seipel, B., Konzeptfindung, Auslegung und Konstruktion eines Verriegelungsmechanismus auf Basis von Formgedächtnislegierungen zur Optimierung des Crashverhaltens von Kraftfahrzeugtüren, diploma thesis, Fraunhofer Institute LBF, Darmstadt, Germany, (2004).

[2] Zimmerman, E., Muntean, V., Generic Fe-Simulation model for the layout of pre-crash actuators in side impact, Transfac Conference San Sebastian, Spain, (2006).

[3] Zimmerman, E., Muntean, V., Pre-crash actuator to improve car structural performance in side impact, EAEVC Conference Budapest, Hungary, (2007).

[4] Zimmerman, E. Dias, J., Pre-Crash Acuators for Side Impact, Presentation, Haus der Technik Essen, Germany; (2006).

[5] CAE - tools manuals: Radioss explicit solver, Hyperworks and EasiCrashRAD pre/post processing tools;

[6] www.destatis.de/ Statistisches Bundesamt Germany.

Eric Zimmerman, Vlad Muntean
Faurecia Interior Systems Germany
57584 Scheuerfeld
Germany
eric.zimmerman@faurecia.com
vlad.muntean@faurecia.com

Tobias Melz, Björn Seipel, Thorsten Koch
Fraunhofer-Institute for Structural Durability and System Reliability LBF
64289 Darmstadt
Germany
tobias.melz@lbf.fraunhofer.de
bjoern.seipel@lbf.fraunhofer.de
thorsten.koch@lbf.fraunhofer.de

Keywords: Aprosys, pre-crash system, actuator, high-speed, reversible, SMA

On the Feasibility of a Bistatic Radar for Detecting Bad Road Conditions

S. R. Esparza, O. Calderón, L. Landazábal
M. A. Deluque, J. V. Balbastre, E. de los Reyes
ITACA Research Institute

Abstract

In this paper a study on the feasibility of a bi static radar system for bad road conditions, including water, snow and ice, is presented. First of all, the most suitable frequency is chosen among all the frequencies available for such a system from a regulatory point of view. Then, the best mathematical model for predicting the electromagnetic behaviour of different layers of water, snow or ice over an asphalt basis is presented. The numerical study is focused on the reflexion attenuation (the difference between the direct and the reflected signal levels). Results for linearly polarized waves are presented and some useful design guidelines are presented.

1   Introduction

Traffic accidents are a huge problem from the social and the economic points of view. The European Traffic Safety Council presented on June 2007 a report supported by the European Commission [1] containing the most relevant figures related with road safety in Europe. According to this report, in Europe there is more than 1.000.000 traffic accidents each year in which over 40.000 people die and many more result injured. Traffic administrations all around Europe develop policies with the aim of reducing those figures. Campaigns devoted to reduce the average speed, the driver's alcohol consumption or the use of mobile phones while driving are implemented year after year. Very promising results have been obtained in highways and other major roads, but not in secondary roads which are narrower, worse paved and signaled and where bad weather conditions can severely affect the road safety.

Cars manufacturers include active safety elements in the most expensive brands and models, but the majority of the European citizens cannot pay the price of those high-tech items. Nevertheless, governments are somewhat responsible of citizen safety and they deploy sensors that detect bad road conditions, but these sensors are expensive and have some operational problems: they must be installed near a power supply line, they cover a rather small area (no more than 20x20 cm) and they are frequently removed by the effect of trucks and snowblowers.

In this paper the architecture of a bi-static pulsed radar system which is able to detect water, snow or ice spots over the asphalt pavement is presented. The system architecture is introduced in section 2, where the most relevant system parameters are presented. In section 3 the most suitable frequency is chosen both from the operational, regulatory and economical point of view. In section 4 the mathematical models available for characterizing the road surface are described and the most suiTab. one for this specific application is identified and then described with some detail in section 5. The numerical results obtained using the models described in section 5 are presented in section 6, while the conclusion relevant to the system design are described in section 7.

2 System Architecture

The system considered in this paper is described in Fig. 1. It is based on two antennas placed one in front the other covering all the road lanes. One of the antennas is connected to the transmitter circuit and the other one is connected to the receiver circuit. This architecture was already described in [2], where continuous wave (CW) bi-static clutter measurements were reported at L band. In this case, in order to get a better space resolution, a pulsed radar will be used instead of a CW one. The transmitter emits one pulse of width τ seconds each T seconds. The pulse width depends on the desired resolution ($\tau = 2\delta_z \cos\theta / c$, being δ_z the specified resolution, c the speed of light and θ the angle defined in Fig. 1). The power detected by the receiver antenna is then given by:

$$P_R = \frac{P_T G_T(\theta_1) G_R(\theta_2) \lambda^2 \sigma_c L}{2(4\pi)^3 R_1^2 R_2^2} \tag{1}$$

where P_T is the power radiated by the transmitter, $G_T(\theta_1)$ and $G_R(\theta_2)$ are the corresponding antenna gain, λ is the operating wavelength, σ_c is the road radar cross section, L is the attenuation due to signal absorption across the propagation path and R_1, θ_1 and R_2, θ_2 are defined in Fig. 1.

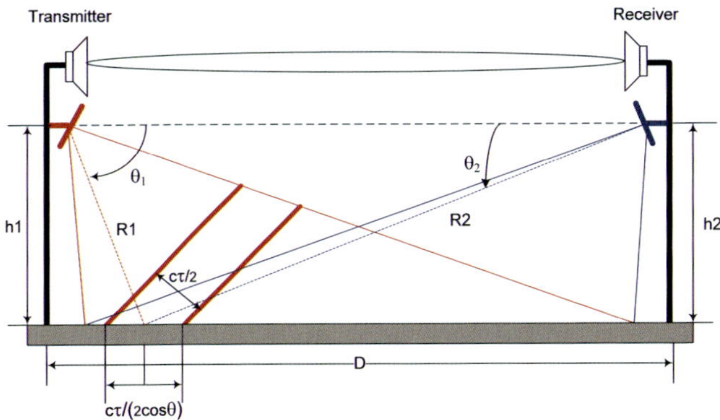

Fig. 1.   Bi static radar system architecture

The road radar cross section, σ_c, is computed from the corresponding surface density σ_C^0 :

$$\sigma_{C,pq} = \sigma_{C,pq}^0 A_{il} = 4\pi R_2^2 \left| R_{p,q} \right|^2 \tag{2}$$

where $R_{p,q}$ is the reflection coefficient on the road surface and the subscripts p,q stands for the different polarization combinations of the transmitter-receiver antennas. Thus, the received power is given by:

$$P_R = \frac{P_T G_T(\theta_1) G_R(\theta_2) \lambda^2 \left| R_{p,q} \right|^2 L}{(4\pi)^3 R_1^2} \tag{2}$$

As stated by the previous equation, the received power depends on the transmission losses, L. In order to derive an estimator independent from the weather conditions, two very directive antennas are built in the system. If both antennas are identical, the received power is given by:

$$P_{R,0} = \frac{P_{T,0} G_0^2 \lambda^2 L}{(4\pi)^3 D^2} \tag{2}$$

Therefore, the Reflection Attenuation can be definded as:

$$A_R = 4\pi \, \frac{P_R}{P_{R,0}} \, \frac{G_T(\theta_1)G_R(\theta_2)}{G_0^2} \, \frac{D^2}{R_1^2} \left| R_{p,q} \right|^2 \qquad (4)$$

All the parameters in the previous equation are known but the reflection coefficient, which depends on the particular combination of materials illuminated by the transmitter antenna (dry asphalt, water, snow, ice and all the possible combinations). Thus, this equation defines a valid estimator to identify whether there is some particular condition affecting road safety.

3　Frequency Selection

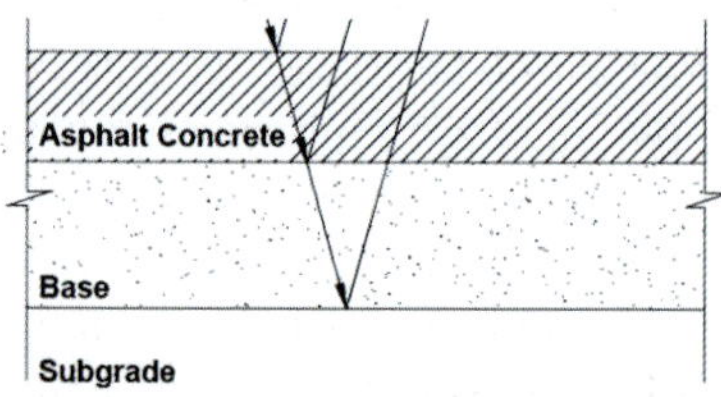

Fig. 2.　Road pavement structure

Road pavement is made of the superposition of several layers, as shown in Fig. 2. From top to bottom, the different materials present in any road are asphalt, a base made of sand and rocks and a subgrade of rocks. Over the outer layer may exist some additional layers of water, snow or ice when the weather conditions are especially bad. The electromagnetic permittivity of all those materials is listed in the following Tab. [3-8].

	2.45 GHz	9.9 GHz	12 GHz	18 GHz	27 GHz	64 GHz
Asphalt	5-j0.2	5-j0.2	5-j0.2	5-j0.2	5-j0.2	5-j0.2
Water	77.2-j9.2	62.6-j29.9	57-j33	45-j37	30-j34	11-j20
Snow	1.8-j0.15	1.7-j0.22	1.65-j0.21	1.63-0.2j	1.52-j0.18	1.4-j0.15
Ice	$3.16\text{-}j2\cdot10^{-4}$	$3.16\text{-}j6\cdot10^{-4}$	$3.16\text{-}j8\cdot10^{-4}$	$3.16\text{-}j8\cdot10^{-4}$	$3.16\text{-}j2\cdot10^{-3}$	$3.16\text{-}j4\cdot10^{-3}$

Tab. 1.　Complex dielectric constants of the different materials considered in the study at the available frequencies.

Electromagnetic waves can penetrate into a material a given distance, known as skin depth. The skin depth depends on both the material properties and the wave frequency. For the materials and frequencies considered in this work, the skin depth is listed in Tab. 2.

	2.45 GHz	9.9 GHz	12 GHz	18 GHz	27 GHz	64 GHz
Asphalt	0.06	0.02	0.015	0.012	0.01	0.005
Water	0.02	0.004	0.003	0.0025	0.002	0.0012
Snow	0.09	0.04	0.03	0.02	0.015	0.008
Ice	0.07	0.03	0.02	0.01	0.008	0.003

Tab. 2.   Skin depth (in meters) in the different materials considered in the study at the available frequencies.

Data included in Tab. 2 permit to extract some useful preliminary conclusions. First of all, since the asphalt layer usually is 10 cm depth, it can be considered semi-infinite and thus the effect of both the base and the subgrade will be discarded in all the models discussed later on this paper, no matter the frequency used. Moreover, the skin depth information is critical for choosing the most suitable frequency for the application. Although some hydroplaning phenomena have been reported for water layers thinner than a few millimetres, severe hydroplaning occurs when the water layer is wider than 8-10 mm [9]. Therefore, the most suitable frequency for the bad road condition radar sensor among all the considered in this work is 2.45 GHz. This frequency is also able to provide us with relevant information concerning the snow and ice scenarios and is, thus, the frequency chosen for this application. Obviously, the antenna size is much bigger at 2.45 GHz than at higher frequencies, but for roadside systems this is not as critical as it is for on vehicle systems. Moreover, the lower the frequency the devices and subsystems are cheaper.

4   Rough Versus Smooth Models

Since the system performance depends on the reflection of electromagnetic waves on the road surface, for an appropriate design it is necessary to develop a model describing the interaction between the high frequency electric field and the surfaces considered in this work. There is an extensive literature on the subject of scattering of electromagnetic waves from arbitrary surfaces. Depending on the surface roughness, the most relevant methods used to solve the scattering problem are [10-12]:

- ▶ To use the laws of the Geometric Optics (reflection and refraction). This approximation is valid for flat surfaces or when the curvature radius is very large compared with the wavelength and the surface roughness is negligible.
- ▶ The Kirchoff method, valid when both the rms value of the surface roughness and the average surface curvature are larger than the wave-

length.

▶ The small perturbation method, valid when both the correlation length and the rms value of the surface roughness are smaller than the wavelength.

Generally speaking, any surface is somewhat rough, but it can be considered flat when the roughness meets the following relationship, known as the Rayleigh criterion [10]:

$$h_R < \frac{\lambda}{8\sin\theta}$$

(5)

Fig. 3 shows the values of the Rayleigh criterion for waves between 1 and 10 GHz and incidence angles between 10 and 90 degrees and it can be seen that at 2.45 GHz it is greater than 4 cm.

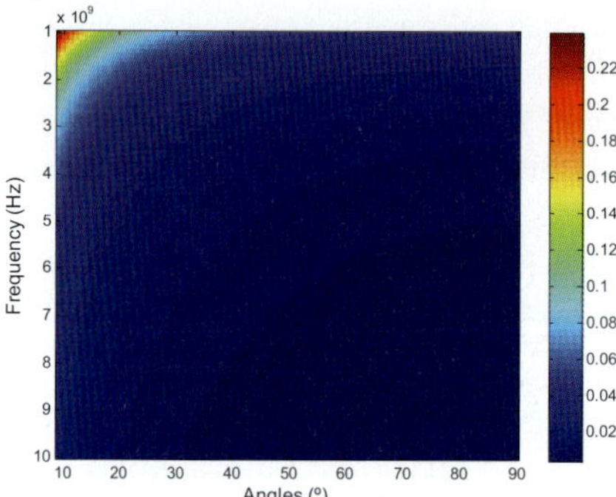

Fig. 3.   Rayleigh criterion in metres for frequencies between 1 to 10 GHz and incidence angles between 10 to 90°

The roughness of the surfaces used in pavements is a zero mean Gaussian random variable with rms value given in Tab. 3 [13].

	h_{rms}
concrete	*0.2 mm*
Smooth asphalt	*0.34 mm*
Rough asphalt	*0.9 mm*

Tab. 3.   Values of rms roughness of different kind of surfaces used in roads

The probability that the asphalt roughness is bigger than the Rayleigh criterion is given by [14].

$$P\left(h > h_R\right) = \int_{h_R}^{\infty} \frac{1}{h_{rms}} e^{\frac{h^2}{2h_{rms}^2}} \, dh = \frac{1}{2\sqrt{2}h_{rms}} \mathrm{erfc}\left(h_R\right) \qquad (6)$$

and it is absolutely negligible. Therefore, all the surfaces involved in this work can be considered smooth and thus analyzed using the simplest model among all the listed above: the Geometric Optics.

5 Geometric Optics and Transmission Line Model

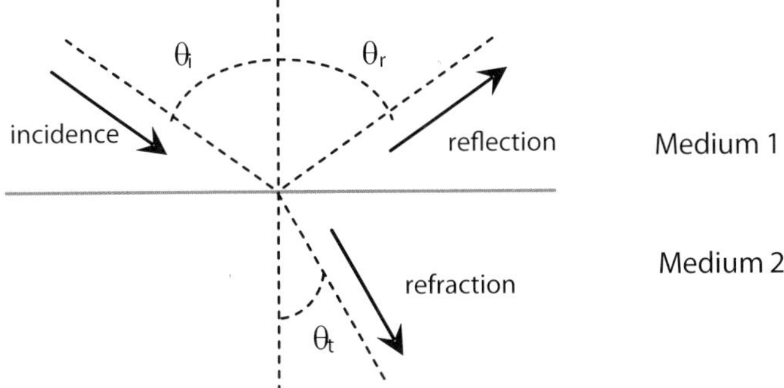

Fig. 3.   GO reflexion and transmission (Snell) laws

The Geometric Optics (GO) explains the reflection and refraction phenomena from surfaces separating two homogeneous media using two simple expressions that can be straightforwardly derived by applying the Maxwell Equations boundary conditions to the plane wave solution of the vector wave .equation at the medium discontinuity. The two GO basic equations are graphically summarized in Fig. 3 and they can be stated as follows:

- ▶ Reflection law: both the incident and reflected vector waves (the unit vector pointing to the wave propagation direction) are contained on the incidence plane (that defined by the wave vector of the incident wave and the unit normal vector to the separation surface at the incidence point). Moreover, the angle between the vector wave of the incident wave and the unit normal vector to the separation surface is the same than the angle between the latter and the vector wave of the reflected wave.

- ▶ Diffraction law: the angle between the vector wave of the incident wave and the unit normal vector to the separation surface and the angle between the latter and the vector wave of the transmitted wave is given by the Snell law:

$$\sqrt{\varepsilon_1\mu_1}\,\sin(\theta_i) = \sqrt{\varepsilon_2\mu_2}\,\sin(\theta_t) \tag{7}$$

where ε and μ are the electrical permittivity and the magnetic permeability of the corresponding medium, respectively.

In this work only linearly polarized waves are considered. The analysis is therefore performed for two orthogonal linear polarizations: Transverse Electric or TE (the electric field of the incident wave is normal to the plane of incidence) and Transverse Magnetic or TM (the magnetic field of the incident wave is normal to the plane of incidence whereas the electric field lies on it). The amplitude of the electric field reflected is related to the incident one by the Fresnel reflection coefficient, which is given by [15]:

$$R = \frac{Z(0) - \eta_0'}{Z(0) + \eta_0'} \tag{8}$$

To take into account the effect of the water, snow or ice layer on the surface road, the impedance is computed using the transmission line model [15]:

$$Z(0) = \eta'\frac{Z(d)\cos k'(d) + j\eta'\,senk'(d)}{\eta'\cos k'(d) + jZ(d)senk'(d)} \tag{9}$$

The above equations are polarization dependent. Indeed, $\eta' = \eta/\cos\theta$ for TE polarization and $\eta' = \eta\cos\theta$ for TM polarization, whereas $k' = k\cos\theta$ for both polarizations. In the previous expressions k stands for the material wavenumber ($k = 2\pi f\sqrt{\mu\varepsilon}$ being f the operating frequency), η for the material intrinsic impedance ($\eta = \sqrt{\mu/\varepsilon}$) and d for the medium depth.

6 Numerical Results

Although the polarization dependence, for perfectly smooth surfaces there is not polarization coupling, i.e. there are not crossed polarization reflected waves. Hence, the following figures show the attenuation reflection on the depth-angle plane for different materials covering the surface road and for the two orthogonal polarizations.

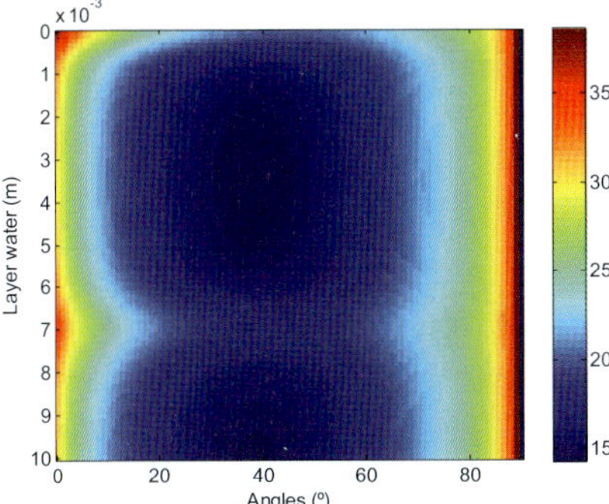

Fig. 5.   AR for water at 2.45 GHz, TE polarization

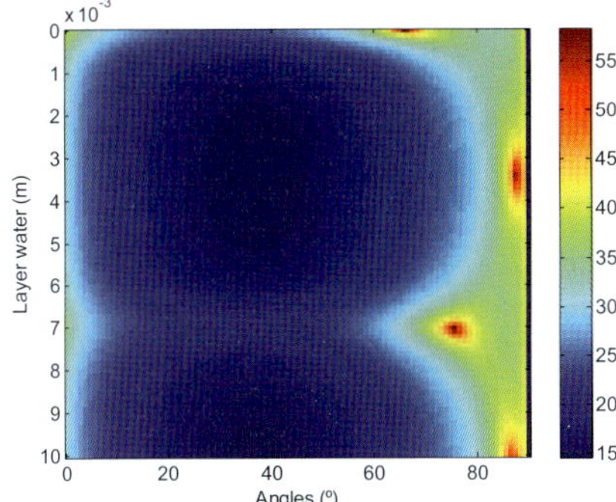

Fig. 6.   AR for water at 2.45 GHz, TM polarization

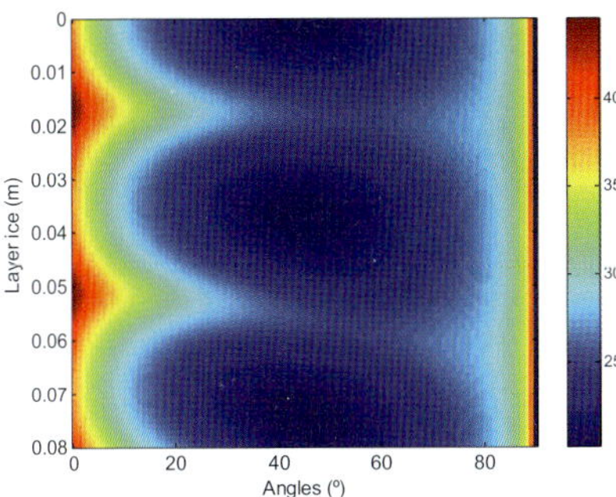

Fig. 7.   AR for ice at 2.45 GHz, TE polarization

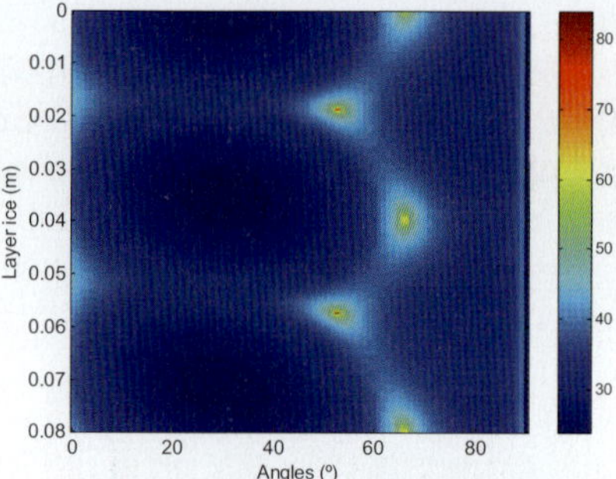

Fig. 8.   AR for ice at 2.45 GHz, TM polarization

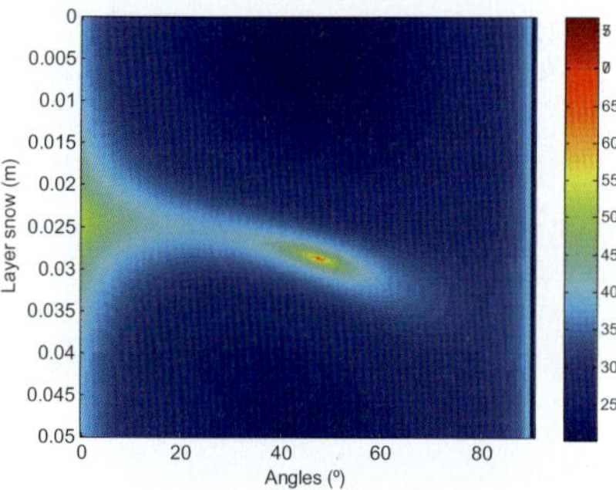

Fig. 9.   AR for snow at 2.45 GHz, TE polarization

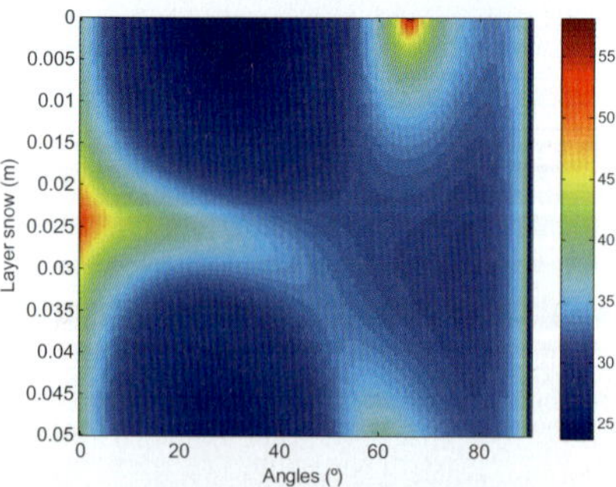

Fig. 10.   AR for snow at 2.45 GHz, TM polarization

The previous figures show that there is relevant difference between the reflection from a road surface covered by different kind of materials (water, snow and ice). This difference is enough to classify the surfaces depending on the material present over the asphalt layer. Since the impedance is a periodic function of the medium depth for a given angle, the response is somewhat periodic, thus the unambiguous detection area must be identified if a depth estimation will be performed on the basis of the attenuation information. Generally speaking the TE response is clearer and presents a larger unambiguous area.

7 Conclusions

The bi static radar architecture proposed in this paper provides an estimator valid for detecting the presence on the road surface of different materials that can affect the traffic safety (water, snow or ice). This estimator is independent from transmission losses. The system response is somewhat periodic because the impedance is also periodic, and hence the unambiguous area must be identified. In this sense, results for TE polarization are clearer and show a larger unambiguous area (which includes a range of material depth relevant from the road traffic point of view). Therefore, the system proved to be feasible and further investigation must be carried out in order to design a practical prototype.

8 Acknowledgements

This work has been supported by the Spanish Ministry of Industry, Tourism and Commerce project IAP 560410-2008-59.

References

[1] http://ec.europa.eu/transport/roadsafety/publications/projectfiles/supreme_en.htm "Summary and Publication of best Practices in Road Safety in the Member States of the European Union and in Norway and Switzerland" (SUPREME), European Traffic Safety Council (ETSC).

[2] Radar Series, Vol 5. Ed. David D. Barton, The Artech Radar Library, 1977.

[3] Edward J. Jaselskis, Jonas Grigas and Algirdas Brilingas, "Dielectric Properties of Asphalt Pavement" Journal of materials in civil engineering, September 2003.

[4] Christian Mätzler and Urs Wegmüller. "Dielectric properties of fresh-water ice at microwave frequencies". Journal of Physics D: Applied Physics.Vol. 20. April 1987.

[5] Gary Koh. "Dielectric Properties of Ice and Snow at 26.5 to 40 GHz". Geoscience and Remote Sensing Symposium International 1992.

[6]   ITU-R P.527-3. "Electrical Characteristics of the Surface of the Earth". 1992.

[7]   Martti T. Hallikainen et al. "Dielectric Properties of Snow in the 3 to 37 GHz Range". IEEE Trans. on Antennas and Propagation, Vol. 34. Nov. 1986.

[8]   Martti E. Tiuri et al. "The Complex Dielectric Constatnt of snow at Microwave Frequencies" IEEE Journal of Oceanic Engineering, Vol 9, No 5, Dec. 1984.

[9]   "Aquaplaning. Development of a Risk Pond Modelfrom Road Surface Measurements, Sara Nygårdhs", PhD Dissertation, Linköping University, Sweden, 2003.

[10] Petr Beckmann and André Spizzichino. "The Scattering of electromagnetic waves from rough surfaces". Editorial Artech House, Inc. 1987

[11] Fawwaz T. Ulaby et al. "Microwave Remote Sensing – Active and Passive". Vol. II.Editorial Addison-Wesley Publishing Company, Inc. 1982.

[12] Fawwaz T. Ulaby et al. "Radar Polarimetry for Geoscience Applications". Editorial Artech House, Inc. 1990.

[13] Eric S. Li and Kamal Sarabandi, "Low Grazing Incidence Millimetre-Wave Scattering Models and Measurements for Various Road Surfaces", IEEE Trans. on Antennas and Propagation, Vol. 47, May 1999

[14] "Probability, Random Variables, and Stochastic Processes", Athanasios Papoulis, McGraw-Hill Kogakusha, Tokyo, 9th edition,

[15] "Electrodinámica para Ingenieros", Luis Nuño et *al*, Intertécnica, Valencia, 2005

Silvia Rocío Esparza Becerra
ITACA Research Institute
Universidad Politécnica de Valencia
Camino de Vera s/n
46022 Valencia
Spain
sesparza@itaca.upv.es
ocalderon@itaca.uvp.es
leilandu@itaca.uvp.es
milangela@itaca.uvp.es
jbalbast@itaca.upv.es
ereyes@dcom.uvp.es

Keywords: bad road conditions detection, bi-static radar, linearly polarized TE waves

Cooperative Intersection Safety – The EU project INTER-SAFE-2

B. Roessler, K. Fuerstenberg, IBEO Automobile Sensor GmbH

Abstract

Today most so called 'black spots' have been eliminated from the road networks. However, intersections can still be regarded as black spots. Depending on the region and country, from 30% to 60% of all injury accidents and up to one third of the fatalities occur at intersections. This is due mainly to the fact that accident scenarios at intersections are among the most complex ones, since different categories of road user interact in these limited areas with crossing trajectories.

1   Introduction

About 30% to 60% (depending on the country) of all injury accidents and about 16% to 36% of the fatalities are intersection related [1]. In addition, accident scenarios at intersections are amongst the most complex (different type of road users, various orientations and speeds).

The INTERSAFE-2 project aims to develop and demonstrate a Cooperative Intersection Safety System (CISS) that is able to significantly reduce injury and fatal accidents at intersections.

Vehicles equipped with communication means and onboard sensor systems will cooperate with the road side infrastructure in order to achieve a comprehensive system that contributes to the EU-25 and "zero accident" vision as well as to a significant improvement of efficiency in traffic flow and thus reduce fuel consumption in urban areas. By networking state-of-the-art technologies for sensors, infrastructure systems, communications, digital map contents and new accurate positioning techniques, INTERSAFE-2 aims to bring Intersection Safety Systems much closer to market introduction.

2   System Description

The novel Cooperative Intersection Safety System combines warning and intervention functions demonstrated on three vehicles: two passenger cars and one heavy goods vehicle. Proof of concept with extended testing and user trials will be shown based on these demonstrators. Furthermore, a simulator will be used for additional R&D.

The functions use novel cooperative scenario interpretation and risk assessment algorithms.

The cooperative sensor data fusion is based on:
- state-of-the-art and advanced on-board sensors for object recognition and relative localisation,
- a standard navigation map

and information supplied over a communication link from
- other road users via V2V if the other vehicle is so equipped
- infrastructure sensors and traffic lights via V2I if the infrastructure is so equipped

to observe the complex intersection environment. Fig. 1 shows a sketch of the INTERSAFE-2 system.

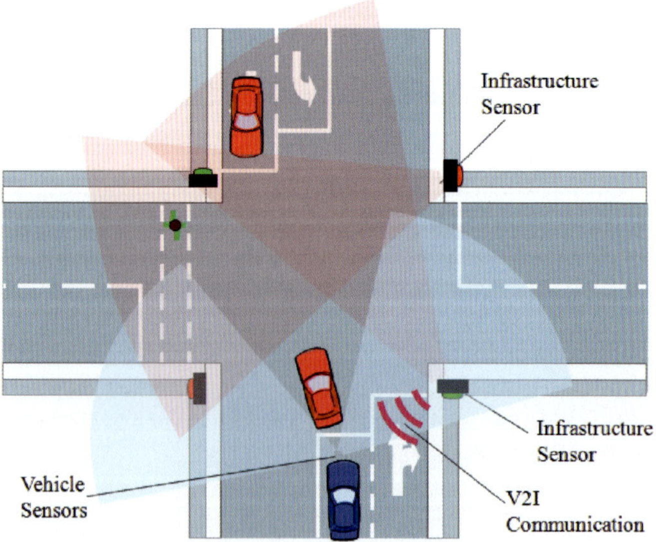

Fig. 1.   The INTERSAFE-2 system

3 Technical Objectives

In the following the technical objectives that are in the focus of the INTERSAFE-2 project are described.

3.1 Demonstrators

Three demonstrator vehicles will be set up which introduce new functionalities ranging from warning systems to active vehicle intervention.

The vehicles will be equipped with
- ▶ on-board sensors that are able to perceive the environment and
- ▶ communication means that complement the overall environment perception from the infrastructural side.

The vehicle support functions comprise:
1. Right turning assistance for heavy goods vehicles and passenger cars to prevent accidents especially with VRUs at intersections
2. Left turning assistance to prevent left-turn related fatal accident types
3. Crossing assistance to prevent accidents that may occur when trying to cross a road with priority at an intersection
4. Traffic light assistance to prevent red light running accidents and to improve traffic flow.

3.2 Bidirectional V2X Communication and Cooperative Sensor Data Fusion

Secure V2V and V2I communication and advanced sensor data fusion techniques will result in reliable perception of environmental data and scene interpretation. Using data not only from vehicle sensors but also from additional data sources (like traffic lights or infrastructure sensors) an advanced and more reliable interpretation of the intersection scene can be realised. Furthermore, real-time traffic management applications become possible. This will contribute on the one hand to traffic safety and on the other hand, to less turbulent traffic flow and therefore, to decreased energy consumption and reduced emissions. Another result will be that standardisation of V2X communication will be enhanced by the cooperative approach developed in this project.

3.3 Infrastructure Monitoring

An overall intersection environment perception even under challenging conditions like occluded field of view will be analysed and the improvement will be demonstrated. One intersection will be equipped with infrastructure sensors like Laserscanners and video cameras. This will enhance the perception area for the Intersection Safety System at wide intersections.

3.4 Relative Intersection Localisation

Based on state-of-the-art technologies, accurate intersection localisation will be performed by the demonstrator vehicles in order to have a basis for further situation analysis. The focus on state-of-the-art in sensor technologies and standard map data enables a more rapid market introduction of such Intersection Safety Systems.

3.5 Intersection Object Tracking and Classification

Objects within the intersection will be reliably classified and tracked both from the demonstrator vehicles and from the equipped intersection. Based on relative intersection localisation and on digital maps, a background elimination process will be performed. Thus, the focus is on road users present in the foreground at the intersection, which will be robustly tracked and classified.

3.6 Cooperative Intersection Scenario Interpretation, Risk Assessment, Warning and Intervention Strategies

Based on the above-mentioned techniques and on improved sensor data fusion, a comprehensive scenario interpretation becomes possible for the demonstrator vehicles. The cooperative environmental model will be established in the demonstrator vehicles based on information from the on-board sensors and from the infrastructure transferred via V2X communication. Thus, a cooperative scenario interpretation and risk assessment can be performed in the demonstrator vehicles. Once the situation is clearly analysed, the warning or intervention strategies can be accomplished.

4   Strategy

Existing accident analysis (e.g. INTERSAFE) has identified the importance of two accident categories in particular:

1. Collisions with oncoming traffic while turning left and
2. Collisions with crossing traffic while turning into, or crossing over an intersection

Thus, development within INTERSAFE-2 will focus on these scenarios as well as others identified during a renewed accident review.

The perception is based on three pillars:

- ▶ V2V and V2I communication
- ▶ on-board sensors, like Laserscanners, Video or Radar in combination with a digital map
- ▶ infrastructure sensors, like Laserscanner or Video and data like the traffic light status.

Based on the information present in the vehicles from

- ▶ the on-board perception systems
- ▶ infrastructure monitoring devices via V2I communication
- ▶ other vehicles via V2V communication

Cooperative risk assessment, warning or intervention strategies will be developed and evaluated. The development starts in the driving simulator in order to be safe, cost- and time-effective. The results are then transferred to the demonstrators to be adapted under real driving conditions.

4.1   Benefit

V2V/V2I communication as a standalone system would be able to deal with the majority of the above mentioned scenarios, if every vehicle and every intersection were equipped. The equipment rate is a crucial topic for a stand-alone V2V communication system. An Intersection Safety system that is based solely on V2V communication will not increase the traffic safety significantly, if just a few vehicles are equipped. As long as there is no initial benefit for the drivers who equip their vehicles first with a communication module the penetration rate and thus the safety benefit will be quite low (Fig. 2).

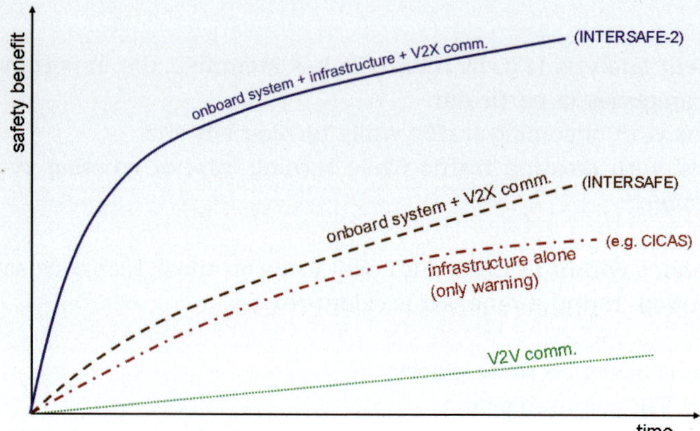

Fig. 2.   System benefit for different Intersection Safety systems

Therefore, there is a need for a cost-effective solution.

By introducing on-board sensors there will be an immediate benefit for every driver who buys such a system. However, on-board systems alone will address just a subset of the relevant situations. Equipping a small number of (the most severe) intersections with infrastructure sensors and V2I communication will provide an initial safety benefit for those who equip their vehicles with V2X communication. Thus, a comprehensive cooperative approach will be cost-effective, address the relevant scenarios and act as an enabler for the introduction of V2X communication.

5   State-of-the-Art and Progress beyond

The subproject INTERSAFE (Intersection Safety) of the European Integrated Project PReVENT aimed to increase the safety at intersections. The running period of the project was from 2004 to 2007. The objectives of INTERSAFE were

- ▶ Development of an Intersection Driver Warning System based on bidirectional V2I communication and on-board sensors.
- ▶ Demonstration of intersection driver assistance functions based on sensors and communication for relevant scenarios in demonstrator vehicles.
- ▶ Demonstration of advanced intersection driver assistance functions in a driving simulator.

INTERSAFE was based on two parallel approaches:

Basic System Approach

The basic system approach started with an accident analysis for intersection related accidents. Extraction of the most relevant scenarios led to the sensor requirements for a basic Intersection Safety System. Based on these requirements a demonstrator vehicle was equipped with suitable state-of-the-art sensors that fulfilled the requirements. The goal was to demonstrate driver assistance at intersections.

Advanced System Approach

The advanced system approach also started with the accident analysis like in the basic system approach. The basis for the function development was the assumption of an ideal sensor system. This was realized in an advanced driving simulator. The results were sensor requirements like range, coverage angles or accuracy for an advanced Intersection Safety System.

5.1   INTERSAFE-2 Progress

Due to its extended cooperative approach INTERSAFE-2 is able to cope with even more critical scenarios at intersections than the predecessor projects INTERSAFE or SAFESPOT. Fig. 3 shows a scenario observed with the INTERSAFE system. A building masks the line of sight between the equipped vehicle, the oncoming vehicle and the person on the pedestrian crossing. Fig. 4 shows the same scenario for the INTERSAFE-2 system. With its infrastructure sensors and V2I communication facilities, the system is able to detect all road users present at the intersection and thus assist the driver.

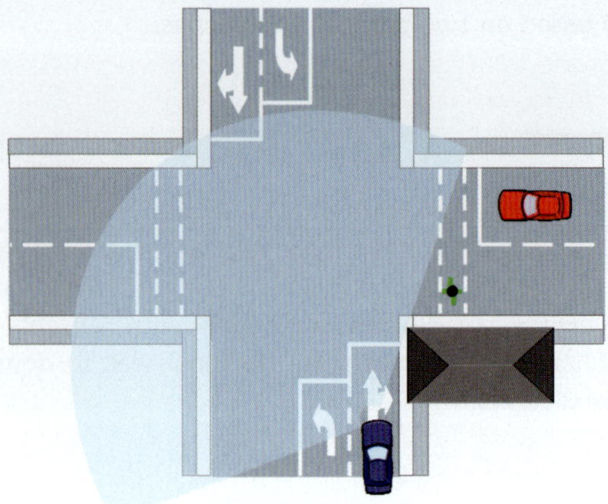

Fig. 3.   Field of view realised by the INTERSAFE project

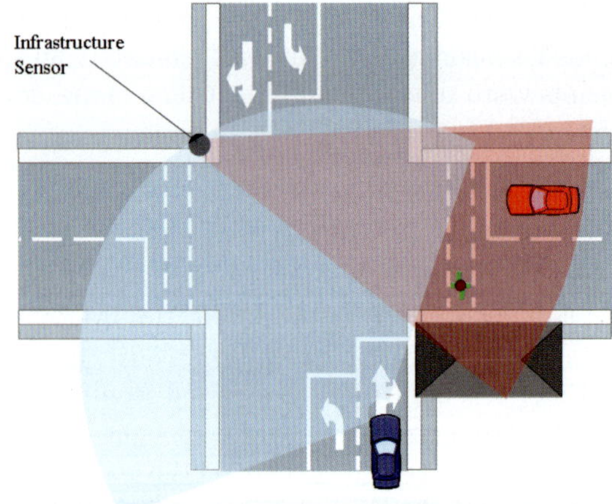

Fig. 4.   Field of view realised by the INTERSAFE-2 project

6   Impact

The INTERSAFE-2 projects has different ITS impacts both on traffic safety, on V2X communication as well as on cleaner environment.

6.1　Impact on Traffic Safety

As a result, the deployment of the INTERSAFE-2 system could provide a positive safety impact of 80% with respect to injuries and fatalities for intersection related accidents. Thus a total safety benefit of up to 40% of all injury accidents and up to 20% of all fatalities in Europe is possible.

6.2　Impact on Communication

The utilization of V2X communication for CISS at a small number of equipped intersections would boost the overall market penetration of communication in vehicles since the benefit for those who buy first could be experienced at every equipped intersection (see Fig. 5).

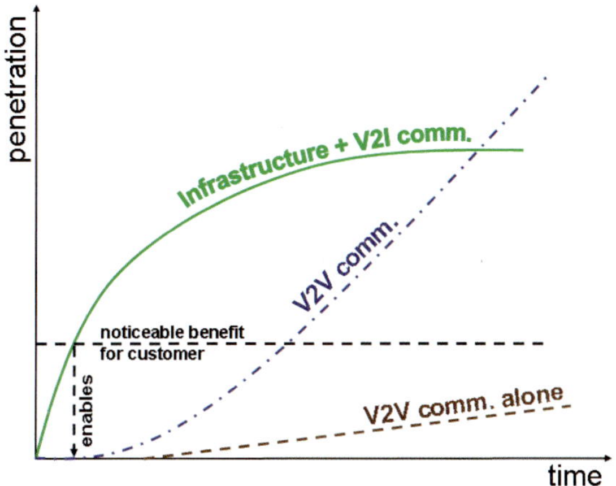

Fig. 5.　Possible boost of the penetration rate for V2V communication

6.3　Cleaner Environment

The deployment of the INTERSAFE-2 project would affect the environment by allowing a better traffic flow with more parameters than today. In addition the speed of the vehicles can be adapted when approaching an intersection due to the recommendations of the system or extended ACC knowing the traffic signal status in advance.

The greatest environmental effects are achieved, however, by avoiding most of the intersection related accidents. Since the traffic flow at intersection – which is usually the bottleneck of urban road transport - is significantly improved.

The project has impacts on the environment also in an indirect way, allowing an optimisation of fleet management systems, and implementing in the near-future warning and speed management systems such as variable, situational speed advisory communication and limits or in-vehicle speed management.

7 Conclusion

INTERSAFE-2 [2] is the follow-up project of the quite successful EU funded project PReVENT-INTERSAFE. It started in June 2008 and will end in May 2011. During this period a comprehensive Cooperative Intersection Safety System will be developed which is able to significantly reduce injury and fatal accidents at intersections. By using state-of-the-art technologies a much closer market introduction of Intersection Safety is aspired.

The INTERSAFE-2 consortium wants to thank the European commission for the support in this project.

References

[1]   PReVENT-INTERSAFE consortium, D40.4 Requirements for intersection safety applications, 2005.
[2]   INTERSAFE-2 website: www.intersafe-2.eu

Bernd Roessler, Kay Fuerstenberg
IBEO Automobile Sensor GmbH
Merkurring 20
22143 Hamburg
Germany
bernd.roessler@ibeo-as.com
kay.fuerstenberg@ibeo-as.com

Keywords: intersection, cooperative driver assistance systems

Intersection Safety for Heavy Goods Vehicles

M. Ahrholdt, Gr. Grubb, E. Agardt, Volvo Technology Corporation

Abstract

Traffic of heavy goods vehicles is an important component of transport in today's cities. Much progress has been made to make it as safe and efficient as possible. In this paper, safety of heavy goods traffic at intersections is investigated. Special regard is given to the safety of vulnerable road users (VRUs), such as pedestrians and cyclist.

In the project INTERSAFE-2, solutions are being developed to reduce accident risks at intersections. One situation identified to be particularly dangerous for VRUs is the right turning of the heavy goods vehicle. This scenario will be specifically addressed by he intersection safety application developed on the Volvo Technology demonstrator vehicle in INTERSAFE-2. This paper describes the requirements for resolving the right-turning accident scenario, and will illustrate the resulting function concept of the demonstrator.

1   Introduction

Driver assistance systems are an important factor for increasing traffic safety even for heavy goods vehicles. Functions like Adaptive Cruise Control (ACC) or Lane Departure Warning (LDW) have been established some years ago and provide good support to the truck driver, mainly in highway and even rural environments.

In urban traffic, however, with complex traffic situations and a high traffic density, additional driver support systems are needed. Within the European research initiative INTERSAFE-2 [1], Volvo is investigating how to support truck drivers in urban intersection situations. Urban traffic of heavy goods vehicles is a vital factor in today's cities to ensure distribution of supplies or collection of waste. A lot of efforts are currently undertaken to make this traffic as safe and eco-friendly as possible.

A special focus for urban heavy traffic is to deal with the limited visibility in the very near vicinity around the host vehicle. Volvo has previously — as a result of the IP PReVENT sub-project APALACI [2] — demonstrated a system observing the area directly in front of the truck, inhibiting the start of the vehicle in case there is a pedestrian present immediately in front of the vehicle [3].

The INTERSAFE-2 project is a European research initiative, started in June 2008 and will be ongoing until May 2011. It consists of partners from 6 European countries, amongst them 3 vehicle OEMs, 3 suppliers, 1 traffic signal manufacturer and 4 research bodies.

Within INTERSAFE-2, the safety of vulnerable road users (VRUs) in intersection situations will be addressed. For a truck turning right exists a large blind spot on the right side of the vehicle, that is not directly visible to the driver, and often cannot be well observed through mirrors as well.

This paper will give an overview of the addressed intersection scenario, motivate the work by evaluation of accidentology studies, describe the particular situation of blind spots at heavy goods vehicles, give a scope of previous work in the field, and describe the safety application to be developed in INTERSAFE-2 in an overview.

1.1 Addressed Scenario

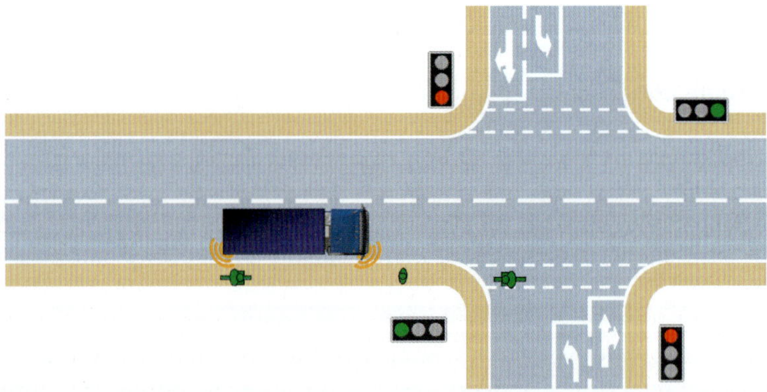

Fig. 1.   Turning Scenario

The vehicle safety function for trucks being developed in INTERSAFE-2 concentrates on one specific scenario, right turning in intersections.

For the driver of a truck, objects and road users are not well visible on the right side of the vehicle, which has caused a number of serious accidents. The issue with this type of accidents is that even small collisions easily lead to the death of a vulnerable road user due to unequal distribution of masses.

Typical attributes of the addressed situations can be characterized as:
- ▶ The truck is turning right in an urban environment
- ▶ The vehicle speed is low.
- ▶ The work load of the truck driver is at a quite high level, both during and before the turning manoeuvre
- ▶ The VRU is not visible in the mirrors, alternatively not observed by the driver.
- ▶ The VRU is not aware that the truck driver does not observe him.
- ▶ The VRU may even have the right of way.

This type of accidents has previously been addressed, first in the 1990ies by improving passive safety through mandatory deflector frames, reducing the risk of the VRU falling in front of the vehicles rear axis. Then in 2003 and 2005, regulations for mirrors and visibility have been toughened, see section 1.3.

1.2   Accidentology and User Studies

A close investigation of truck accidents is done by the Volvo Accident Research Team [4]. The analysis ranks intersection accidents with turning vehicles with 20 % of casualties among the most relevant scenarios with VRUs, see Table 1. This table is based on the investigation of 127 fatal accidents between vulnerable road users and heavy trucks in urban environment during 1992-2003.

A user interview study has been made in the INTERSAFE-2 project in cooperation with the Institut für Kraftfahrzeuge (ika) of the RWTH Aachen. Out of 44 truck drivers, 72 % of the interviewees stated overlooking pedestrians in right turns and 85 % overlooking cyclists as the two most critical situations when driving in intersections. More results from this user survey are available in [5].

Unprotected	Relative share	Description	Traffic environment	Speed	Typical cause
1. Truck- unprotected collision, truck front vs. unprotected when taking off	10%	Collision with unprotected, frontal part of truck in low speed manoeuvring or starting from stationary e.g. at crossroads or zebra crossings. (mostly pedestrians)	Urban areas, daylight.	Low	- Limited visibility; front of cab, right or left side of cab. - Limited driver knowledge of blind spots. - Lack of communication with other road user. - Driver stressed, inattentive or distracted.
2. Truck- unprotected collision, truck vs. un-protected when reversing	20%	Collision with unprotected, rear parts of truck/ trailer in low speed reversing. Distribution trucks when delivering goods/ garbage collectors. (mostly pedestrians)	Urban areas, daylight. Most often elderly people, but also children	Low	- Limited visibility rear of truck. - External acoustic warning signal not enough. - Working routines not good enough. - Lack of knowledge. - Driver stressed, inattentive or distracted.
3. Truck- unprotected collision, cross road collision	20%	Collision with unprotected at intersection, moderate or high speed. (pedestrians, bicycles, mopeds)	Urban areas. The driver is often surprised and not prepared for the sudden situation.	High	Other road user: - Lack of judgement. - Misjudgement of speed of truck. - Drugs. - Truck speed above allowed.
4. Truck- unprotected collision, truck side or front vs. unprotected	20%	Collision with unprotected, most often right turn. (pedestrians, bicycles, mopeds)	Urban areas, low speed, narrow city streets, often a parallel bicycle lane or a zebra crossing.	Low	- Limited visibility side of truck. - Lack of knowledge about the blind spots. - Driver stressed, inattentive or distracted
5. Truck- unprotected collision, lane driving	10%	Collision with unprotected. Lane driving, e.g. lane change, merge, cut in, not keep in lane. (bicycles, mopeds, motorcycles)	Urban areas.	Medium	- Lack of visibility - Driver stressed, inattentive or distracted.
6 Other truck- unprotected collision,	20%				

Tab. 1.   Truck accidents causing injuries to unprotected road users. [4]

1.3 Visibility

Visibility is a crucial topic for trucks, particularly in dense and urban environments. To the close front and to the right, the driver has a very limited direct sight. Mirrors are used to give additional fields of view. Originally, they were regulated in the EC directive 71/127/EC, which has been improved with extended field of view requirements in 2003 and 2005 by directive 2003/97/EC [6] and 2005/27/EC [7], respectively . They have been used to construct the visibility map shown in Fig. 2. The visibility and blind spot regions are defined for the ground plate as depicted.

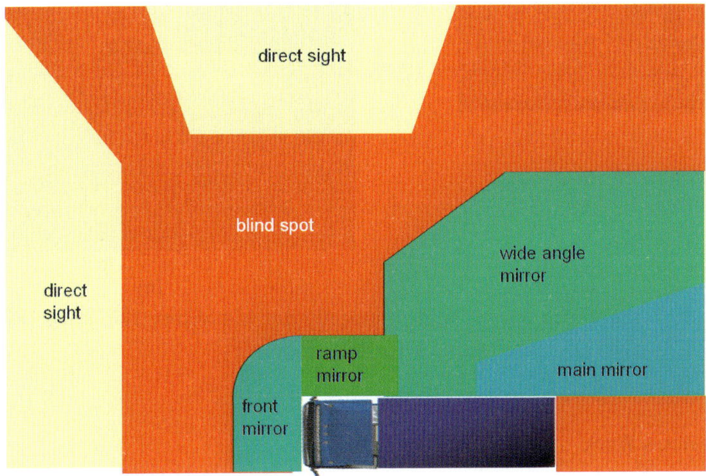

Fig. 2.   Blind spot and areas observable by mirrors (approximately correct scale)

Fig. 3(a) shows the set of three mirrors used on the right hand side of the vehicle. However, the efficiency of indirect sight through mirrors is limited: Firstly, mirror functionality depends significantly on the correct adjustment of the mirrors. Particularly the ramp and wide angle mirror are sensitive to small mis-adjustments, which will lead to a decreased observation area, see Fig. 3(b). Secondly, all mirrors create an additional blind spot themselves. Particularly along the vehicle's A pillar, the set of mirrors easily can cover a car-size object in some distance. And thirdly, the driver has to focus on a quite complex traffic situation in urban intersections, such that he might overlook objects in the mirror, particularly since the objects may be highly distorted in wide-angle mirrors. Note that the legal requirements of a heavy goods vehicle include a set of six mirrors in total (left, left wide angle, right, right wide angle, ramp and front). There also is a proposal of improving the field of view by another wide angle mirror, so-called Dobli mirror [8]. However, opinions have been two-fold, the concerns on additional mirrors stated above apply even here.

Improved mirrors have reduced the blind spot, but not totally eliminated it, and objects in the mirror may still be overlooked. Therefore, in this paper, we focus on systems for also warning the driver in critical turning situations.

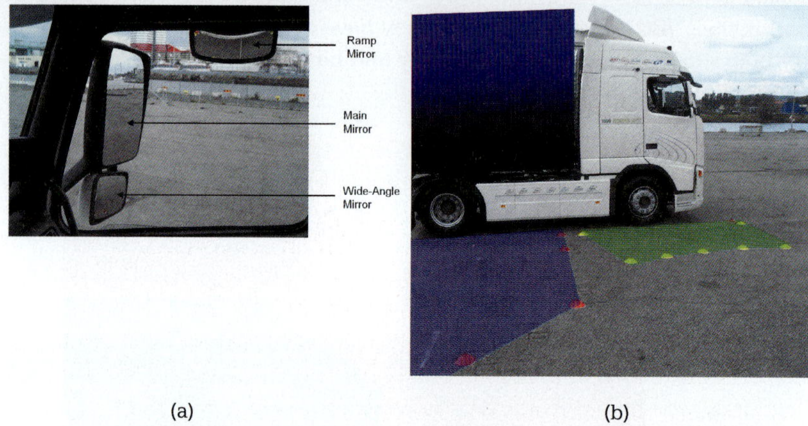

(a)   (b)

Fig. 3.   Right side mirrors and practically derived observation areas (with typically misadjusted mirror settings).

1.4 Previous Work

Safety of pedestrians during truck turning manoeuvres has been investigated previously based on a radar approach within the German program MoTiV (1996-2000) [9], based on a first generation laser scanner, and within the European initiative PROTECTOR (2000-2003) [9], based on 24 GHz radar sensors. For the work described in this paper, radar sensors have not been considered suitable, due to the limited detection capabilities in the very close range in terms of resolution and angular estimation.

In [10], a system has been proposed for the use of an array of ultrasonic sensors to monitor the close environment in front of and next to the vehicle. However, this solution is currently limited to work only when the host vehicle is stationary.

Within the APALACI project in FP6 IP PReVENT, pedestrian safety in the close vicinity of trucks has been investigated for the start-inhibit functionality. Vulnerable road users have been detected in the blind spot area in front of the truck by a combination of short-range radar and stereo vision [2] [3]. In the approach demonstrated by Volvo Technology, the vehicle was inhibited from driving away when a VRU was standing in the blind spot directly ahead of the vehicle.

The previous INTERSAFE project [11] has addressed different aspects of intersection safety, but not covered the particular aspects of heavy goods vehicles during turning manoeuvres.

2 Safety Application

Particularly addressing accidents between vulnerable road users and heavy goods vehicles, Volvo Technology is currently together with partners of the INTERSAFE-2 project developing a driver assistance function to prevent the type of accident described in section 1.1.

At the time of writing of this contribution, definition of the requirements, functionality and the perception system have been concluded. However, integration, development of the system and final tests remain, such that no final results can be presented here.

2.1 Perception System

The perception system to be developed in INTERSAFE-2 comprises a set of different components:

▶ A front and side vicinity monitoring system, consisting of a set of laser scanners and ultrasonic sensors.
▶ A forward vision module to recognize intersection land marks and road markings, and thus localize the host vehicle with respect to the intersection.
▶ A wireless communication link between traffic light and vehicle, to inform the vehicle system about signal phases and push-button requests of pedestrians or cyclists, using the proposed IEEE 802.11p standard for vehicle-to-vehicle/infrastructure (V2X) communication.
▶ An ADAS map, so-called e-horizon, comprising information such as intersection location, number and angle of connecting roads, presence of traffic lights and right-of-way regulations.

An overview of the perception system is shown in Fig. 4. The system is built up scalably, such that functionality shall also be achieved in intersections not equipped with a communication device or with missing map information.

A crucial point for the functionality is to cover the close vicinity of the vehicle to detect both moving and stationary objects. This is why laser scanner technology is employed to be able to classify objects. Ultrasonic sensors enable a wider lateral coverage area.

For system development and tests, one public intersection in Gothenburg will be equipped with a V2X communication unit.

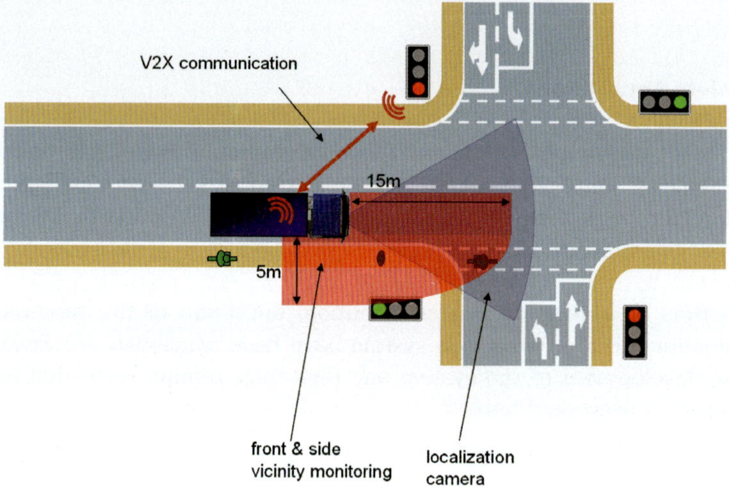

Fig. 4.   Perception system field of view.

2.2   Situation Assessment

During a right turning manoeuvre, the environment on the front and right side will be constantly monitored by the sensor system. Stationary and moving objects are distinguished, but both are taken into account for the risk assessment. Objects will be classified to distinguish between VRU or infrastructure elements. Particular regard is given to moving objects, such as VRU. From the path of the host vehicle and the motion of the VRU (or other object) the risk of collision will be calculated.

2.3   Driver Interface

The risk of collision with a VRU or other object will be constantly monitored, and if critical, a warning will be issued to the driver. Several positions exist for the HMI design for issuing a warning, which will be examined as the project progresses. An overview of strategies and guidelines is given in [5]. (However, the primary project focus is on developing the sensing and assessment system, rather than the HMI).

3 Conclusion

This contribution describes the intersection safety system currently under development in the INTERSAFE-2 project. Intersection safety is a very important factor for heavy goods vehicles, particularly in urban environments together with vulnerable road users.

In this paper, the current situation is analyzed with particular regard to visibility and blind spots. Accidentology and user interviews have identified this topic to be among the statistically relevant accident scenarios between trucks and VRU.

This contribution describes the INTERSAFE-2 approach to tackle this issue, with a modular perception system, including both laser scanner, ultrasonic, localization, communication and electronic horizon. Future presentations will report on the progress of the system development.

Acknowledgement

The work described in this contribution has been co-funded by the European Commission, in the project INTERSAFE-2 within the Seventh Framework Programme.

References

[1]   INTERSAFE-2 project web site http://www.intersafe-2.eu/

[2]   APALACI project web site http://www.prevent-ip.org/en/prevent_subprojects/vulnerable_road_users_collision_mitigation/apalaci/

[3]   Massimo Bertozzi, et al., Stereo Vision-Based Start-Inhibit for Heavy Goods Vehicles, IV'06 IEEE Intelligent Vehicles symposium, Tokyo, Japan, 13 - 16 June 2006.

[4]   Volvo 3P Accident Research Safety Report 2007.

[5]   INTERSAFE-2 Deliverable D3.1 "User needs and requirements for INTERSAFE-2 assistant system", available via http://www.intersafe-2.eu/

[6]   European Commission Directive 2003/97/EC

[7]   European Commission Directive 2005/27/EC

[8]   DOBLI company web site http://www.dobli.com

[9]   Jens Sauerbrei – MAN Abbiegeassistent: Ein System zur Unfallvermeidung beim Rechtsabbiegen von Lkw, Tagung Aktive Sicherheit durch Fahrerassistenz, Garching 11-12 March 2004.

[10]   Karl Viktor Schaller – Driver assistance systems from the point of view of a vehicle manufactuer, Tagung Aktive Sicherheit durch Fahrerassistenz, Garching 7-8 April 2008.

[11]   INTERSAFE project web site http://www.prevent-ip.org/en/prevent_subprojects/ intersection_safety/intersafe/

Malte Ahrholdt, Grant Grubb, Erik Agardt
Volvo Technology Corporation
Intelligent Vehicle Technologies
40508 Gothenburg
Sweden
malte.ahrholdt@volvo.com
grant.grubb@volvo.com
erik.agardt@volvo.com

Keywords: intersection, heavy goods vehicle, INTERSAFE-2

V2X Communication and Intersection Safety

L. Le, A. Festag, R. Baldessari, W. Zhang, NEC Laboratories Europe

Abstract

Vehicle-to-vehicle and vehicle-to-infrastructure communication (V2X communication) has great potential to increase road and passenger safety, and has been considered an important part of future Intelligent Transportation Systems (ITS). Several R&D projects around the world have been investigating various aspects of V2X communication. Some of these projects focus on specific issues of V2X communication for intersection safety (communication-based intersection safety) because intersections are the most complex driving environments where injury and fatal accidents occur frequently. In this paper, we discuss the technical details of V2X communication and discuss how it can be used to improve intersection safety.

1 Introduction

In recent years, V2X communication has attained significant attention from both academia and industry because it has great potential to increase road and passenger safety. Several R&D projects around the world have been working on different aspects of V2X communication. Some of these projects have been investigating specific issues of V2X communication for intersection safety (communication-based intersection safety) because intersections are the most complex driving environments where injury and fatal accidents occur frequently. Research activities for communication-based intersection safety have been conducted in different research projects in Europe, Japan, and in the U.S. [1, 2, 3, 4]. However, many questions related to communication-based for intersection safety remain open. For example, previous work implemented proprietary solutions for communication-based intersection safety as a proof of concept but did not consider important standards currently being defined in the Car-to-Car Communication Consortium (C2C-CC) [5] and the ETSI Technical Committee for Intelligent Transport Systems (ETSI TC ITS). Further, solutions for communication-based intersection safety developed so far do not consider multi-hop forwarding, a networking technology that can provide important benefits for road safety by significantly extending a driver's vision and increasing his reaction time. In this paper, we investigate an integrated approach for communication-based intersection safety by considering it in the big picture of V2X communication. The rest of our paper is organized as follows. We review relevant related work in Section 2. We provide the background for V2X communication in Section 3. We discuss how V2X communication can improve intersection safety in Section 4. We summarize our paper in Section 5.

2 Related Work

Thanks to its importance, intersection safety has been investigated in several research projects around the world. In this section, we review a number of important activities for intersection safety in Europe, Japan, and the U.S. We pay special attention to communication-based intersection safety.

2.1 Communication-Based Intersection Safety in Europe

In Europe, specific issues related to communication-based intersection safety were first addressed in the pioneering project INTERSAFE which combined sensor and communication technologies to increase intersection safety [1, 7, 8]. INTERSAFE's goal was to develop an *Intersection Assistant* that can reduce or even eliminate fatal accidents at intersections. An *Intersection Assistant* can provide intersection safety in two main ways. First, the *Intersection Assistant* can help a vehicle detect others in its neighborhood by means of sensors and bidirectional wireless communication based on the IEEE 802.11p standard. When a potential collision is detected, a driver can be warned to stop for traffic from other directions. Second, a traffic light controller can also be equipped with sensors and communication devices. In this way, the traffic light controller communicates with approaching vehicles via bidirectional wireless communication and informs them about the traffic light's status, road conditions and potential hazards detected by sensors installed at an intersection.

2.2 Communication-Based Intersection Safety in Japan

In Japan, the Driving Safety Support Systems (DSSS) have been investigated by the National Policy Agency and the Universal Traffic Management Society of Japan (UTMS) [4, 9]. DSSS strives to prevent accidents by providing drivers with warning about potential danger at intersections. Main target scenarios for DSSS are stop sign violation, red light violation, turning accidents, crossing-path accidents, rear-end collision, and collision with pedestrians. For DSSS, UTMS has been developing vehicle-infrastructure cooperative systems and conducting operational tests in four different test sites: Tochigi, Aichi, Kanagawa, and Hiroshima [3, 10].

The roadside infrastructure of DSSS consists of an infrared beacon and a Dedicated Short Range Communication (DSRC) beacon. The infrared beacon is placed before an intersection while the DSRC beacon is installed near an intersection. The infrared beacon is periodically broadcast by the roadside infrastructure and realizes two main functions. First, it delivers static information such as road alignment, distance to the intersection, and traffic regulation. Second, it informs approaching vehicles about their specific geographic location and their lane number (infrared beacon is particularly suited for this purpose since its communication range is limited within a few meters and thus provides good accuracy for localization). The DSRC beacon is broadcast periodically and provides dynamic information at an intersection such as position and speed of pedestrians or other vehicles as detected by roadside sensors. DSRC beacon can provide relevant messages for specific lanes at an intersection. In this case, a vehicle's onboard unit (OBU) can perform message filtering using the vehicle's lane number as provided by the infrared beacon. The OBU performs a risk analysis based on information received from the roadside infrastructure. If imminent danger is detected, OBU delivers an acoustic or a visual warning signal.

Initial system evaluation for DSSS has been conducted and received positive feedback from test subjects. The evaluation also demonstrated that considerable speed reduction could be achieved for vehicles approaching an intersection. Further cooperative experiments between DSSS and Advanced Safety Vehicle (ASV) are currently being considered for future large-scale ITS operational tests.

2.3 Communication-Based Intersection Safety in the U.S.

Intersection safety is addressed in the Cooperative Intersection Collision Avoidance Systems Initiative (CICAS) in the U.S. [2,11]. CICAS implements critical safety applications combining different ITS technologies to reduce intersection accidents by providing real-time warnings both in the vehicle and on the infrastructure. The ITS technologies used in CICAS include in-vehicle positioning, roadside sensors, intersection maps, and two-way wireless communication. For wireless communication, CICAS leverages the DSRC technology developed in the Vehicle Infrastructure Integration program (VII). Four main safety applications are developed in CICAS.

CICAS-Violation Warning System (CICAS-V). This application allows the infrastructure to send status information of the traffic light to approaching vehicles using DSRC. Based on this information and in-vehicle GPS, CICAS-Vestimates the risk that the vehicle will violate a traffic light. If this risk is sufficiently high, CICAS-V provides a warning to the driver. Two important objects contained in CICAS-V messages are SPAT (signal phase and timing) and GID (geometric intersection description). SPAT informs an approaching vehicle about the traffic light's status and its remaining time. GID provides geospatial encoding and reference points of an intersection.

CICAS-Stop Sign Assist (CICAS-SSA). This application uses sensors installed at an intersection to help drivers in deciding when they can proceed onto or across a high-speed road after stopping at a rural road stop sign. CICAS-SSA provides drivers with assistance either via animated display sign or wireless communication.

CICAS-Signalized Left-Turn Assist (CICAS-SLTA). This application uses infrastructure sensors and wireless communication (building from CICAS-V) to assist drivers in making turning maneuvers at an intersection. The application takes oncoming traffic, pedestrians, and other obstacles into consideration.

CICAS-Traffic Signal Adaptation (CICAS-TSA). This application combines infrastructure sensors and wireless communication (building from CICAS-V) to detect a dangerous situation when a vehicle violates a red light and can potentially collide with other vehicles. In this case, CICAS-TSA triggers a red light in all directions to protect drivers from an imminent danger. Further, when a vehicle detects a dangerous situation, it can also send a warning message to the infrastructure to trigger an all-red traffic light and prevent a chain reaction of accidents.

2.4 Summary of Related Work

A number of pioneering projects in Europe, Japan, and the U.S. have addressed intersection safety. However, many questions related to communication-based intersection safety still remain open. While these projects laid the groundwork for communication-based intersection safety, they did not address several important issues of V2X communication such as robustness, reliability, and scalability. Further, these projects did not consider multi-hop forwarding, an important V2X networking technology that can provide significant benefits for road safety by extending a driver's vision and increasing his reaction time. We plan to address these issues in the ongoing EU project INTERSAFE-2 [6].

3 V2X Communication

The core networking concept for V2X Communication is *Geocast*, an ad hoc routing protocol that provides wireless multi-hop communication over short-range wireless radio without the need of a coordinating infrastructure as in cellular networks. The basic idea of *Geocast* was originally proposed for mobile ad hoc networks [12] and has been further developed for other systems, i.e., wireless sensor networks and vehicular ad hoc networks.

In principle, *Geocast* provides data dissemination in ad hoc networks by using geographical positions of nodes. For multi-hop communication, nodes forward data packets on behalf of each other. *Geocast* is attractive in vehicular environments for two reasons. First, it works well in highly mobile network where network topology changes frequently. Second, *Geocast* offers flexible support for heterogeneous application requirements, including applications for road safety, traffic efficiency and infotainment. In particular, *Geocast* provides periodic transmission of safety status messages at high rate, rapid multi-hop dissemination of packets in geographical regions for emergency warnings, and unicast packet transport.

Geocast provides two basic and strongly coupled functions: *geographical addressing* and *geographical forwarding*. With *geographical addressing*, *Geocast* can address a node by a position or address multiple nodes in a geographical region (*geo-address*). For *geographical forwarding*, *Geocast* assumes that every node has a partial view of the network topology in its neighborhood and that every packet carries a *geo-address*, i.e., the geographical position or geographical area as the destination. When a node receives a data packet, it uses the *geo-address* in the data packet and its own view of the network topology to make an autonomous forwarding decision. Thus, packets are forwarded "on the fly" and nodes do not need to set up and maintain routes. In order to achieve this, *Geocast* assumes that every network node knows its geographical position, e.g. by GPS or another positioning systems, and maintains a location table of geographical positions of other nodes as soft state.

Geocast has three core protocol components: *beaconing, location service* and *forwarding.*
- ▶ *Beacons*, also referred to as *heartbeats*, are small packets that each node broadcasts periodically to inform other nodes about its ID, its speed, its heading, and its current geographical position.
- ▶ Nodes can cooperatively provide *location service* that resolves a node's ID to its current position on a query/reply basis.
- ▶ *Forwarding* basically means relaying a packet towards the destination.

These three protocol components are combined to support *Geocast* forwarding. There are four types of *Geocast* forwarding: *Geographical Unicast (GeoUnicast)*, *Geographical Broadcast (GeoBroadcast)*, *GeoAnycast*, and *Topologically-Scoped Broadcast (TSB)*.

▶ *Geographical Unicast (GeoUnicast)* (illustrated in Fig. 2) offers packet relay between two nodes via one or multiple wireless hops. When a node wants to send a unicast packet, it first determines the destination position (by means of location table look-up or the location service) and forwards the data packet to the next node in the direction of the destination. This node in turn re-forwards the packet along the path until the packet reaches the destination. Greedy forwarding is an algorithm for where a node selects the next hop based on position information such that the packet is forwarded in the geographical direction of the destination. An illustration of greedy forwarding is provided in Fig. 1.

▶ *Geographical Broadcast (GeoBroadcast)* (illustrated in Fig. 3) distributes data packets by flooding, where nodes re-broadcast a packet if they are located in the geographical region specified by the packet. Packet is forwarded only once: If a node receives a duplicate packet that it has already received, it will drop packet.

▶ *GeoAnycast* is similar to *GeoBroadcast* but addresses any node in a geographical area.

▶ *Topologically-Scoped Broadcast (TSB)* (illustrated in Fig. 4) offers re-broadcasting of a data packet from a source to all nodes that can be reached in certain number of hops (all nodes in an n-hop neighborhood). Single-hop broadcast is a special case of TSB that is used to send *heartbeats* including application data payload.

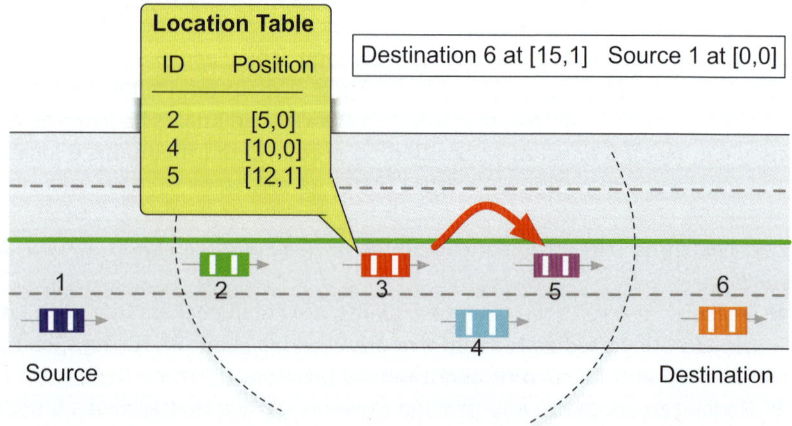

Fig. 1.   Illustration of greedy forwarding: Node 3 selects node 5 as the next hop.

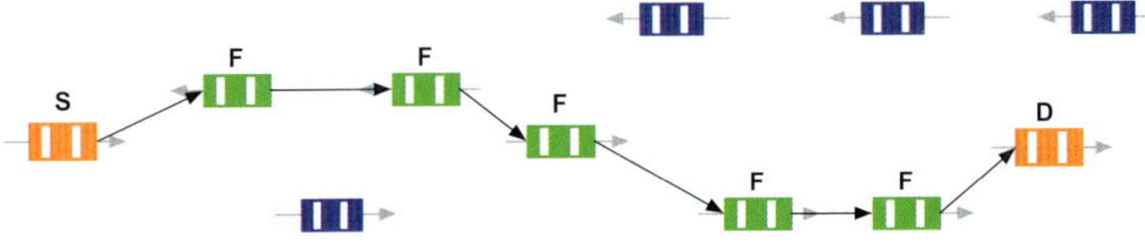

Fig. 2.   Illustration of GeoUnicast.

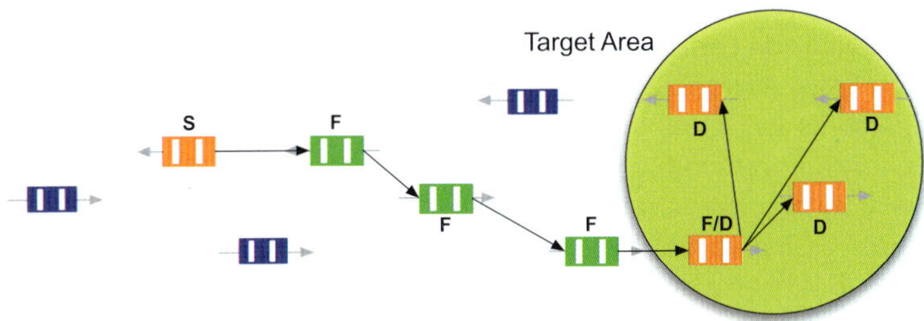

Fig. 3.   Illustration of GeoBroadcast.

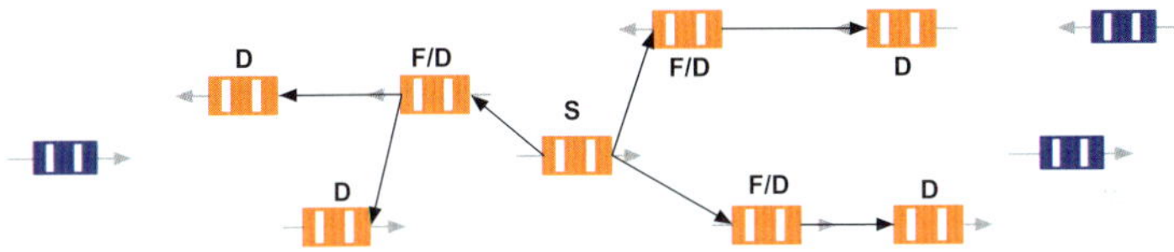

Fig. 4. Illustration of Topologically-Scoped Broadcast

In *Geocast*, a node usually processes all data packets that it receives on the wireless links to keep track of all nodes in its surrounding (this requires nodes to operate their network interfaces in promiscuous mode). Since each data packet contains the source's and previous forwarder's positions, a receiving node can update its location table accordingly. A *Geocast* packet header contains fields for node identifier, position and timestamp for source, sender, and destination, and more. The originator of a packet is referred to as source, and the last forwarder as sender. Fields in a *Geocast* header are classified as immutable and mutable: *Immutable fields* are not changed in the forwarding process. On the other hand, *mutable fields* are updated by forwarders. This allows a forwarder to change some header fields on the fly, e.g. in case it has more recent information in its location table about a given destination.

4 Applications and Use Cases

In this section, we discuss several use cases for communication-based intersection safety that can be supported by V2X communication. Some of these use cases have been identified by the pioneering projects reviewed in Section 2. While the use cases discussed here are not meant to be comprehensive, we believe that they are representative for communication-based intersection safety.

Prevention of traffic light violation. In this use case, an infrastructure-based intersection assistant can use V2X communication to inform approaching vehicles about the traffic light's status and the remaining time until the status changes. Since a traffic light system is inherently complex and involves many inputs such as inductive loops and push buttons for pedestrians, traffic light can change unpredictably. For this reason, real-time V2X communication is necessary to provide vehicles with accurate and up-to-date signal phase of a traffic light. Given a traffic light's status, a vehicle can deliver a warning to a driver when a potential traffic light violation is detected. Further, the vehicle can also adjust its velocity to achieve optimized fuel consumption. This use case can be supported efficiently by *GeoBroadcast* or *Topologically-Scoped Broadcast* forwarding.

Prevention of turning and crossing-path collision. This use case assists drivers in their turning or crossing-path maneuvers at an intersection. Sensors installed at an intersection can detect objects and vehicles and construct an overview of the intersection. This view can be broadcast at a regular interval on the wireless channel to inform a driver about the presence of other road users at an intersection. Further, V2X communication can be used as an enabling technology for cooperative fusion of sensor data acquired in vehicles and at the infrastructure side. Cooperative data fusion provides a driver with a better vision of an intersection and helps detect other road users that the driver would overlook due to obstacles, distraction, or bad weather. Special protection for vulnerable road users (VRUs) can be obtained. This use case can be supported efficiently by *GeoBroadcast* or *Topologically-Scoped Broadcast forwarding.*

Prevention of rear-end collision. This use case prevents an accident from happening when a vehicle reduces its velocity abruptly at an intersection and other vehicles behind it do not have sufficient time to react. In this use case, multi-hop V2X communication can distribute a message within an intersection's surrounding area to warn other drivers. The warning message can be triggered by a vehicle's braking system or infrastructure-based sensors. Further, the traffic light controller can use V2X communication to inform drivers about the recommended driving speed before they enter an intersection. With this information, drivers can avoid reducing their speed abruptly at an intersection. This use case can be supported efficiently by *GeoBroadcast* or *Topologically-Scoped Broadcast* forwarding.

Traffic signal adaptation for emergency warning and prioritized road users. This use case can provide a dynamic traffic signal adaptation when an accident occurs at an intersection. In this case, an intersection assistant can broadcast an alert on the wireless communication channel. Further, in case a vehicle causes an accident, it can also send an emergency message to the infrastructure to trigger an all-red traffic light and prevent other vehicles from entering the intersection until the situation becomes clear. In emergency scenarios, an intersection can also support prioritized road users, e.g., emergency vehicles by giving them the green light in their direction. This use case can be supported efficiently by *GeoBroadcast* or *Topologically-Scoped Broadcast* forwarding.

Traffic efficiency. In this use case, RSUs can monitor road conditions and traffic density at intersections and provide a backend traffic management center with real-time information. Using this information, the traffic management center can obtain a global view of road systems and can compute alternate routes for vehicles. The traffic management center sends this information back to RSUs to inform the drivers. Drivers can use this information to optimize their route selection according to their needs. This use case can be supported efficiently by *GeoBroadcast* or *Topologically-Scoped Broadcast* forwarding.

We note that while *GeoBroadcast* or *Topologically-Scoped Broadcast* forwarding provide efficient support for most use cases, *GeoUnicast* and *GeoAnycast* forwarding might be useful for future use cases of communication-based intersection safety that are currently not considered.

5   Summary

V2X communication has been considered a key ITS technology due to the fact that short-range wireless communication technology has become mature, inexpensive, and widely available. Using V2X communication, a vehicle can communicate with other vehicles in its neighborhood in order to support safety applications such as cooperative collision warning. For intersection safety, V2X communication can be used as an enabling technology to combine traffic light system, in-vehicle sensors, and infrastructure-based sensors. In this paper, we provide technical background for V2X communication. We review state-of-the-art of communication-based intersection safety and highlight several use cases where V2X communication can provide considerable benefits for intersection safety. We plan to address these use cases and open issues related to communication-based intersection safety in the ongoing EU project INTERSAFE-2 [6].

References

[1]   INTERSAFE. http://www.prevent-ip.org/intersafe.

[2]   Cooperative Intersection Collision Avoidance Systems. http://www.its.dot.gov/cicas/.

[3]   S. Itagaki, et al., Outline of safety support systems for intersections and NEC's activity. NEC Technical Journal, 3(1), March 2008.

[4]   Universal Traffic Management Society of Japan. http://www.utms.or.jp.

[5]   Car-to-Car Communication Consortium. http://www.car-to-car.org.

[6]   INTERSAFE-2. http://www.intersafe-2.eu.

[7]   A. Benmimoun, et al., Communication-based intersection assistance. IEEE Intelligent Vehicles Symposium, June 2005.

[8]   K. Fuerstenberg. A new European approach for intersection safety - the EC-Project INTERSAFE. IEEE Conference on Intelligent Transportation Systems, September 2005.

[9]   M. Fukushima, et al., Progress of V-I Cooperative Safety Support System in Kanagawa, Japan. ITS World Congress. November 2008.

[10] M. Yamamoto, et al., Development of Vehicle-Infrastructure Cooperative Systems – Field Operational Test of Hiroshima DSSS. ITS World Congress. November 2008.

[11] C. Goodwin, et al., A Comparison of U.S. and European Cooperative System Architectures. ITS World Congress. November 2008.

[12] B. Karp and H.T. Kung. GPSR: Greedy Perimeter Stateless Routing for Wireless Networks. In Proceedings of the 6th ACM/IEEE International Conference on Mobile Computing and Networking (MobiCom), pages 243–254, Boston, MA, USA, August 2000.

Long Le, Andreas Festag, Roberto Baldessari, Wenhui Zhang
NEC Laboratories Europe
Kurfürsten-Anlage 36
D-69115 Heidelberg
Germany
le@nw.neclab.eu
festag@nw.neclab.eu
baldessari@nw.neclab.eu
zhang@nw.neclab.eu

Keywords: V2X communication, intersection

Utilization of Optical Road Surface Condition Detection around Intersections

M. Kutila, M. Jokela, VTT Technical Research Centre
B. Rössler, Ibeo Automobile Sensors GmbH
J. Weingart, Signalbau Huber GmbH

Abstract

This paper presents experimental results regarding road condition monitoring by machine vision techniques. The system will be further developed by VTT in the European INTERSAFE-2 project, focused on applications at road intersections. Knowing the presence of adverse conditions such as icy or wet roads, important effects can be obtained for intersection safety by means of more effective driver assistance functions. A very good reliability up to 93% has been found for the detection of icy roads, while lower values around 60% have been measured in the case of wet surfaces, due to a higher sensitivity to environmental conditions, especially outdoor light. Future steps will include an advanced classification algorithm and the implementation of active lighting. Integration of the vision equipment into a cooperative system for intersection safety is being investigated.

1   Introduction

Intersection safety is a challenging subject due to the complexity of the heterogeneous environment. But it is as well one of the most important areas under discussion with respect to the huge number of accidents which occur on intersections. In this context, the *Cooperative Intersection Safety* project (INTERSAFE-2) was established within the European Seventh Framework Program to develop advanced driver assistance systems and contribute to a significant increase of safety for this type of road structures.

The project develops key technologies to monitor dangerous situations related to traffic and to warn drivers so that they can adapt vehicle speed accordingly. Various techniques are studied, such as in-vehicle camera vision, laser scanners and radar sensors, in order to detect the different road users at the intersection. This allows vehicles to extend via a mutual communication the driver's ability to observe other vehicles, obstacles and pedestrians. Moreover, the vehicles will communicate with the intersection infrastructure, receiving specific sensor data. In this case, the infrastructure data include road friction measurements based on a vision system and the vehicle positions based on a laser scanner.

The reason for introducing the friction measurement is the importance of this parameter to correctly implement driver assistance functions. Slippery roads due to ice, snow or water are involved in more than 3800 fatalities annually in Europe (EU-18) which is approx. 12 % of total amount [1]; in Nordic countries, such as Finland and Sweden, a rate of 14 % has been found. In traffic junctions, adverse weather conditions (including fog) were identified to be present in 14,8 % of the accidents [2]. Therefore, detecting a decreased friction level in the intersection area is a key point to reduce fatalities and provide suitable warnings. In that respect, the project contributes to bring Europe closer to the vision of reducing by half the number of fatalities from 2000 to 2010.

This paper first describes the road-state monitoring system, based on machine vision and developed by VTT (called IcOR), and then discusses how this system is addressing the needs of cooperative intersection driving. Many vehicles today include devices able to detect slippery road conditions, using an analysis of tyre rotational speeds. These systems are dedicated to active safety in order to prevent tyre skidding. However, there has been recently additional interest in the automotive sector to utilise environmental sensors, able to detect a possibly reduced friction level. On the other hand, the road operators have installed traffic and road condition detectors in order to analyse the road quality and e.g. if salting is needed. Within INTERSAFE-2, the road state monitoring will be infrastructure based, and therefore vehicles do not need any additional sensor installation (Fig. 1).

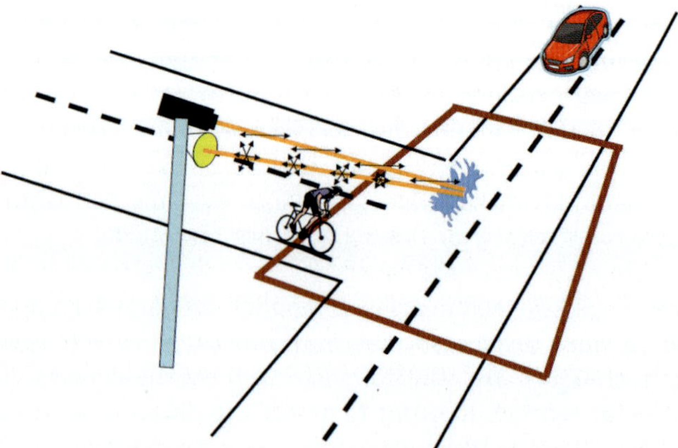

Fig. 1.   Road state monitoring in the intersection infrastructure

The road side equipment will include "a smart traffic pole" which is constituted by traffic lights and includes a controller to process the captured traffic data. The traffic light adaptation and driver warning provision via the in-vehicle HMI will support fluent, environment friendly and safe driving when crossing an intersection. The most important infrastructure elements are the monitoring facilities which are intended to track vehicles and the road conditions. The intersection environment will be equipped with the laser scanners, which provide accurate relative position of the vehicles and pedestrians in the intersection and a camera vision technique for executing advanced polarisation analysis which monitors the actual road conditions.

VTT has investigated two types of camera based techniques able to monitor the road conditions. The first one uses the near infrared band above 1200 nm, where the reflectance of snow changes radically and the response of ice is high. This approach was proven to be very reliable, but unfortunately involves a high cost for the camera (>5000 Euro). Considering therefore the interest for a cost effective deployment after the project, we have focused our efforts on the second technique, which is based on the analysis of light polarisation and image graininess. This method can be implemented by a low cost device (< 2000 Euro). A first prototype was already investigated in the previous SAFESPOT-EU-FP6 project, with promising results. Ongoing work within INTERSAFE-2 is now focused on improving the performance for road state detection by a new classification algorithm and by implementing active lighting. A further goal is the system integration within a monitoring platform, part of the infrastructure, which is a key element for cooperative safety systems.

2 State of the Art Knowledge

The following remarks intend to highlight some recent research work in the domain of road status monitoring by optical techniques.

Casselgren et al. [3] inspected the light behaviour on asphalt covered by water, ice or snow. They performed exhaustive investigations concerning the changes of light intensity with varying incident angle and the spectrum of light. The wavelength band in their experience was from 1100 to 1700 nm. They observed that the snow reflection ratio drops consistently at wavelengths above 1400 nm, and as a conclusion they proposed two different bands in the NIR region (below and above 1400 nm) to classify road conditions. Their intention is to build a photo-detector for automotive applications, called RoadEye [4]. The advantage of this approach is that the equipment is accurate and relatively cheap, but the drawback is the measuring area limited to one spot on the road.

Yamada et al [5] proposed to utilise changes in the intensities of different polarisations to estimate the road surface conditions. In addition to the light polarisation analysis, this work proposed to improve the performance of the system using texture analysis. The authors achieved good results in the classification of road conditions. Compared to their system, our approach is characterised by an improved processing scheme and adaptation to a co-operative traffic monitoring environment. Moreover, the system diagnostics have been introduced, which is an important feature when considering application oriented developments.

Yusheng and Higgins-Luthman [6] are owners of a patent application dealing with black ice detection, based on the analysis of polarisation. Their application covers the measurement principle of using polarisation filters and active lighting to detect icy puddles in front of a vehicle. However, the system performance seems quite limited, since it relies only on the polarisation measurements and does not include a graininess analysis.

The Finnish company Vaisala Oyj has a product which measures different road conditions like water, ice, snow and can even estimate a friction level [7]. Their DSC111 camera system uses a spectrometric measuring principle and thus, probably operates in the same wavelength bands as the RoadEye system [4]. Vaisala are also proposing a solution based on thermal imaging (CST111) for detecting road conditions. The system uses very advanced but somewhat expensive techniques, whereas our approach is to keep the cost at a low level.

Many projects funded by the European Union have worked in the area of the road state and friction monitoring. Two such examples are FRICTION and SAFESPOT. The first one developed an on-board technology in order to predict the available friction level [4]. On the other hand, the Safespot project developed a technology which can be used also by roadside systems [8] to detect reduced friction levels. VTT has developed a camera system in both projects. Despite some similarities, specific solutions were needed for the two environments, which involve totally different operating conditions. New aspects to be considered for the intersection application are the complexity and dynamics of the scenarios, the geometry, and the interaction between road systems and vehicle systems.

3   Measurement Principle

3.1   Light Polarization Changes

Light reflection from a non-metallic surface (asphalt, ice or a wet patch) reduces vertically polarized electo-magnetic radiation compared to horizontal one. The reflection coefficient depends on the surface type and characteristics. A film of ice or water on the road causes incident light to be partially reflected and partially refracted into the medium. However, the non-metallic substances cause changes to the light polarisation; these phenomena can be utilised to detect whether the road surface is icy/wet or the reflection is due to pure asphalt.

Most of the light is reflected in the specular direction when impinging on a road surface covered by ice or water. Therefore, the characteristics of the measurement strongly depend on the relative position of the illumination source and the viewing camera. When the light source comes from the opposite side, the reflection from an icy or wet surface is stronger. However, when the illumination device is integrated into the camera, most of light is reflected away, and only a small fraction is scattered back from the asphalt to the detector (see. Fig. 2). The transmitted beam in the media (ice/water) is slightly polarised, and therefore the influence is not as strong as in the case when light is coming from the opposite direction. Our first version based on the light reflected towards camera from opposite direction but the future INTERSAFE-2 approach will include lighting equipment itself and hence, the measurement principle will change.

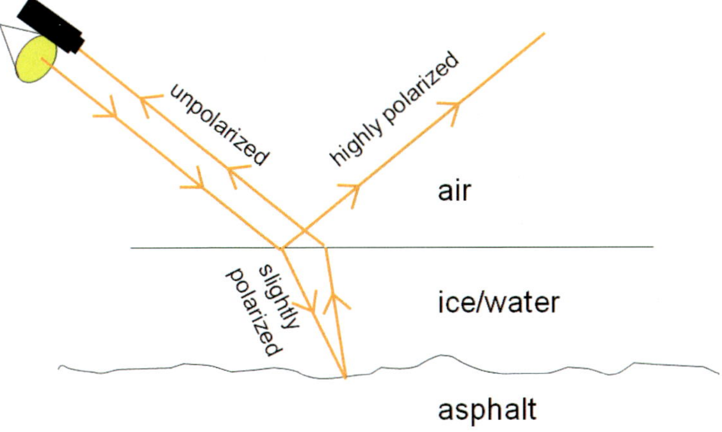

Fig. 2.   Light behaviour in the interface of two media

3.2 Graininess Analysis

Graininess analysis has originally been proposed to analyse paintings [9]. However, the methodology is useful also to distinguish road conditions since in general an icy road is smoother than a snow surface, or dry asphalt, which typically has a granular structure.

The key idea of this method is to treat the image with low-pass filtering, which makes it more blurry. Calculating the contrast difference between the original and the low-pass filtered images can provide information on the amount of "the small elements" in the picture. The low-pass filtering has been executed by a Wiener filter, which assumes that the additive white noise has a Gaussian distribution. The mathematical expression for the Wiener filter is:

$$G = \frac{H^*}{|H|^2 + \dfrac{P_n}{P_g}} \tag{1}$$

where H^* is the complex conjugate of a Fourier transform of the point-spread function. P_n is the noise power and P_g is the power spectrum of the model image, which is calculated by taking the Fourier transform of the autocorrelated signal.

Blurriness is measured by estimating the total amount of contrast in the image, considering rows and columns in a first step, and then a sum of these two cases. The contrast (C) is defined as the difference of the adjacent pixels aligned horizontally or vertically (see. Fig. 3).

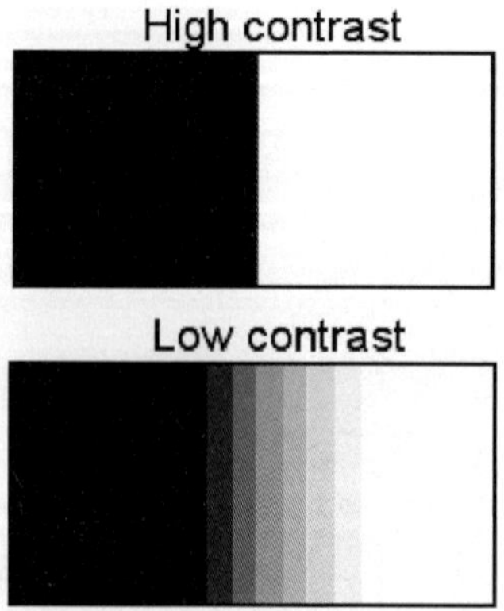

Fig. 3.   Contrast is a measure of sharpness of the changes.

The overall contrast is then the sum of differences computed on rows and columns respectively:

$$C_{total} = C_{row} + C_{column} \tag{2}$$

Finally, the graininess of the image S is obtained from a comparison of contrasts in the different images, as follows:

$$S = \frac{C_{original} - C_{filtered}}{C_{original}} \tag{3}$$

4 The System Description

4.1 Equipment

The current IcOR system uses two cameras installed side by side; this provides an ideal configuration when considering spatial and temporal synchronisation between the captured images. The cameras are of the CMOS type (containing the Micron MT9V032 imager element) and there are horizontal and vertical polarisation filters in front of the optics. The cameras acquire monochrome images with a maximum resolution of 640 x 480 pixels. They are enclosed in a robust housing which is heated, water proof and meets the IP67 standard (Fig. 4). The polarisation filters are selected so that their eliminate most of the ambient light and only a narrow spectral band is passed to the detectors.

Fig. 4. The camera housing of the VTT's IcOR system

Fig. 5. The computing unit of the road side monitoring system in the Swedish test site of SAFESPOT-EU-FP6.

The road state monitoring algorithms run in the computer (Fig. 5) located inside the "cabin" of the road-side unit. The cameras are connected to the road side unit using two Firewire cables. In the future, the system will be modified so that only one wire is needed. The output is available via Ethernet connection or optionally a CAN interface.

4.2 Software

The software is designed not only for the detection task itself, but also as a tool for analyzing and optimising system performance offline. The program provides a classification of the road surface, which can be *icy, wet, snowy* or *dry asphalt*; moreover a classification confidence level is available. A specific output *unknown* is given when the confidence level is not sufficient for a reliable result.

The user interface of the system shows (Fig. 6) the two images acquired by the cameras with horizontal and vertical polarisation. The operator may select to visualise the analyzed image, where the ROI area and difference image is shown, instead of the vertically polarized image. Since the software can be utilised also for applications where only one camera is needed (e.g. for spectral analysis), the operator may also configure whether one or two cameras are connected.

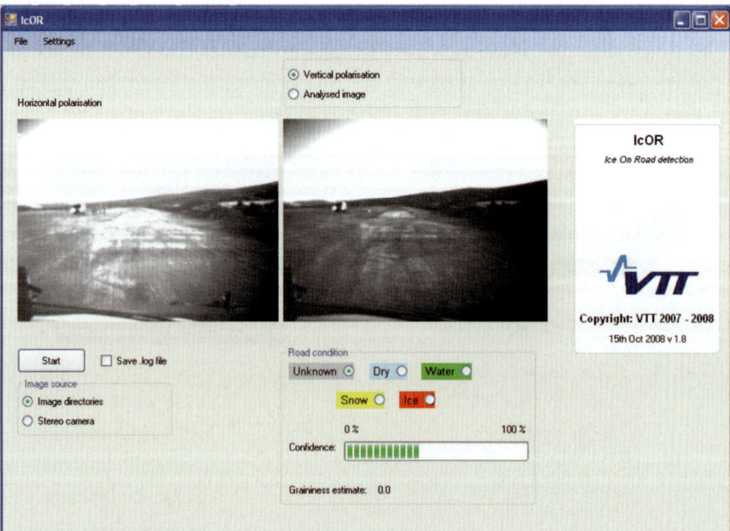

Fig. 6.   User interface of the IcOR program

Additional features of the IcOR system include the ability to read images from the computer's data base. This enables an offline investigation of the optimal detection parameters, as well as tuning of the software by means of recorded images from real conditions.

The detection algorithm looks for the presence of ice or water from a pre-defined region of interest (ROI), which covers the lanes in the static infrastructure installation. This ROI is determined manually during the installation phase and it depends on the height and the field of view of the camera.

5 Preliminary Test Results

5.1 Data gathering Description

During the project FRICTION two different sites were implemented to gather essential test data: Arjeplog in Sweden and Aachen in Germany. A car and a truck were employed for different tests: in both cases, the camera pair was installed inside the vehicle, behind the windshield, to avoid difficulties coming from dirt on the optics or low temperatures of the camera. The tests have been performed with installing the IcOR system to instrumented test vehicles which enabled comprehensive tests by arranging different road conditions on the equivalent environmental and lighting conditions. (Fig 7). Technically, the on-board system corresponds with the infrastructure system and therefore, performance rates correlate completely with the infrastructure sub-system.

Fig. 7.   Audi used in the Aachen tests in a wet curve

Data from winter conditions was gathered in Arjeplog in March 2008 both from public roads and a private test track. The gathered data cover snowy, slushy, icy and wet road surfaces. Due to the rainy and cloudy weather only a small amount of dry road data was recorded and only from a heated part of the road on the test track. The dark weather made the lighting conditions rather poor.

In Aachen the recordings concentrated mainly on the summer and extremely adverse conditions, such as icy curves and excess water on the road. The tests were done on a closed test track, which made it possible to create the extreme conditions without jeopardising the traffic. The weather conditions were quite the opposite of the ones in Arjeplog being sunny and dry during the tests. Hence, lots of dry asphalt data were also recorded.

5.2 Results

The winter tests in Sweden provided promising results concerning system performance and the detection of icy, wet or snowy road conditions. Comparable results concerning dry asphalt were not available, due to the specific weather conditions. As Fig. 8 shows, the winter scenarios are well separated in the "map", where the polarisation difference is plotted in the horizontal axis and the image graininess in the vertical axis. Fig. 9 gives some preliminary indications related to improved road conditions with partially dry asphalt. As expected, the polarisation difference for dry asphalt decreases compared to the icy or snowy road conditions.

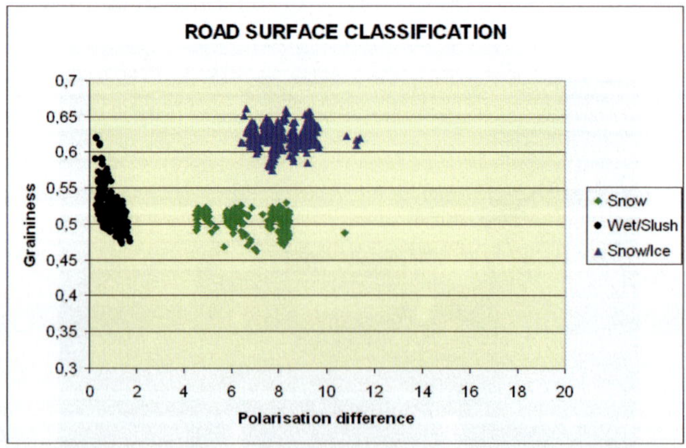

Fig. 8.   Road condition detection results in the test track

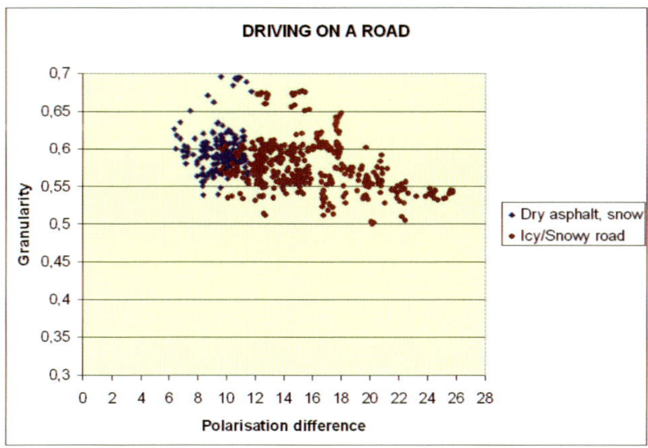

Fig. 9.   Detecting ice from partially snowy and dry asphalt

Fig. 10 indicates that separating dry and wet road surfaces can be rather diffcult according to the data captured with the passenger car in Germany. This was somehow expected, since even with the human eye it is sometimes difficult to observe a water bed on a road. On the other hand, variations in the polarisation difference are more spread compared to the wet road case. From this observation, we could assume that the system might be used when the wiper are activated, to distinguish if this is due to a wet road or just to cleaning the windscreen. The dark (night time) conditions are another aspect of the existing system requiring proper consideration (see. Fig. 11). In these cases, the graininess analysis behaves as expected, being higher than for the dry asphalt compared to icy roads, but the polarisation difference appears to be too small in order to perform a reliable measurement in outdoor conditions. An active lighting technique would obviously improve this situation.

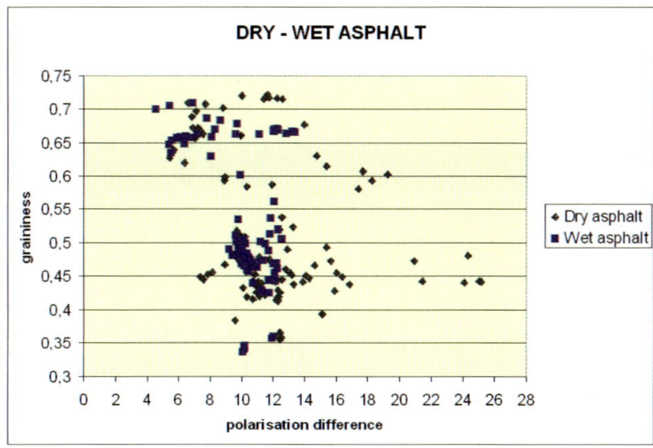

Fig. 10.   Results of wet road detection

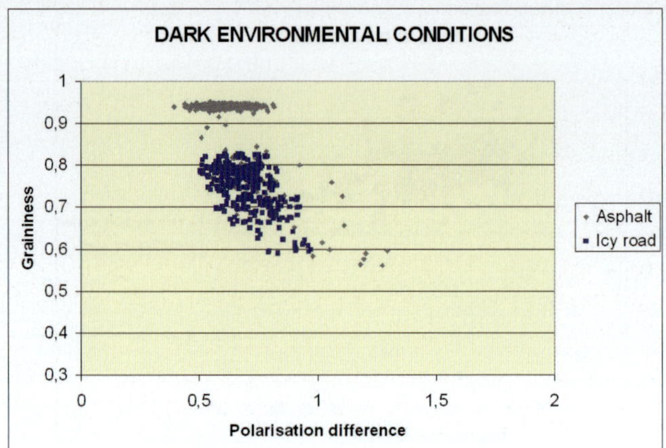

Fig. 11.   Icy road detection in a dark environment

The following two tables (Tab. 1 and Tab. 2) provide some overall data concerning performance of the IcOR system. As the data show, ice detection performs very well when sufficient illumination is present: this aspect was the main goal of this research. The achieved accuracy is 93,4 %, which is judged perfectly adequate for the application requirements at intersections, especially when considering that additional data are available in the road side unit to further improve the detection rate. If one considers instead the water detection, a lower accuracy of 61,1 % has been obtained. This value indicates that additional work is needed for this aspect; an improvement of approximately 15% would be needed to meet the expectations.

Ice presence on a road				
			Event occurrence reality	
			Dry road	Icy road
Event occurrence detection	*Dry road*		218	54
	Icy road		2	386
Detection rates, Ice presence				
Correct alarm rate	87,7 %			
Missed alarm rate	12,3 %			
Correct non-alarm rate	99,1 %			
False alarm rate	0,9 %			
Accuracy	93,4 %			

Tab. 1.   Performance rates of detecting ice on road surface

Water presence on a road			
		Event occurrence reality	
		Dry road	**Wet road**
Event occurrence detection	**Dry road**	*33*	*33*
	Wet road	*63*	*236*
Detection rates, Water presence			
Correct alarm rate	*87,7 %*		
Missed alarm rate	*12,3 %*		
Correct non-alarm rate	*34,4 %*		
False alarm rate	*65,6 %*		
Accuracy	*61,1 %*		

Tab. 2.   Performance rates of detecting water on road surface

6   Integration of the System to the INTERSAFE-2 Platform

6.1   Requirements concerning the INTERSAFE-2 Infrastructure

In the INTERSAFE-2 project, there will be laser scanners detecting dynamic objects and road users (e.g.: pedestrians, bicycles and animals) on the intersection. Position, size and speed of these objects will be measured relative to the intersection coordinate frame. Moreover, the road surface conditions will be detected in the vicinity lanes of the intersection, in order to warn drivers before entering the intersection, especially if they proceed with a too high speed. The infrastructure platform may use a predefined map of the intersection, including data on all the static objects like lanes, guard-rails, poles, buildings, etc.

Considering these object detection sensors, the coverage area will depend on the speed of the vehicles when entering to the intersection and amount of vehicles/bicycles/pedestrians in monitored area. Practically, this leads to a longitudinal detection range of approximately 50-100 m before the stop line of the intersection. Each installed sensor should cover at least one and preferably more lanes, and the detection rate should exceed 5 Hz to be reactive enough. For the road surface condition monitoring, the surveillance area can be selected more freely and the timing requirements are not so strict, so that a

1 Hz refresh frequency would be sufficient. The maximum longitudinal range will be from 10 to 20 m from the sensor.

The detection system should use either 12 VDC or 230 VAC power supplies. Since the system is intended to run stand-alone, a self-diagnosis is desired. Housings have to withstand all normal outside weather conditions throughout the year. This implies also the use of heated camera housings, especially at northern latitudes. The sensing system should be able to operate also without specific maintenance like cleaning.

6.2 Laser Scanner Setup in Intersection

The current Ibeo Lux Laserscanner (see. Fig. 12) combines a 4-channel laser range finder with a scanning mechanism, and robust design suitable for integration into vehicles as well as at the infrastructure.

Technical data:

▶ Scan Frequency: 25Hz
▶ Horizontal angle: 100° field of view
▶ Range: 0,3 m to 200 m
▶ Resolution: range 4 cm, angle 0,1° to 1°
▶ parallel and simultaneous scanning layers
▶ vertical field of view 3.2°
▶ up to three reflection measurements per laser impulse
▶ Size: H85 x W128 x D83
▶ Weight: 900 g (less than 1 kg)
▶ built-in processing
▶ laser class 1
▶ Ethernet- and CAN-Interface

Fig 12. The Ibeo Lux laserscanner

The laserscanner will be used in order to monitor road users in the intersection from the demonstrator vehicles and the infrastructure side. The road users (like vehicles, pedestrians, bicyclists) present at the intersection are tracked and classified based on the 3D laserscanner measurements. The added value of the infrastructure integration is the extended field of view that can be covered. Fig. 13 shows a sketch of the INTERSAFE-2 system with the Laserscanner setup in the vehicle alone (left image) and combined in the vehicle and at the intersection infrastructure (right image). If the field of view of the host vehicle is occluded by an object, the laserscanners at the infrastructure can be used to monitor this occluded part.

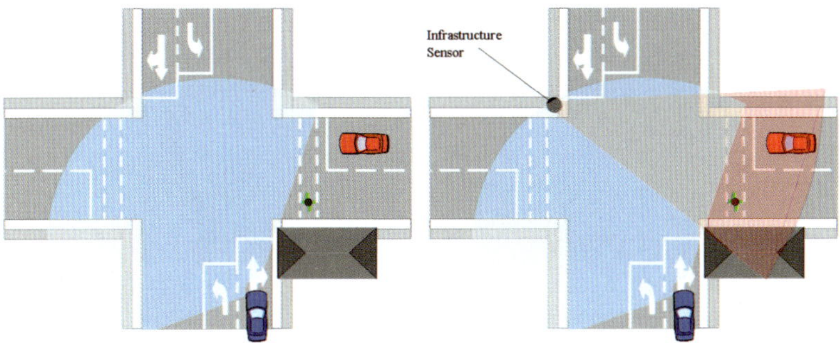

Fig. 13.   Laserscanner integration at the intersection infrastructure

The number of Laserscanners at the intersection can vary with the size and structure of the intersection. A suitable installation will be found that is able to cover the whole test intersection. The information from the infrastructure sensors will be transferred by a wireless data link to the approaching vehicles.

6.3   Traffic Light Controller

At the demonstrator sites of the INTERSAFE-2 project, the powerful ACTROS traffic controller (see. Fig 14) of Signalbau Huber will be used. This modular and flexible controller is providing interfaces like Ethernet and CAN, which makes it easy to integrate new components and the INTERSAFE-2 roadside features to modern traffic control systems.

The traffic controller not only supplies the main power and housing for the extended roadside equipment (like laser scanner and environmental sensors), but also is able to process the data in real time. This process includes the fusion of data from the different sensors, and also the joint analysis with its internal information concerning the intersection geometry and layout. The processed information is transferred via the wireless V2I communication channel, which is developed to meet the most recent communication recommendations on upcoming standards. A part of this communication will embrace also intelligent methods to increase the accuracy of vehicles positioning, using an integrated GPS receiver.

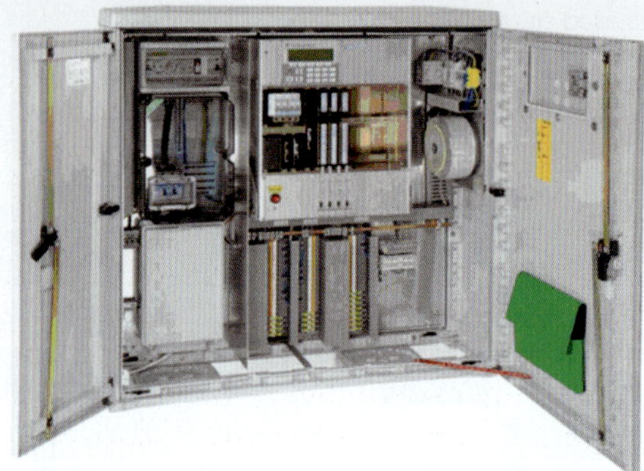

Fig. 14.   The ACTROS traffic controller which will include the data processing capacity of the INTERSAFE-2 infrastructure sub-system

For installation and final tests at public intersections the certified safety concept with defined fail-safe technology of the ACTROS controller will ensure the needed reliability for all involved users in the INTERSAFE-2 demonstrations.

7   Conclusions

Machine vision techniques appear a very promising approach to detect the conditions of the road surface, especially the presence of ice, water or snow. Preliminary experiments on vehicles are preparing the work to be done in the INTERSAFE-2 project, which is addressing applications for intersection safety, based on a road-side platform.

Performance of the system for ice detection fully corresponds with the assumptions made when the development was initiated. The icy road sections could be identified with 93 % detection accuracy, which is a satisfactory level, especially when the system is compared to other existing devices, which are five times more expensive. The IcOR system was also benchmarked against behaviour of the RoadEye system and clear correspondence between these devices was observed.

The detection capability for a water bed is at the moment not as good as required: an accuracy level around 61 % has been achieved, and new approaches are under investigation to improve this value by about an additional 15 %. Since the main reason for this lower capability is a poor signal-to-noise ratio, techniques based on averaging will be studied among others.

One positive achievement concerning the road monitoring system is the good performance in high intensity conditions (glares, sun light, etc.). On the other hand, the use of averaging methods (or even active light) will also improve the operation in a dark environment (e.g. night time), especially when street lamps are not present.

The initial idea of the road state monitoring device was development of a cost effective system compared to in-market products. The price of the developed monitoring equipment is approx. 4 kEur whereas the existing products cost 5 times more. This enables wider deployment of the system since the investment budgets of road operators allows more products to be bought. In addition to the cost effective solution, another important aspect of the on-going development is to provide a system which is originally designed to support cooperative vehicle and traffic solutions. A key aspect is the interface configuration, as well as the compatibility with multiple sensor sources and their related cross links. In particular, the solution under study allows using additional relevant data, like those provided by passing vehicles.

Considering the challenging requirements of systems for intersection safety, the vision system will be a key element, together with the object data captured by the laser scanners and the static information and actuation capability of the traffic light controller, in order to provide a comprehensive understanding of safety risks at the intersection. The knowledge of road status is expected to play a fundamental role to improve driver assistance functions.

8   Future Work

The current application version is affected by the external and variable illumination. This is a common problem for camera based systems. In this case, the problem is not easy to solve since the changes in polarisation level are small, and therefore any additional filtering of ambient light would also decrease the signal-to-noise ratio significantly. Instead of additional filters, the idea for future work is to perform averaging of images in a 10 − 20 sec time window, in order to stabilise the environment conditions. At the moment, it is under investigation if the narrow NIR-band polarisation filter would eliminate the influence of glares, strong sun light etc. Moreover, an active lighting will be implemented to the system in order to stabilise lighting conditions.

Another area where improvements are expected from the on-going work is the classification algorithm. The current algorithm is not flexible enough to determine clusters of different road conditions. Thus, the new classifier should include a graphical tool to determine the optimal boundaries of different clusters (ice, water, snow, dry asphalt). They can be even adapted, according to changing environment conditions, and semi-automatically by increasing size of the cluster if variation of the measures significantly changes. Look-up tables could also be used to adapt the setups and tune the processing according for instance to the time and season.

Finally, the potentiality of a colour camera to enrich the information using the spectral response is also an interesting option. For the human eye, snowy/icy and wet/dry asphalt are easily distinguished according to the colour response. This approach requires however an illumination with full spectral coverage and not just an emission in the infrared band.

Acknowledgements

We would like to express gratitude to the European Commission for funding the INTERSAFE-2 initiative and the FRICTION and SAFESPOT projects. We also thank the partners of the aforementioned projects for interesting discussions related to the design of the road state monitoring system and the architecture of the co-operative traffic safety platform. Finally, we would like to special thanks prof. Giancarlo Alessandretti due to reviewing and fruitful comments/remarks concerning content and readability of this article.

References

[1] SafetyNet. Annual Statistical Report 2007. Deliverable No: 1.16. Data based on Community database on Accidents on the Roads in Europe – CARE. Available in [http://ec.europa.eu/transport/road_safety/observatory/statistics/reports_graphics_en.htm]. cited on 8th Jan 2009. 2007.

[2] SafetyNet. Traffic Safety Basic Facts 2007 – Junctions. Data based on Community database on Accidents on the Roads in Europe – CARE. available in [http://ec.europa.eu/transport/road_safety/observatory/statistics/reports_graphics_en.htm]. cited on 8th Jan 2009. 2007.

[3] Casselgren, J. Sjödahl, M. & LeBlanc, J. Angular spectral response from covered asphalt. *Applied Optics*. Vol. 46, Issue 20. pp. 4277-4288. Society of America. U.S.A. 2007.

[4] Haas, T., et al., J. Fusion of vehicle and environment sensing in Friction project. *Proceedings of the 17th Aachen Colloquium "Automobile and Engine Technology" congress.* Germany, Aachen. 6-8 Oct 2008.

[5] Yamada, M., et al., Discrimination of the Road Condition Toward Understanding of Vehicle Driving Environments. *IEEE Transaction on Intelligent Transportation Systems.* Vol. 2. No. 1 pp. 26-31. 2001.

[6] Yuesheng, L. & Higgins-Luthman, M. Black ice detection and warning system. Patent application. Application number: US2007948086A. 2007.

[7] Vaisala Oyj web pages. [www.vaisala.com]. cited on 15th January 2008.

[8] Kutila, M., et al., S. Optical road-state monitoring for infrastructure-side co-operative traffic safety systems. *Proceedings of 2008 IEEE Intelligent Vehicles Symposium (IV'08).* the Netherlands, Eindhoven. 4-6 June 2008.

[9] ART SPY. Graininess Detection. Available in [www.owlnet.rice.edu/~elec301/ Projects02/artSpy]. Cited on 16th Jan 2009.

Matti Kutila, Maria Jokela
VTT Technical Research Centre of Finland
Tekniikankatu 1
33720 Tampere
Finland
matti.kutila@vtt.fi
maria.jokela@vtt.fi

Bernd Rössler
Ibeo Automobile Sensor GmbH
Merkurring 20
22143 Hamburg
Germany
berd.roessler@ibea-ass.com

Jürgen Weingart
Signalbau Huber GmbH
Kelterstraße 67
72669 Unterensingen
Germany
juergen.weingart@signalbau-huber.de

Keywords: camera, machine vision, traffic safety, intersection, co-operative traffic, road condition, ice, water, snow

Stereovision-Based Sensor for Intersection Assistance

S. Nedevschi, R. Danescu, T. Marita, F. Oniga, C. Pocol, S. Bota
Technical University of Cluj-Napoca
M.-M. Meinecke, M. A. Obojski, Volkswagen AG

Abstract

The intersection scenario imposes radical changes in the physical setup and in the processing algorithms of a stereo sensor. Due to the need for a wider field of view, which comes with distortions and reduced depth accuracy, increased accuracy in calibration and dense stereo reconstruction is required. The stereo matching process has to be performed on rectified images, by a dedicated stereo board, to free processor time for the high-level algorithms. In order to cope with the complex nature of the intersection, the proposed solution perceives the environment in two modes: a structured approach, for the scenarios where the road geometry is estimated from lane delimiters, and an unstructured approach, where the road geometry is estimated from elevation maps. The structured mode provides the parameters of the lane, and the position, size, speed and class of the static and dynamic objects, while the unstructured mode provides an occupancy grid having the cells labeled as free space, obstacle areas, curbs and isles.

1   Introduction

Intersections are the most complex traffic situations, as they can be both confusing and dangerous. The standard vehicle trajectories and the standard road geometries that are covered by most of the driving assistance systems do not apply in most of the intersections, but, on the other hand, most of the traffic accidents happen there. Due to the demanding nature of the scenario, the research community is increasingly focused on solving the perception and acting problems related to the intersection. The European research project INTERSAFE, part of the broader project PReVENT (http://prevent.ertico.web-house.net/en/prevent_subprojects/intersection_safety/intersafe/), had the goal to develop a system that was able to perceive the relative vehicle localisation, path prediction of other objects and communication with traffic lights, in order to warn the driver and simulate active measures. A new joint research project,

INTERSAFE-2 (www.intersafe-2.eu), aims at developing a system for cooperative sensor data fusion, based on state of the art passive and active on-board sensors, navigation maps, information from other traffic participants and from intelligent infrastructure, in order to generate a complete and accurate description of the complex intersection environment.

The dense stereovision sensor is maybe the sensor that provides the highest amount of usable information, as it combines the visual data with the dense 3D information that can be deduced through precise inference from the binocular view. A reliable stereovision sensor for urban driving assistance has been developed by our team [1]. However, the intersection scenario has some specific demands from a stereo sensor, demands that impose changes in the physical setup and in the software algorithms.

A wide field of view is essential for detecting and monitoring the relevant objects at an intersection, but that means lower focal length and consequently reduced working depth for stereo algorithms, and increased lens distortions in the images. The calibration algorithms, combined with an accurate image rectification step, must ensure that the quality of the reconstruction is not affected. The stereovision correlation process has to be performed on rectified images, by a dedicated stereo board to free processor time for the high-level algorithms. Even there are some methods reported in the literature for dense stereo systems calibration [2], [3], their accuracy is not well assessed. Therefore, a calibration method derived from previous approaches used for high precision and far range stereo-vision is proposed [4], [5], [6]. Camera parameters are estimated on the full-size images. For the extrinsic parameters calibration lenses distortion correction is performed in advance. This ensures the highest possible accuracy of the estimated parameters (which depends on the accuracy of the detected calibration features). The full-size images are further rectified, un-distorted and down sampled in a single step using reverse mapping and bilinear interpolation [7] in order to minimize the noise introduced by the digital rectification and image correction.

Due to the complex nature of the intersection, a solution for perceiving the environment in two modes is proposed: a structured approach, for the scenarios where the road geometry is estimated from lane delimiters, and an unstructured approach, where the road geometry is estimated from elevation maps.

Lane detection is one of the basic functions of any sensorial system dedicated to driving assistance and safety. The intersection scenario is the most difficult one for a lane tracking system, as many of the assumptions that the classical solutions rely on are not valid anymore. The good condition of the lane markings and the continuous nature of the road geometry on the highways lead to the first lane detection solutions, based on Kalman filtering, one of the first ones being presented in [8]. Since then, the Kalman filter was used, with some changes, for more difficult conditions. The solution presented in [9] used a backtracking system to select the measurement features in such a way that the false clues such as shadows or cracks in the road could be overcome (and thus the system could work with simpler delimiters, not only with markings). 3D lane estimation algorithms using stereo information in 3D space is presented in [10]. The solution presented in [11] uses the Kalman filter along with an elaborate selection of lane delimiters in the form of lines, in order to overcome the large quantity of possible lane delimiting cues and the short viewing distance that define the urban environment. All the Kalman filter solutions expect, more or less, a continuously varying road situation, and the discontinuities are usually handled by reinitializing the tracking process. The solution presented in [12] handles some particular case of road discontinuities by using two instances of the road model, but it is clear that the Kalman filter is not the best choice for tracking discontinuous roads.

Since the particle filter has been introduced in visual tracking under the name of condensation [13], several lane tracking solutions have been based on it. In [14] a lane detector based on a condensation framework, which uses lane marking points as measurement features is presented. Each point in the image receives a score based on the distance to the nearest lane marking, and these scores are used to compute the matching score of each particle. The system uses partitioned sampling (two-step sampling and measurement using subsets of the state space, achieving a multiresolution effect), importance sampling, and initialization samples (completely random samples from the whole parameter space) which cope faster with lane discontinuities. In [15] we find a lane detection system that uses the particle filtering framework to fuse multiple image cues (color, edges, Laplacian of Gaussian). This solution also uses initialization samples for faster lane relocation, and additional sampling around the best weighted particles for improvement of accuracy. The much simpler way in which a particle filter handles the measurement information allows the use of a wider range of cues. Such is the case of the lane detector for country roads, presented in [16], where the image space is divided into road and non-road areas and each pixel in these areas contribute to the final weight by its intensity, color, edge and texture information. The work presented in [17] also shows the value of the particle filtering technique for heterogeneous cue fusion, when image information is fused with GPS and map information for long distance lane estimation.

In [18] a lane tracking technique based on stereovision and particle filtering for handling discontinuous urban scenarios is described. For the intersection applications, the proposed lane detection system based on stereovision and particle filtering had to be extended and reorganized, so that it can handle multiple lane models and multiple measurement cues. These changes are designed to increase the robustness of lane tracking through the intersection, and give the system capability to handle various lane geometries. Due to the modular and configurable nature of the tracking system, we aim to extend it to detect other painted road structures, such as arrows and pedestrian crossing signs.

In structured environments, the obstacles are modeled as cuboids, having position, size, orientation and speed. From the top view of the dense stereo data, the cuboid reconstruction algorithms analyze the local density and vicinity of the 3D points, and determine the occupied areas which are then fragmented into obstacles with cuboidal shape: without concavities and only with 90° convexities. The orientation of the obstacles also is determined in order to get a very good fitting of the cuboidal model to the obstacles in the scene ahead and, consequently, to minimize the free space which is encompassed by the cuboids. The cuboids are tracked using a Kalman filter-based approach, in order to reduce the measurement noise and improve the stability of the results, and to measure the object speed. An experimental tracking algorithm, based directly on 3D raw data, is also proposed in this paper.

An intersection assistance system is required to make decisions based on all the relevant factors present at the intersection environment. One important factor is the class of obstacles involved, either mobile or belonging to the intersection infrastructure. Relevant object classes include vehicles, pedestrians, bicyclists, curbs, pillars, trees and traffic signs. Discriminating between these object classes provides opportunities for implementation of comfort functions and active safety functions in intelligent vehicles.

Pedestrians are especially vulnerable traffic participants. Therefore, recognizing the pedestrian class is extremely important, and a lot of scientific research has been conducted in this area. Classification methods are traditionally limited to 2D image intensity information. A contour matching approach is described in [19]. Another technique using Adaboost with illumination independent features is described in [20]. There are some classification methods that use 3D information directly like [21] and [22]. The use of template matching in conjunction with 3D information has not been extensively explored. Some approaches aimed at pedestrian detection have used dense 3D information, but only as a validation method [23]. Another important feature for walking pedestrian detection is their walking pattern. There are a number of works, e.g. [24], [25] which have used motion cues in pedestrian detection. Other types of information such as IR images have also been used, as described [26].

In this paper we present a multi-feature, multi-class approach to classification. We use 2D intensity image information, 3D information and motion in order to obtain a reliable classification. Our detected classes include pedestrians, cars and poles. The architecture of our classifier allows us to easily incorporate more feature and more classes.

Intersections in urban areas often have no lane markings or these are occluded by crowded traffic. The lanes cannot be detected in such particular scenes, thus the obstacle/road separation becomes difficult. An alternative/complementary method must be used to detect elevated areas (obstacles), regions where the ego vehicle cannot be driven into. Complementary, the obstacle-free road areas can be considered as drivable. Starting from the representation of the dense 3D data as an elevation map, an occupancy grid which defines the drivable area, traffic isles and obstacles is proposed and implemented.

Digital elevation maps (DEMs) are less used to represent and process 3D data from stereovision, but they offer a compact way to represent 3D data, suitable for real-time processing. A complex method for building the digital elevation map of a terrain (for a planetary rover) is proposed in [27]: local planar surfaces are used to filter the height of each DEM cell, and the stereo correlation confidence for each 3D point is included in the filtering process. In [28] the elevation map is built straightforward from the disparity map. The authors avoid using a 3D representation of the reconstructed points by projecting a vertical 3D line for each DEM cell onto the left, disparity, and right image. Based on these projections, the disparity of the point associated with the cell is selected and possible occlusions are detected. The obtained DEM is used for a global path-planning algorithm.

The road, traffic isle and obstacle detection algorithm presented in [29] uses digital elevation maps to represent 3D dense stereo data. This method is suitable for intersection scenarios, since it does not rely on high-gradient road features (lane markings) for the road surface detection. However, if the road surface is not detectable (poor texture resulting in few 3D points) an alternative to detect curbs (traffic isle borders) was designed and presented in [30]. A combination of these two approaches, more robust and suitable for intersection scenes, is presented in this paper.

The output provided by the stereo sensor supports intersection driving assistance systems in multiple ways: detection of the approaching intersection, intersection sensorial modeling, relative localization of the ego vehicle, precise alignment of the intersection sensorial model and the corresponding intersection map, obstacle detection, tracking and classification, collision prevention and collision mitigation.

2 Stereo System Setup and Calibration

2.1 Stereo System Architecture

The proposed stereovision system architecture (Fig. 1) is derived from our previous experience gained with urban driving assistance application based on dense-stereovision [1]. Adjustments were made in order to fit the requirements of intersection assistance applications. E.g. a very large horizontal field of view is compulsory for such applications which imply short focal length lenses. Short focal lenses are introducing pronounced radial lens distortions which should be corrected before the stereo reconstruction. The intersection is mainly an unstructured road environment (sometimes, there are no lane delimiters, only curbs) and therefore a dense 3D points' map is required for providing rich information.

The architecture has a versatile and hybrid structure: some modules can be implemented on dedicated hardware boards or as software libraries in a conventional PC architecture.

Two monochrome (grayscale) imagers mounted in a quasi-canonical configuration are used. The image sensors can be integrated in a dedicated stereo-head or built from stand-alone cameras mounted rigidly on a stereo-rig. The stereo cameras (Fig. 1- module 1) can be interfaced (Fig. 1- module 2) trough a framegrabber, a dedicated hardware interface or a common PC interface (Firewire, USB). Image acquisition is controlled by a software interface (Fig. 1– module 3) specific for the used cameras. The quality of the acquired images is controlled by tuning the Gain and Exposure parameters of the cameras through an auto-adaptation to the lighting condition module (Fig. 1– module 4).

The stereo images are acquired at the full resolution of the image sensors. They are further rectified (to meet the canonical configuration), corrected (to eliminate the lens distortions) and down-sampled (to meet the processing capacity of the hardware dense-stereo reconstruction engine – e.g. 512 pixels in width). This process (Fig. 1 – module 5) can be performed on dedicated hardware or implemented in software. The software implementation of this process is reduced to an image warping approach performed in a single step using reverse mapping and bilinear interpolation [7] in order to minimize the noise introduced by the digital rectification and image correction. The implementation is optimized for processing speed-up using MMX instructions and lookup tables.

The rectified (canonical) down-sampled images are fed to the stereo reconstruction module (Fig. 1 – module 6) which is implemented in hardware. From the depth map provided by the stereo-engine the 3D coordinates of reconstructed points (Fig. 1 – module 7) are computed knowing the camera parameters and are further provided to the processing modules (Fig. 1 – module 8).

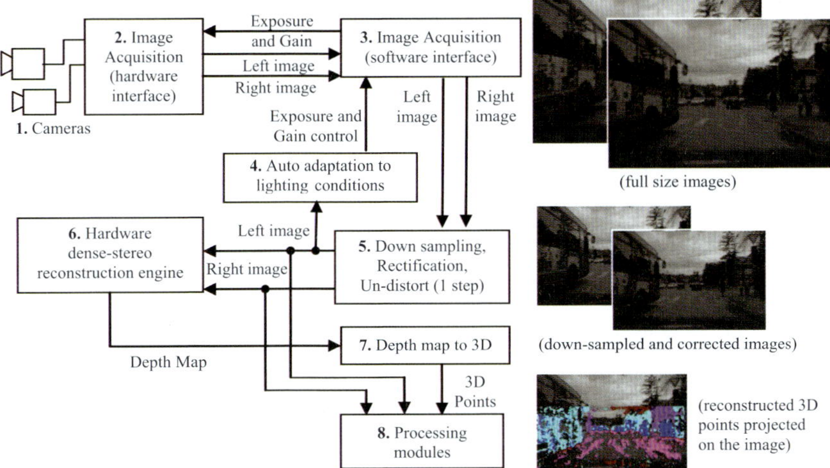

Fig. 1.   The stereovision system architecture

2.2   Stereo Cameras Setup

The tuning of the image acquisition system to specific application requirements can be done by choosing the proper parameters of the stereo cameras: focal length of the lenses, imager resolution and size, baseline length.

Intersection assistance applications require a very large horizontal field of view (HFOV) but in the same time an accurate 3D reconstruction in a depth range from 0 m up to 20 … 25 m, measured from the front of the ego car. Spherical lenses (with HFOV > 90°) do not comply with this detection range. Therefore pinhole lenses with short focal lenses were chosen. Having the focal length as a known constant, the detection range is adjusted by choosing the proper baseline length (the horizontal displacement between the two cameras). The lowest limit of the depth range is computed with the formula bellow, while the upper limit is determined experimentally.

$$Z_{MIN} = \frac{b \cdot f}{d_{MAX}} \approx 0 \tag{1}$$

Z_{MIN}	lower limit of the detection range
b	is the baseline (horizontal displacement between the cameras) [mm]
f	focal length [number of pixels]
d	horizontal disparity between reconstructed features [pixels]. The values of d_{Max} is stereo engine dependent.

Two setups are proposed (Tab. 1):

1.   The first one uses an integrated stereo-head along with a dedicated board that performs the image acquisition along with the automatic exposure and gain control, image down-sampling, correction and rectification, and dense stereo reconstruction (Fig. 1 - modules 2, 4, 5 and 6). This setup has the advantage of using a compact size, cheap stereo head but provides medium quality images due to the CMOS imager and miniature lenses.

2.   The second one uses a stereo-head built from stand-alone cameras with 1.3 megapixel CCD sensors and high quality lenses, mounted rigidly on a stereo-rig. The cameras are interfaced with the PC with a dedicated frame grabber trough the standard CameraLink interface. Functionalities of modules 4, 5 (Fig. 1), are implemented in software libraries, while the dense stereo-reconstruction (Fig. 1 - module 6) is performed on the same dedicated hardware board. This second setup has the disadvantage of larger stereo-head size and higher system price, but provides the best image quality.

System no.	1	2
Full size image [pixels]	752x480	1376x1030
Imager size (inch)	1/3	2/3
Pixel size [mm]	6.00	6.45
Lens focal [mm]	4.00	4.80
HFOV [deg]	75	90
Focal - full size image [pixels]	618	744
Focal – down-sampled images [pixels]	421	276
Baseline - b [mm]	220	340
Max. disparity - d_{Max} [pixels]	52	52
$Depth_{Min}$ (camera) [m]	1.78	1.81
$Depth_{Min}$ (car) [m]	-0.02	0.01
$Depth_{Max}$ (car) [m]	33.68	34.14
$Depth_{Max}$ (car) (75% $Depth_{MAX}$) [m]	25.26	25.60

Tab. 1.   Comparison of the two stereo setup parameters

2.3 Camera Calibration

In order to reconstruct and measure the 3D environment, the cameras must be calibrated. The calibration process estimates the camera's intrinsic and extrinsic parameters. Both intrinsic and extrinsic parameters are estimated on the full-size images. For the extrinsic parameters calibration lenses distortion correction is performed in advance. This ensures the highest possible accuracy of the estimated parameters (which depends on the accuracy of the detected calibration features/control points).

The intrinsic parameters of each camera are calibrated individually using the Bouguet's [31] method. The accuracy of the obtained intrinsic parameters depends on the quality of the calibration object (size, planarity, accuracy of the control points' positions) and on the calibration methodology used (number of views, number of control points on each view, coverage ratio of each of the view by the calibration object etc.). If the proper conditions are fulfilled, the uncertainties of the estimated parameters are below one pixel for the focal length and principal point.

For the extrinsic parameters estimation an original method dedicated for high range stereo-vision was developed [4]. The estimated parameters are the T_L and T_R translation vectors and R_L and R_R rotation matrices of each camera. The extrinsic parameters calibration method assures not only very accurate absolute extrinsic parameters of each camera individually but also very accurate relative extrinsic parameters, which allows a precise estimation of the epipolar lines (near 0 pixel drift) [5] with beneficial effects for the stereo matching process. Therefore the method is suited for the calibration of any stereovision system used for far or mid range 3D reconstruction like in vision based driving assistance systems.

3 Structured Approach for Environment Description

3.1 Lane Detection

The lane tracking system for intersection safety applications was designed in a modular fashion, the equations for the vertical and horizontal profile being easily configurable, as lanes at the intersections have varying geometries. The measurement process is independent on the 3D model, as long as sets of 3D points for the delimiters are available.

We have designed a general-purpose particle filtering framework that is the basis for the lane tracking system, and that can be easily extended to cover multiple situations (Fig. 2). The *abstract particle* class contains the matrices and vectors that can describe any process, and provides interface for specific methods of initialization, measurement, drift and specifics. The *particle set class* aggregates particles regardless of their specifics, and implements the Resample – Drift – Measure – Estimate cycle by calling individual particle methods. This class also provides abstract interface for specific calculations, such as computing the trajectory curvature from CAN sensor data.

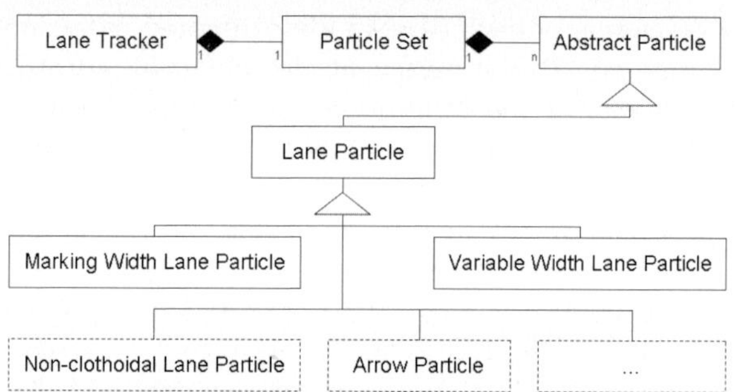

Fig. 2.   Base hierarchy for developing lane detection solutions

Building a lane detection and tracking system based on this framework means extending the Abstract Particle class with a Lane Particle class, that will implement the lane specifics, such as projecting the 3D lane model into the measurement data space, or the measurement itself, which means weighting the lane hypothesis particle against the measurement data.

The model-independent measurement process relies on projecting chains of points for each lane delimiter in the image space, and use the distance transform for the measurement cues to compute a distance between the cues and the lane projection. Measurement cues can be lane markings, image edges that are located on the road surface, or projections of the curb points detected using dense stereo analysis. For each delimiter we use two rows of points, one row that is located on the delimiter (the delimiter points) and one row located inside the lane surface, near the delimiter (the inner points). Ideally, we would like to have a minimum distance for the delimiter points (they should fall on the delimiting features) and a maximum distance for the inner points. The two distances are combined, and a probability measure that is higher as the distance is lower is computed. A more detailed explanation of the measurement process is given in [18].

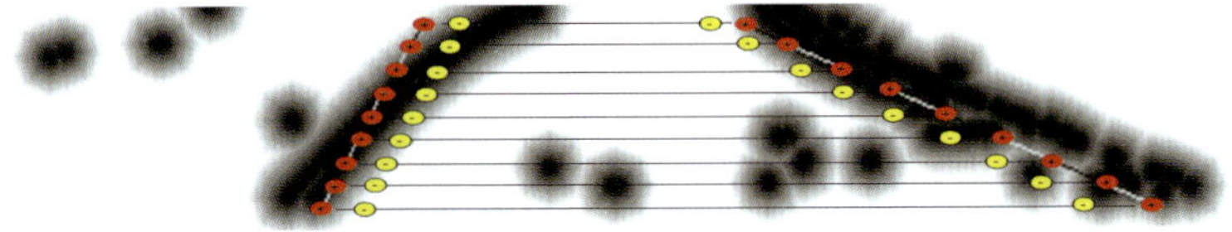

Fig. 3.   Boundary and inner points. For the boundary points, the distance is added, and for inner points the distance is subtracted

Using a new 3D lane model implies that we modify only the initialization of the vectors, the building of the state transition matrices, and the factory method. A specific function GetLaneXY that retrieves the X and Y positions given Z and the model is used to decouple the measurement from the 3D model. Implementing the system based on the new model, by inheritance of the CLaneParticle class and modification of the specifics of the model takes about 100 lines of code, mainly declarations.

We have implemented two extensions from the classical lane description (as used in [18]): modeling the lane delimiter's marking width and modeling a variable lane width. For modeling the lane marking width, the lane state vector had to be enhanced with two more parameters, and the measurement process had to include an additional cue – a horizontal gradient image computed with a perspective-aware kernel, which allows for a smooth transition towards the maximum. The gradient image itself was used as a distance image. As a side effect, estimating the marking width in this manner improved the lane estimation stability in cases when the estimation was difficult (such as when only one delimiter is present).

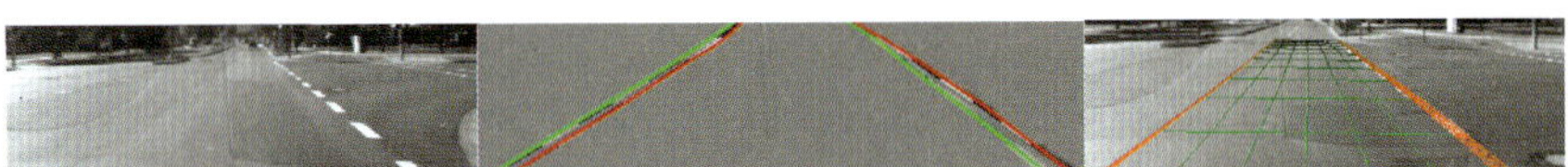

Fig. 4.   Estimation of the lane marking width

Modeling and estimating a lane with variable width required much less than modeling the lane marking width, because the measurement system could be reused completely. Only the functions describing the 3D model needed to be overloaded.

Fig. 5.   Variable width model

The lane estimation framework based on particle filters will be extended in the future towards detecting target lanes (lanes we are not on yet), and this will require changes in the model's restrictions and changes in the measurement process (we have to rely more on horizontal features). Also, a future use will be to detect "Capped" lanes, which are lanes shorter than the detection distance, ending with a horizontal line, that usually mark the beginning of the intersection. They require a change in the model (addition of the length) and in the measurement (addition of the ending cue).

3.2   Obstacle Detection and Tracking

Starting from the 3D points situated above the 3D profile of the lane, obstacles described by cuboids, having position, orientation and size, are detected. The confident fitting of cuboids to obstacles is achieved in several steps. By analyzing the vicinity and the density of the 3D points, the occupied areas are located. An occupied area consists of one or more cuboidal obstacles that are close to each other. By applying shape criteria, the occupied areas may get fragmented into cuboidal parts. The orientation of the obstacles on the road surface is extracted as well.

The 3D points above the height of the ego vehicle are discarded (Fig. 6.b). It is supposed that the obstacles do not overlap each other on the vertical direction. In other words, on a top view (Fig. 6.c) the obstacles are disjoint. Consequently, in what follows the elevation of the 3D points will be ignored and all the processing is done on the top view.

Due to the perspective effect of the camera, further obstacles appear smaller in images, providing fewer pixels, and therefore, less and sparser 3D reconstructed points in the 3D space. On the other hand, the error of the depth reconstruction increases with the distance too, which, contributes to the 3D points sparseness as well. To counteract the problem of the points' density, a schema to divide the Cartesian top view space into tiles with constant density is proposed (Fig. 6.c). The horizontal field of view of the camera is divided into polar slices with constant aperture, trying to keep a constant density on the X-axis. The depth range is divided into intervals, the length of each interval being bigger and bigger as the distance grows, trying to keep a constant density on the Z-axis.

A specially compressed space is created, as a matrix (Fig. 7.a). The cells in the compressed space correspond to the trapezoidal tiles of the Cartesian space. The compressed space is, in fact, a bi-dimensional histogram, each cell counting the number of 3D points found in the corresponding trapezoidal tile. For each 3D point, a corresponding cell C (*Row, Column*) in the compressed space is computed, and the cell value is incremented.

The column number is found as:

$$Column = ImageColumn/c \tag{2}$$

where *ImageColumn* is the left image column of the 3D point and c is the number of adjacent image columns grouped into a polar slice as shown in Fig. 6c ($c = 6$).

The depth transformation, from the Z coordinate of the Cartesian space into the *Row* coordinate of the compressed space has a formula obtained using the following reasoning:

a) The Cartesian interval corresponding to the first row of the compressed spaces is:

$$[Z_0 \ldots Z_0 + IntervalLength(Z_0)] = [Z_0 \ldots Z_1] \tag{3}$$

where $Z_0 = Z_{min}$, the minimum distance available through stereo reconstruction. The length of the interval beginning at a certain Z is

$$IntervalLength(Z) = k*Z \tag{4}$$

k being empirically chosen). Thus

$$Z_0 + IntervalLength(Z_0) = Z_0 + k*Z_0 = Z_0*(1+k) = Z_1 \tag{5}$$

b) The Cartesian interval corresponding to the nth row of the compressed spaces is

$$[Z_n \dots Z_n + IntervalLength(Z_n)]$$
$$Z_n = Z_0 * (1+k)^n \qquad (6)$$

The above equation can be proven by mathematical induction.

c) For a certain 3D point, having depth Z, the ith interval it belongs to is

$$[Z_i \dots Z_i + IntervalLength(Z_i)] = [Z_i \dots Z_{i+1}].$$

From equation (6), we can find the row number as the index of the point's interval

$$Row = i = \left[\log_{1+k} \frac{Z}{Z_0} \right] \qquad (7)$$

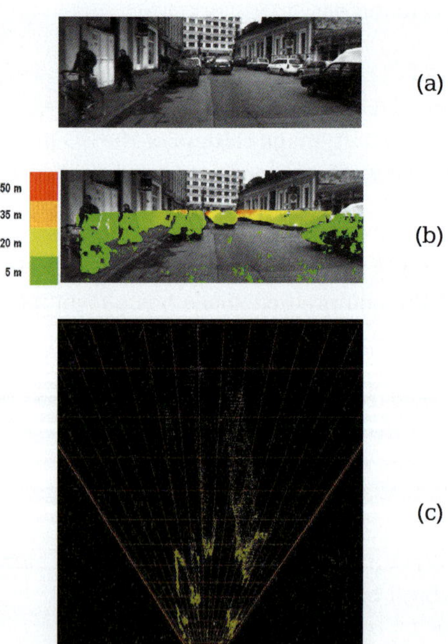

(a)

(b)

(c)

Fig. 6.   Division into tiles. a) Gray scale image, b) 3D points – perspective view, c) 3D points – top view; the tiles are here considerably larger for visibility purpose; wrong reconstructed points can be seen in random places; reconstruction error is visible as well

The histogram cells that have a significant number of points indicate the presence of obstacle areas. On these cells, a labeling algorithm is applied, and the resulted clusters of cells represent occupied areas (Fig. 7.b). The small occupied areas are discarded in this phase.

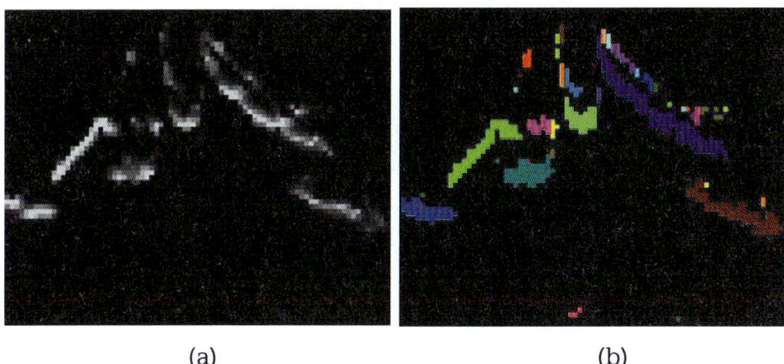

(a)   (b)

Fig. 7.   The compressed space (for scene in Fig. 6) – a bi-dimensional histogram counting 3D points.
Occupied areas are identified by cell labeling

The occupied areas may contain several obstacles and by consequence they may have miscellaneous shapes. The obstacle tracking algorithms as well as the driving assistance applications need the individual cuboidal obstacles. Therefore the fragmentation of the occupied areas into the individual cuboidal obstacles is required.

An indication that an area requires fragmentation is the presence of concavities. In order to detect the concavities, the envelope of the cells of an occupied area is drawn, and then for each side of the envelope, the gap between the side and the occupied cells is determined. If the gap is significant, we consider that two or more obstacles are present, and the deepest point of the concavity gives the column where the division will be performed. The two sub-parts are subject to be divided again and again as long they have concavities.

In Fig. 8.c the bottom side of the envelope for the cells in Fig. 8.b delimits a significant concavity. For each new sub-part, the envelope of the cells has been calculated again (and painted as well), but without revealing big concavities for new divisions to be performed.

By reconsidering the coordinates (including the height) of the 3D points that have filled the cells of an obstacle, the limits of the circumscribing box are determined. Boxes are shown in Fig. 8.d (perspective view) and Fig. 8.e (top view).

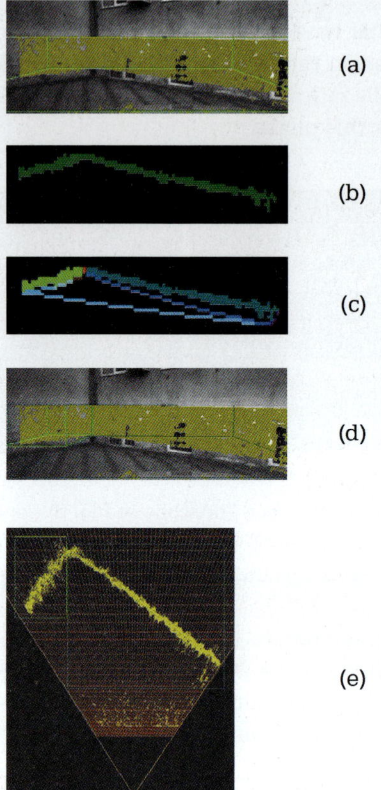

Fig. 8.   Fragmentation of occupied areas into primitive obstacles. a) an occupied area, b) the labeling in the compresses space, c) sides of the envelope and the two primitive obstacles of the occupied area – compressed space, d) the two primitive obstacles – perspective view, e) the two primitive obstacles – top view

The next step is obstacles orientation estimation. In Fig. 9.a, it can be observed that, even though the real obstacle is oblique oriented, the cuboid encompasses some free space because the cuboid is parallel with the coordinate system. The shape of the cloud of 3D points is modeled by their envelope and an analysis of its visible sides (towards the camera) can estimate the orientation of the obstacle. If the analysis cannot determine a preponderant orientation of these sides, the box remains un-oriented with sides parallel with the axes of the coordinates system.

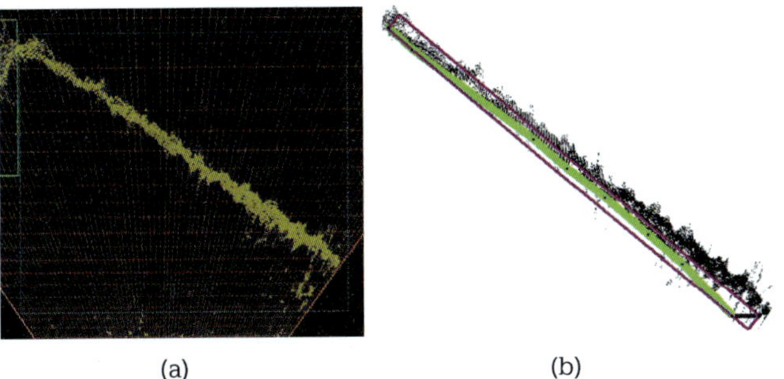

(a)   (b)

Fig. 9.   Obstacle orientation (top view): a) un-oriented box b) the longest chain of visible envelope sides (green) and its surrounding rectangle

Another case when an occupied area must be fragmented is signaled by the following observation. A real obstacle is accurately detected when the space between the sides of its oriented cuboid and the visible shape of the cloud of its 3D points (on the top view) has a small area. When a free space of more than 10 cells (an empirically chosen threshold) between the cuboid corners and the inner cloud appears a fragmentation into two or more sub-obstacles is necessary.

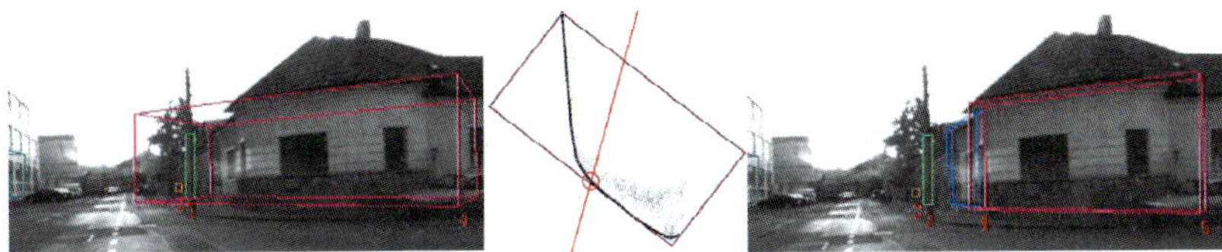

Fig. 10.   Fragmentation of an obstacle that doesn't have a cuboidal shape: a) initial cuboid, b) top view of the initial cuboid and the optical ray of the fragmentation, c) the two sub-obstacles obtained by fragmentation

An example is shown in Fig. 10.a and b, where no single oriented cuboid would fit. The fragmentation is done by splitting along an optical ray that passes through one of the sharp vertices of the unoccupied triangle. The procedure is recursively applied on the newly obtained obstacles if needed. The fragmentation of the large cuboid in Fig. 10.a is shown in Fig. 10.c.

A more detailed description of the obstacle detection system is given in [32]. Some examples, showing obstacle detection in intersection scenarios, are presented in Fig. 11:

Fig. 11. Results on obstacle detection in intersection scenarios

As a major resource for improving the obstacle detection accuracy for complex intersection scenarios the use of color images and optical flow are being studied.

Tracking is the final stage of the stereovision based object detection system. At the end of this step, the sensor outputs the cuboid list containing parameters of position, size, orientation and speed. For tracking, we have two methods: an established solution based on the Kalman filter [33], which uses as measurement the cuboids extracted using the methods described in the previous paragraphs, and an experimental one, based on particle filters, which uses raw 3D obstacle data.

The new solution is based on multiple particle filters, and does not require points grouping, working directly on the raw obstacle data projected on the top view, data which can be provided, for instance, by elevation map processing [29]. This solution is a multiple object tracking scheme that uses a two-level approach. First, a particle filter-based tracker that starts with an initial probability density function that covers the whole possible object parameter state space runs freely on the 3D data, and its particles will tend to cluster around one object. When the clustering is achieved, the particle state is passed to an individual object tracker, that will track the object until it disappears, and the initialization tracker is reset and will start searching for another object.

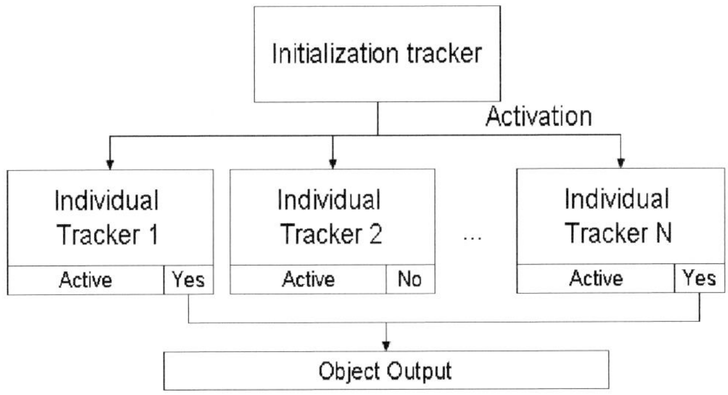

Fig. 12.   Top-level architecture of the particle filter-based tracking system

Although the individual trackers and the initialization tracker are built around similar particle filters, their behavior is different. Fig. 13.a shows the flowchart for the initialization tracker. We'll briefly describe each block in the diagram.

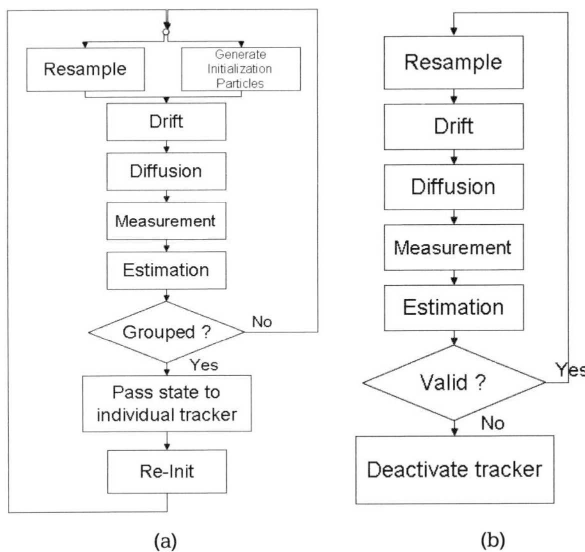

Fig. 13.   a) Flowchart of operation for the initialization tracker, b) Flowchart
of operation for the individual object tracker

Resample – the process that transforms the weighted particle set describing the past probability density into an equivalent set of weightless particles, by random weighted sampling.

Generation of initialization particles – a fraction of the particle set will be sampled from the initial probability distribution, that is, from the whole range of possible object states. These particles help the tracker evade a false hypothesis faster.

Drift – the uniform motion dynamic model is applied to the particles (details in the next section)

Diffusion – a random value is applied to each particle, accounting for the uncerties of state transition.

Measurement – each particle is compared to the measurement data, and weighted according to a fitness score.

Estimation – an estimated object state is computed by weighted average of the particle values. This step also computes a standard deviation value for each estimated parameter.

"Grouped" decision - The initialization tracker uses only the standard deviations of the object position to decide whether the particles are grouped. If they are not grouped, the tracker starts another cycle. If they are grouped, the whole particle distribution is passed to an individual tracker, which is made active, and the initialization tracker is reset.

The operation of the active individual object tracker is depicted in Fig. 13.b. The basic blocks are similar to the initialization tracker, but there are some notable differences. First, there are no initialization particles, as these trackers operate at local level, their target already selected. Second, after estimation the next block is a "valid" condition, which tests the grouping of the particles, the position and size of the estimated object (if grows too big, or too small, or goes out of the field of view, it becomes invalid). If the tracker becomes invalid, it goes into the inactive state, and the tracked object is declared lost.

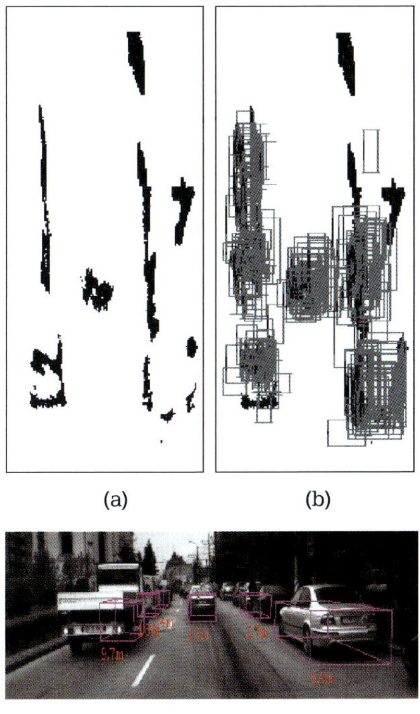

(a)   (b)

(c)

Fig. 14. a) Obstacle areas in 3D top view, b) Particles of the object trackers,
c) Results (estimations) in perspective projection

3.3 Obstacle Classification

Although recognition of various obstacle types is easy for humans (at least in normal visibility conditions) it proves to be an extremely challenging task for machine vision systems. In order to cope with the problem's difficulty we designed a complex, multi-class, multi-feature, multi-stage classification system. All the information supplied by the stereovision sensing approach: intensity images, range information and motion is used by our classifier. The training and assessment of the classifier was performed using an extremely large database of manually labeled stereo-data (approximately 50000 object instances). The architecture of our classifier is presented in Fig. 15.

The input of the classifier is a list of 3D tracked cuboids, the set of reconstructed 3D points and the left intensity image. The first step performed is to create depth masked images, i.e. images that capture the foreground pixels for each object (see Fig. 16). These images contain only the points which are inside the object box, thus eliminating background points. All subsequent feature extractors use only the foreground points.

A number of simple features are extracted in the first step: the 3D box dimensions (width, height & depth) and the lateral and longitudinal speed. For tracked, non-stationary objects, the motion signature feature [34] is extracted and computed. This feature represents the variance of the 3D motion vector fields. It is based on the computation of 2D optical flow vectors for corner points [35] which belong to the object (hence the need for masking) followed by their back-projection in 3D. The motion signature is large for articulated, moving objects such as pedestrians and small for rigid objects such as cars. Another feature related to the motion signature is its periodicity. Periodic motions are more reliable features, because occasionally, an object may have a large, spurious motion signature caused by noise in optical flow computation. An example of using the motion signature and motion periodicity features is presented in (Fig. 17).

All the above features apply to all object classes. The remaining features are more class specific in the sense that they help validating a hypothesized class or discriminate between two similar classes (dichotomizers). Therefore, it is unnecessary to extract these features for all objects. In order to choose which features to extract for which object, a coarse classification is performed, based on the features extracted thus far (object width, height, depth, lateral and longitudinal speed, motion signature and motion periodicity). The result of this coarse classification is a class probability distribution.

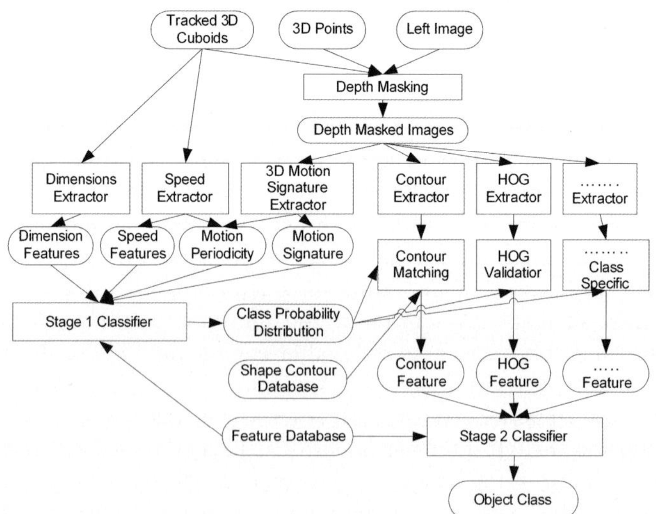

Fig. 15.   Architecture of the object classifier

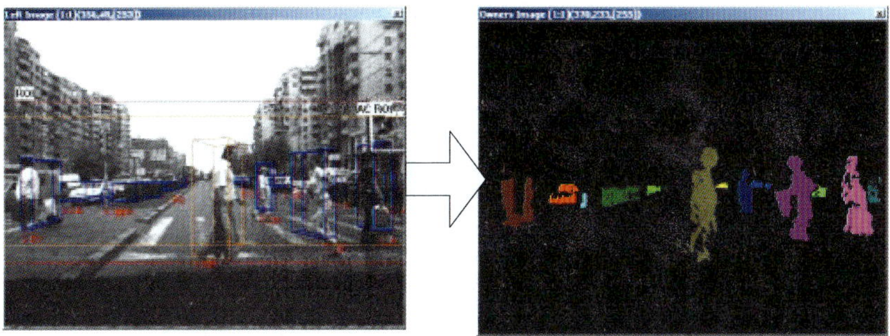

Fig. 16. Depth masking - background pixels are black, color pixels belong
to objects

The remaining features are extracted only if the relevant class or classes for
which they apply have a high enough probability according to the coarse classification. The first such feature is the object's contour, which is matched with
a hierarchy of high-quality template contours extracted using a controlled fixed
background scenario. The matching is performed using distance transform as
presented in [19]. Both the template is matched with the distance transformed
extracted contour and the distance transformed template is matched with the
extracted contour. The two resulting distances are added together and form
the contour matching feature. This ensures a more reliable matching process.
We currently use contour matching only for pedestrians (Fig. 18).

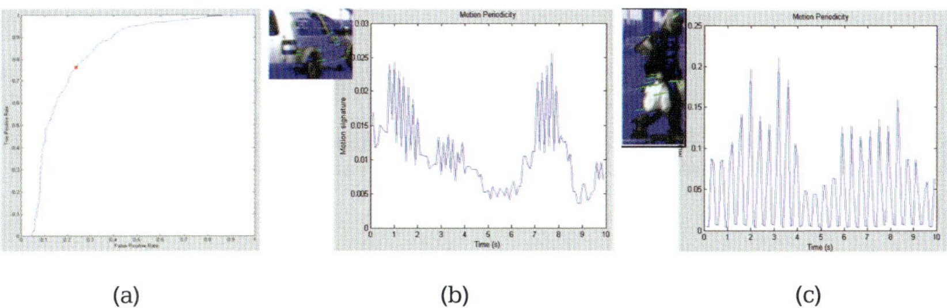

(a)   (b)   (c)

Fig. 17. a) ROC curve showing the relevance of motion signature for pedestrian recognition. b) Motion signature evolution in time for rigid
objects. c) Motion signature evolution in time for pedestrians

Another feature we use is the histogram of oriented gradients [36]. This is
applied for both pedestrians and cars. Because of the flexible architecture of
our classifier, many more features can be added, as suggested in Fig. 15. After
all these features are extracted, they enter the stage 2 (final) classifier, which
picks up the class with the highest probability for each object, and outputs it.

Fig. 18.   Pedestrian contour matching. Green contours are extracted from the images and yellow contours are templates matched with these contours

Both the coarse classifier and the final classifiers were trained using the WEKA environment [37] and are implemented by decision trees. We evaluated our classifier using two scenarios. In the first scenario 66% of the dataset was used for training and the remaining 34% was used for testing. The classes considered were car, pedestrian, pole and other. Out of a number of 3744 of instances, 3110 were correctly classified (83.0662%). The results are presented in Table 2. For the second scenario we performed a 10 fold cross validation using a number of 11013 instances. The "Other" class was eliminated. Out of 11013 instances, a number of 9223 (83.7465 %) instances were correctly classified. The results are summarized in Tab. 2.

TP Rate	FP Rate	Precision	Recall	F-Measure	ROC Area	Class
0.899	0.095	0.823	0.899	0.859	0.966	Car
0.914	0.048	0.872	0.914	0.892	0.974	Pedestrian
0.815	0.009	0.864	0.815	0.839	0.978	Pole
0.706	0.096	0.797	0.706	0.748	0.895	Other

Tab. 2   Classification results for scenario 1

TP Rate	FP Rate	Precision	Recall	F-Measure	ROC Area	Class
0.877	0.071	0.856	0.877	0.866	0.968	Car
0.909	0.048	0.869	0.909	0.889	0.974	Pedestrian
0.814	0.01	0.844	0.814	0.829	0.978	Pole

Tab. 3.   Classification results for scenario 2

These are only preliminary results. The aim is to build a real-time classifier, with an extremely high classification rate of pedestrians, cars, poles, traffic signs, trees etc. that will provide the necessary information for an intersection assistance system.

4 Unstructured Approach for Environment Description

Intersections in urban areas often have no lane markings or these are occluded by crowded traffic. For such scenarios a more general approach has to be used. An occupancy grid, which defines the drivable area, traffic isles and obstacles, should be computed, by using digital elevation maps to represent 3D data.

4.1 Computing the Occupancy Grid from Elevation Maps

The road, traffic isle and obstacle detection algorithm presented in [29] uses digital elevation maps (DEMs) to represent 3D dense stereo data. The DEM is enhanced based on the depth uncertainty and resolution models of the stereo sensor. A RANSAC-approach, combined with region growing, is used for the detection of the optimal road surface. Obstacles and traffic isles are detected by using the road surface and the density of 3D points. This algorithm outputs the road surface parameters and an occupancy grid (Fig. 19.c) with three distinct cell types: road, traffic isles and obstacles.

A region of interest of the 3D space is represented as a digital elevation map (DEM). The DEM is a rectangular grid (matrix) of cells and the value of each cell is proportional to the 3D elevation of the highest point (within the cell). The DEM presents poor connectivity between road points at far depths (due to the perspective effect). Connectivity is improved by propagating (to empty cells) the value of valid cells (with 3D points), along the depth (Fig. 19.b). The propagation amount is computed from the stereo geometry of the system, to compensate the perspective effect. Additional features are stored with each DEM cell, such as the density of 3D points per cell.

(a)

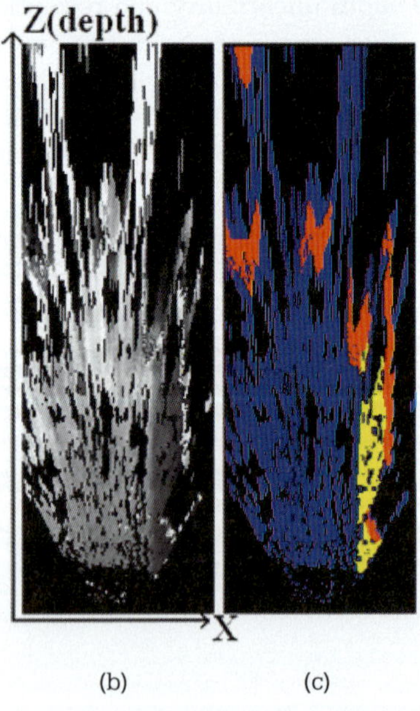

(b)   (c)

Fig. 19.   a) top: A common urban scene, bottom: the occupancy grid projected onto the left image b) The elevation map c) The occupancy grid resulting after classification: road cells with blue, traffic isle cells with yellow, and obstacle cells with red

A quadratic surface is used to model the road, instead of the less general planar model. A RANSAC-approach, combined with region growing, is proposed to compute the parameters of the surface. First, the road model is fitted, using RANSAC, to a small rectangular DEM patch in front of the ego vehicle. A primary road surface is extracted optimally for the selected patch. Next, the primary solution is refined through a region growing process, where the initial region is the set of road inliers of the primary surface, from the initial rectangular patch. New cells are added to the region if they are on the region border and they are inliers of the current road surface. The road surface is recomputed (LSQ fitting to the current region) each time its border expands with 1-2 pixels.

The uncertainty of the stereo sensor is modeled and used to discriminate between road inliers and outliers (Fig. 20). Using a height uncertainty increasing with the depth is more reliable (compared to a constant uncertainty strip) and in accordance with the stereo system characteristics:

Obstacle / road separation is performed based on the detected road surface. Each DEM cell is labeled as drivable if it is closer to the road surface than its estimated height uncertainty or as non-drivable, otherwise. Small clusters of non-drivable cells are rejected based on criteria related to the density of 3D points. Remaining non-drivable clusters are classified into obstacles or traffic isles, based on the density of 3D points.

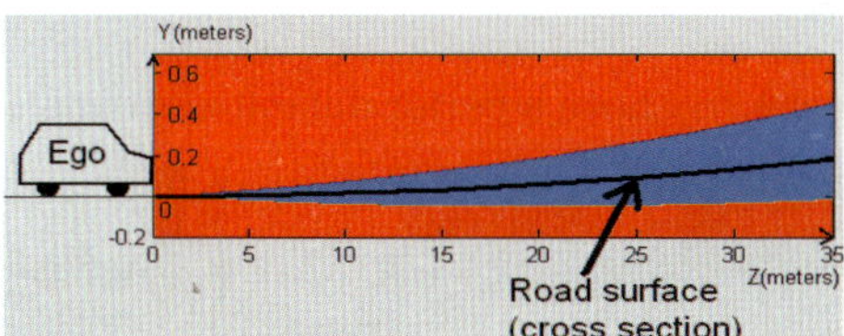

Fig. 20.   Lateral view (a slice from 3D with a constant value for the X coordinate) of the inliers (blue) area around a quadratic road

This algorithm outputs the classified grid (Fig. 19.c) with three distinct cell types: road, traffic isles and obstacles. There are some difficult scenarios, with poor road texture or poor scene illumination, where only few 3D road points are reconstructed through stereovision. The road surface cannot be computed for such scenes, but the algorithm is still able to output an occupancy grid made of drivable and obstacle cells. In such scenes only the density of 3D points is used for discrimination, but the traffic isles cannot be identified.

The curb detection algorithm presented in [30] detects curbs as chains of segments on the DEM. Edge detection is applied to the DEM in order to highlight height variations. A method to reduce significantly the 3D noise from dense stereo was proposed, using a multi-frame persistence map. Temporal filtering was performed on edge points, based on the ego car motion, and only persistent points were validated. The Hough accumulator for lines was built with the persistent edge points. A scheme for extracting straight curbs (as curb segments) and curved curbs (as chains of curb segments) was proposed. Each curb segment was refined using a RANSAC approach to fit optimally the 3D data of the curb (Fig. 21).

Fig. 21.   A curb is detected in front of the ego while turning left in an intersection

Both algorithms for road, traffic isles and obstacles detection [29] and for curb detection [30] use the same elevation map representation as input data, and therefore their results can be integrated into a single occupancy grid. Scenes with poor road texture will have a better delimited drivable area (Fig. 22).

Fig. 22.   The occupancy grid augmented with the detected curbs

The concept of multi-frame persistence map used for curb detection was also implemented for traffic isles, in order to remove false detections. Similar to false curbs, false traffic isles have an important feature: they persist only for a limited number of consecutive frames (mostly for two frames). Furthermore, traffic isles are static scene items. Based on these features, a fast and efficient approach can be used to filter false traffic isles: a multi-frame persistence map is built (the ego motion between frames is taken into consideration) and only traffic isle cells that persist for several frames are validated.

Another improvement that can be done is the enhancement of the occupancy grid, to discriminate between static and dynamic obstacle cells. Again, the persistence map concept can be used. It shows the lifetime for each obstacle cell (for how many consecutive frames it was labeled as an obstacle cell, in a global reference system). Static obstacle cells will tend to have an increased lifetime while dynamic ones will have a shorter lifetime. However, depending on the obstacle size and speed, some of the obstacle cells might overlap cells from the same obstacle along several frames. This can offer false clues about the nature of individual cells. Therefore, the discrimination must be performed at blob level, as all the cells from the same obstacle are treated unitary. If the average persistence of the blob is above a threshold (a desired number of frames with no or negligible blob movement), the blob is considered static, otherwise dynamic.

Fig. 23.   The enhanced grid (with static / dynamic obstacles) projected onto the left image. The front car is the only moving obstacle; the vehicles on the left are waiting for the green light

4.2   Using the Occupancy Grid

The occupancy grid can be used for generating a compact, higher level representation of the environment components. A new representation was proposed and implemented [38], each scene component being represented as a 3D poly-line with additional information about its type (resulted from classification) and dynamic features.

Fig. 24.   Poly-line representation derived from the occupancy grid

The occupancy grid can be integrated over time using the ego car's motion parameters (yaw-rate and speed data), and the static grid cells. A global map is obtained (Fig. 26). The scene static features can be used for alignment with the intersection GPS or aerial maps (Fig. 26).

Since traffic isles (sidewalks) and/or curbs are included in the grid, these can be used as focus of attention for pedestrian detection. Another important issue in intersections is the detection of traffic lights and signs: static obstacle areas can be used for focusing attention for traffic lights / signs detection (Fig. 24).

Another way to use the occupancy grid is the detection (awareness) of the intersection situation (for passive intersections) based on: the configuration of static delimiters in the scene (ending traffic isles/curbs, enlarging drivable area, etc), and the configuration of dynamic obstacles (increased lateral motion).

Curbs can be used as measurement data for the lane detection algorithm, because the lane marking is usually not present near sidewalks, in urban scenes (Fig. 22).

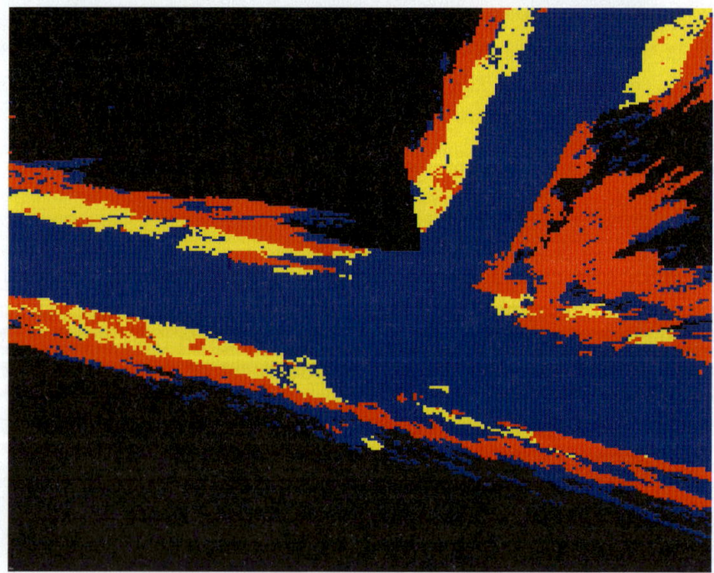

Fig. 25.   The global occupancy grid

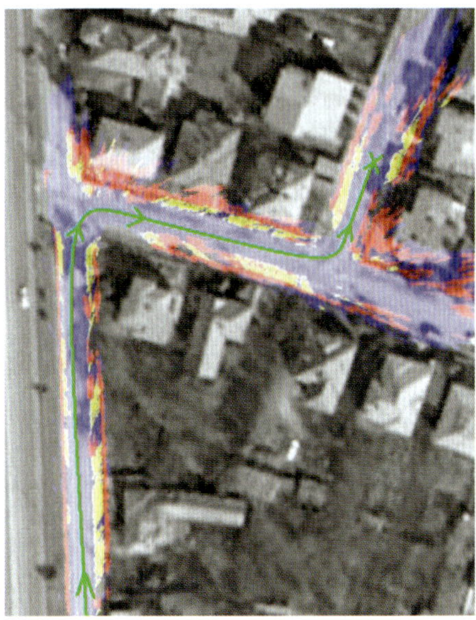

Fig. 26. Global occupancy grid superimposed to the satellite image of the region. The trajectory of the ego car is shown with the green curve

5   Conclusions and Future Work

In this paper we have proposed an on-board stereovision sensor for the perception and description of urban and intersection environments. The sensor has as main features the exploitation of grayscale and stereo 3D information through specialized algorithms that provide structured and unstructured approaches for environment perception.

The system is able to detect the static and the dynamic components of the intersection environment. When the conditions allow it, the road geometry is estimated using a modular and configurable tracker based on particle filter, which allows the use of multiple lane models and the integration of multiple measurement cues. The obstacles are detected and tracked using a cuboid model, and subsequently classified to discriminate their type, and thus their importance in traffic. If no lane delimiters are available, or the geometry is unknown, the road surface and the free space ahead can be estimated in an unstructured fashion from the elevation map's occupancy grid. The obstacles are alternatively detected and described from the non-drivable cells of the occupancy grid. Thus, the system is able to detect, track and recognize most of the components of the traffic environment.

For the future, a fusion process between the structured and the unstructured descriptions is planned, in order to increase the detection robustness. Future work will include estimation of multiple types of road structures, detection and tracking of more object models, and enhancing the object classifier to handle more classes and improve the classification accuracy.

Cooperation between the stereo sensor and other types of data sources, such as the active sensors, or the GPS and map information is also planned for the future.

The stereovision process itself can be improved by using color information, and by experimenting with different camera and lens setups, to find the one that will provide the best coverage for the intersection.

Acknowledgement

This work was supported by Volkswagen AG Germany, Technical University of Cluj-Napoca Romania and recently is supported by the "INTERSAFE-2" EU founded project.

References

[1]   S. Nedevschi, et al., "A Sensor for Urban Driving Assistance Systems Based on Dense Stereovision", Proceedings of 2007 IEEE Intelligent Vehicles Symposium (IV2007), June 13-15, 2006, Istanbul, Turkey, pp 276-283 (2007).

[2]   T. Dang, C. Hoffmann, "Stereo calibration in vehicles", Proceedings of IEEE Intelligent Vehicles Symposium (IV2004), June 14-17, 2004, Parma, Italy, pp.268-272. (2004).

[3]   T. Dang, C. Hoffmann, and C. Stiller, "Self-calibration for Active Automotive Stereo Vision", Proceedings of IEEE Intelligent Vehicles Symposium (IV2006), June 13-15, 2006, Tokyo, Japan, pp.364-369. (2006).

[4]   T. Marita, et al., "Camera Calibration Method for Far Range Stereovision Sensors Used in Vehicles", Proceedings of IEEE Intelligent Vehicles Symposium (IV2006), June 2006, Tokyo, Japan, pp. 356-363. (2006).

[5]   S. Nedevschi, C. Vancea, T. Marita, T. Graf, "On-Line Calibration Method for Stereovision Systems Used in Far Range Detection Vehicle Applications", IEEE

Transactions on Intelligent Transportation Systems, vol.8, no. 4, pp. 651-660, 2007.

[6] T. Marita, et al., "Calibration Accuracy Assessment Methods for Stereovision Sensors Used in Vehicles", in Proceedings of IEEE 3-rd International Conference on Intelligent Computer Communication and Processing (ICCP2007), 6-8 September 2007, Cluj-Napoca, Romania, pp. 111-118. (2007).

[7] C. Vancea, S. Nedevschi, "Analysis of different image rectification approaches for binocular stereovision systems", in Proceedings of IEEE 2nd International Conference on Intelligent Computer Communication and Processing (ICCP 2006), vol. 1, September 1-2, 2006, Cluj-Napoca, Romania, pp. 135-142. (2006).

[8]. E.D. Dickmanns, B.D. Mysliwetz, "Recursive 3-d road and relative ego-state recognition", IEEE Transactions on Pattern Analysis and Machine Intelligence, vol. 14, no.2, pp. 199-213, 1992.

[9] R. Aufrere, et al., "A model-driven approach for real-time road recognition", Machine Vision and Applications, vol 13, no. 2, pp. 95-107, 2001.

[10] S. Nedevschi, et al., "High accuracy stereo vision system for far distance obstacle detection", in Proceedings of IEEE Intelligent Vehicles Symposium (IV 2004), June 2004, Parma, Italy, pp. 292-297. (2004).

[11] R. Danescu, et al., "Lane Geometry Estimation in Urban Environments Using a Stereovision System", in Proceedings of IEEE Intelligent Transportation Systems Conference (ITSC 2007), September 2007, Seattle, USA, pp. 271-276. (2007).

[12] R. Labayrade, et al., "A Multi-Model Lane Detector that Handles Road Singularities", in Proceedings of IEEE Intelligent Transportation Systems Conference, October 2006, Toronto, Canada, pp. 1143-1148. (2006).

[13] M. Isard, A. Blake, "CONDENSATION – conditional density propagation for visual tracking", International Journal of Computer Vision, vol. 29, nr. 1, pp. 5-28, 1998

[14] B. Southall, C.J. Taylor, "Stochastic road shape estimation", in Proceedings of IEEE International Conference on Computer Vision, 2001, Vancouver, Canada, pp. 205-212.

[15] K. Macek, et al., "A Lane Detection Vision Module for Driver Assistance", in Proceedings of IEEE/APS Conference on Mechatronics and Robotics, 2004.

[16] U. Franke, et al., "Lane Recognition on Country Roads", in Proceedings of IEEE Intelligent Vehicles Symposium, June 2007, Istanbul, Turkey, pp.100-104. (2007).

[17] P. Smuda, et al., "Multiple Cue Data Fusion with Particle Filters for Road Course Detection in Vision Systems", in Proceedings of IEEE Intelligent Vehicles Symposium, 2006, Tokyo, Japan, pp. 400-405. (2006).

[18] R. Danescu, et al., "A Stereovision-Based Probabilistic Lane Tracker for Difficult Road Scenarios", in Proceedings of IEEE Intelligent Vehicles Symposium 2008 (IV2008), Eindhoven, The Netherlands, June 4-6, 2008, pp.536-541. (2008).

[19] D. M. Gavrila, "Pedestrian detection from a moving vehicle," in Proceedings of the European Conference on Computer Vision, 2000,pp. 37–49.

[20] A. Khammari, et al., "Vehicle detection combining gradient analysis and AdaBoost classification", in Proceedings of Intelligent Transportation Systems Conference (ITSC 2005), pp. 61-71, 2005. (2005).

[21] D. Huber, et al., "Parts-based 3D object classification", in Proceedings of IEEE Conference on Computer Vision and Pattern Recognition (CVPR), June 2004, vol 2, pp. 82-89. (2004).

[22] R. Osada, et al., "Matching 3D Models with Shape Distributions", Shape Modeling International, Genova, Italy, May 2001.

[23] D. M. Gavrila, "A Bayesian Exemplar-based Approach to Hierarchical Shape Matching", IEEE Transactions on Pattern Analysis and Machine Intelligence, Vol. 28 No. 8, pp. 1408-1421, 2007.

[24] L. Havasi, et al., "Pedestrian Detection Using Derived Third-Order Symmetry of Legs", in Proc. of the Int. Conference on Computer Vision and Graphics, 2004, pp. 733-739. (2004).

[25] A. Shashua, et al., "Pedestrian Detection for Driving Assistance Systems: Single-frame Classification and System Level Performance", in Proceedings of IEEE Intelligent Vehicle Symposium(IV2004), June 2004, Parma, Italy, pp. 1-6. (2004).

[26] M. Bertozzi, et al., "Stereo Vision-based approaches for Pedestrian Detection", in Proceedings of the IEEE Comp. Soc. Conf. on Computer Vision and Pattern Recognition - Workshops, June 2005, San Diego, USA, vol. 3, pp. 16, 2005.

[27] Z. Zhang, "A stereovision system for a planetary rover: Calibration, correlation, registration, and fusion," in Machine Vision and Applications, vol. 10, no. 1, pp. 27-34, 1997.

[28] M. Vergauwen, et al., "A stereo-vision system for support of planetary surface exploration", Machine Vision and Applications, vol. 14, no.1, pp. 5–14, 2003.

[29] F. Oniga, et al.,, "Road Surface and Obstacle Detection Based on Elevation Maps from Dense Stereo", in Proceedings of the 10th International IEEE Conference on Intelligent Transportation Systems (ITSC 2007), Sept. 30 - Oct. 3, 2007, Seattle, Washington, USA, pp. 859-865. (2007).

[30] F. Oniga, et al., "Curb Detection Based on a Multi-Frame Persistence Map for Urban Driving Scenarios", in Proceedings of the 11th International IEEE Conference on Intelligent Transportation Systems (ITSC 2008), Oct. 2008, Beijing, China, pp. 67-72. (2008).

[31] J.Y. Bouguet, Camera Calibration Toolbox for Matlab, www.vision.caltech.edu/bougetj

[32] C. Pocol, et al., "Obstacle Detection Based on Dense Stereovision for Urban ACC Systems", in Proceedings of 5th International Workshop on Intelligent Transportation (WIT 2008), March 18-19, 2008, Hamburg, Germany, pp. 13-18.

[33] R. Danescu, et al., "Stereovision Based Vehicle Tracking in Urban Traffic Environments", in Proceedings of the IEEE Intelligent Transportation Systems Conference (ITSC 2007), October 2007, Seattle, USA, pp.400-404. (2007).

[34] S. Bota, S. Nedevschi - Multi-Feature Walking Pedestrians Detection for Driving Assistance Systems, IET Inteligent Transportation Systems Journal, Volume 2, Issue 2, pp. 92-104, 2008.

[35] J.-Y. Bouguet, "Pyramidal implementation of the Lucas Kanade feature tracker", available: http://mrl.nyu.edu/ bregler/classes/vision spring06/bouget00.pdf.

[36] N. Dalal and B. Triggs, "Histograms of oriented gradients for human detection," in Proceedings of IEEE Computer Society Conference on Computer Vision and Pattern Recognition (CVPR 2005), June 2005, pp. 886–893. (2005).

[37] I. H. Witten, E. Frank "Data Mining: Practical machine learning tools and techniques, 2nd Edition", Morgan Kaufmann, San Francisco, 2007.

[38] S. Nedevschi, et al., "Forward Collision Detection using a Stereo Vision System", Proceedings of IEEE 4th International Conference on Intelligent Computer Communication and Processing (ICCP 2008), August 28-30, 2008, Cluj-Napoca, Romania, pp. 115-122. (2008).

Sergiu Nedevschi, Radu Danescu, Tiberiu Marita
Florin Oniga, Ciprian Pocol , Silviu Bota
Technical University of Cluj-Napoca
Str. C. Daicoviciu 15
400 020 Cluj-Napoca
Romania
sergiu.nedevschi@cs.utcluj.ro
radu danescu@cs.utcluj.ro
tiberiu.marita@cs.utcluj.ro
florin.oniga@cs.utcluj.ro
ciprian.pocol@cs.utcluj.ro
silviu.bota@cs.utcluj.ro

Marc Michael Meinecke, Marian Andrzej Obojski
Volkswagen AG
Brieffach 01117770
38436 Wolfsburg
Germany
marc-michael.meinecke@volkswagen.de
marian-andrzej.obojski@volkswagen.de

Keywords: stereovision, intersection, lane detection, obstacle detection, tracking, classification, elevation maps, occupancy map

Object Recognition based on Infrastructure Laserscanners

M. Sowen, F. Ahlers, Ibeo Automobile Sensor GmbH

Abstract

In the integrated research project SAFESPOT the concept of a Cooperative Pre-Data-Fusion has been implemented as a novel approach for an environmental perception system. This paper describes the approach of eliminating measurements of a Laserscanner which are not part of the environment of interest. It is being developed in the subproject INFRASENS which specifies and develops an infrastructure-based sensing platform.

1 Introduction

SAFESPOT is an integrated research project co-funded by the European Commission Information Society Technologies among the initiatives of the 6th Framework Program. The objective is to understand how intelligent vehicles and intelligent roads can cooperate to produce a breakthrough for road safety. Therefore, the general aim of the project is to create a Safety Margin Assistant, detecting potentially dangerous situations between road users of any kind in advance [1]. The complete system structure has been outlined previously in [2]. As promised there, quantitative results for the performance of the SAFESPOT system had been obtained as the project progressed. These results will be presented in the final paper.

2 System Architecture

The key component of the INFRASENS infrastructure system are Laserscanners installed in infrastructure housings shown in Fig. 1. These units consisting of a Laserscanner and a background label ECU are placed at several spots at an intersection. Each unit gathers a range profile in its field of view and preprocesses the Laserscanner's raw data. The preprocessed data is then send to a central fusion Electronic Control Unit (Fusion ECU) which detects tracks and classifies all objects seen by all Laserscanners and passes the information on to the SAFESPOT system using the Ethernet protocol. In Fig. 2 an example setup is shown with three infrastructure Laserscanner units (LS1-3, red) and one central fusion ECU (blue).

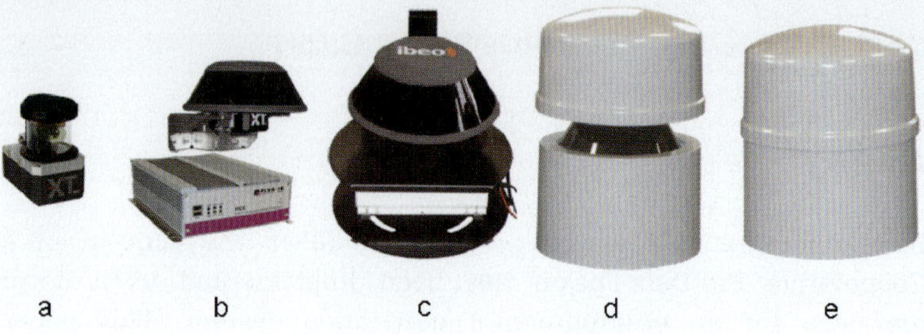

Fig. 1.   Infrastructure sensor components Alasca XT Laserscanner (a), including background label ECU (b), mounted (c), installed in housing (d) and closed for transportation (e).

2.1 Laserscanner Sensor

The Laserscanner observes its environment horizontally (240° field of view) and gathers a range profile of the Laserscanner's vicinity of up to 200 meters. The rotational frequency of the mirror guiding the laser beam is 12.5 Hz. The measurement frequency enables distance measurements every 0.25 degrees horizontally. The applied multi-echo technology ensures proper performance even under adverse weather conditions like rain, snow or any other precipitation. The Laserscanner is connected to a background label ECU via ARCnet interface.

2.2 Background label ECU

All infrastructure housings are equipped with an ECU (Fig. 1b) processing the Laserscanners' raw data. A special grid map (referred to as a background map) of the area visible to the Laserscanner is stored on each ECU. It is used to label all distance measurements from the Laserscanner as valid or background, which distinguishes between measurements relevant to the scenario or not. Typical examples for measurement not relevant and therefore labeled as background are bushes or buildings far off the road as they do not impact the traffic scenario in terms of safety. The labeled scan data is transmitted via Ethernet to a central fusion ECU.

2.3 Fusion and Tracking ECU

Gathering measurements of the connected Laserscanner sensors, the fusion and tracking ECU accumulates the data into a single environmental model. By clustering, tracking and classifying the distance measurements, the fusion ECU outputs UDP protocol based object data specified in the INFRASENS documentation. Those object data represents the road user's current state (position, speed, driving direction etc.), class (pedestrian, car, truck, etc.) and further safety relevant information (e.g. dimensions).

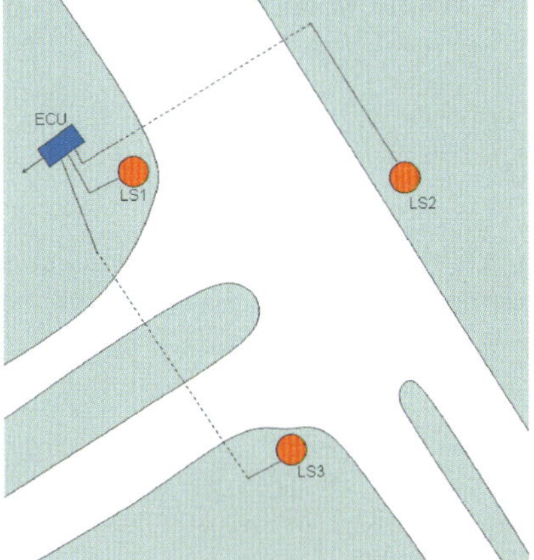

Fig. 2.   Example system architecture with 3 Laserscanners at intersection

3   Background Measurements

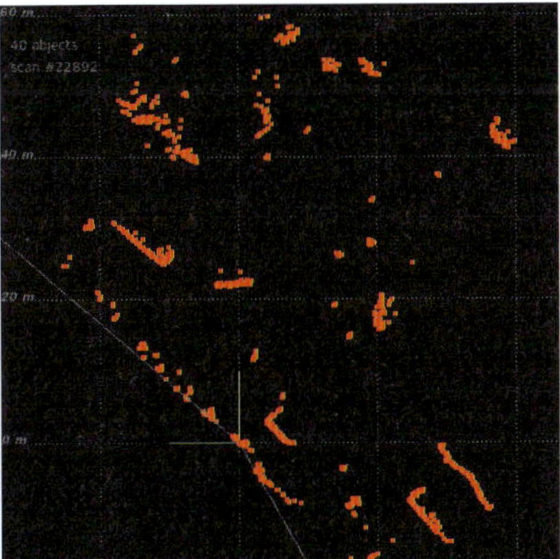

Fig. 3.   Measurements of one Laserscanner at intersection

A Laserscanner based infrastructure sensor measures distances to all objects in the field of view (range profile). In Fig. 3 an example "bird"-view of an intersection is shown. The Laserscanner is located at the origin. With the large field of view of more than 180 degrees, the Laserscanner is able to see all relevant parts of the intersection. This is not only limited to road users, but also objects located in the periphery of the surrounding area. These background measurements are not in the scope of the INFRASENS project and put a higher demand on the processing power of the fusion ECU for the tracking and classification algorithm. To reduce the processing power to a minimum, the algorithms are designed to focus on the measurements of road users only.

4 Background Map

Infrastructure sensors are installed at fixed positions at an intersection; hence the positions of the road and pedestrian crossings relative to the sensors never change. Consequently, it is feasible to create a static map of the surrounding area for each Laserscanner, marking the parts which are of interest and the ones that can be considered as background.

4.1 Building the Background Map

A semi-automatic approach for creating such a background map is to use the OnlineMap algorithm. The OnlineMap subdivides the area of the field of view of one infrastructure Laserscanner into grid cells. For each grid cell an occupancy probability is calculated for a series of measurement runs [3]. Since the area of the road and the pedestrian crossings are only part-time occupied, the occupancy probability is very low. Grid cells which lie in the area of static objects like poles, traffic lights and bushes have a very high probability. The regular OnlineMap algorithm assumes a moving Laserscanner and therefore a constantly changing environment. When building a static background map, further assumptions on the probabilities of each cell can be made. If one cell was considered to be unoccupied once, the cell is considered as free space e.g. drivable area. Therefore the background building algorithm has two phases:

Initialization phase: During this phase the OnlineMap algorithm works in a regular manner. If a cell happens to be occupied the cell's occupancy probability is increased. The probability of the cell is decreased if it is unoccupied. This phase lasts for 200 scans, which results in a time duration of 16 seconds when considering the scan frequency of 12.5 Hz of the Laserscanner.

<u>Stabilizing phase</u>: After 200 scans the background building algorithm switches to the stabilizing mode. A probability threshold of p_{th} = 0.3 is used to keep each cell unoccupied once it has dropped below this value. A car driving across the intersection will not affect a low occupancy probability anymore even though it shortly occupies the cells over which it passes.

The second phase lasts for another 1800 scans which results in a total background map building time of 160 seconds. This time duration appears to be sufficient for most intersections since the intersection usually was cleared for at least some time, due to different traffic light phases.

After the algorithm is done it generates a gray-scale PNG image file. This file is considered to be the automatic generated background map, where each pixel represents a cell. Cells with low occupancy probabilities are marked with a bright pixel and high probabilities with a dark pixel.

The background map should be further enhanced by manually editing the PNG image file using an image editing program. Areas where e.g. parked cars reside during the background map building phases can be painted white to support tracking and classification of objects there as well. But in general, the procedure does not any manual intervention.

A visualisation of the final background map of the intersection of Fig. 3 is shown in Fig. 4.

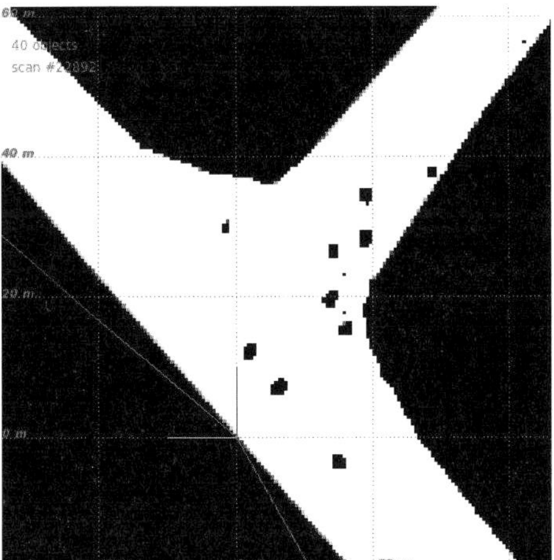

Fig. 4.   Background map of intersection

4.2 Using the Background Map

Once the background map for a Laserscanner is finalized it is stored on the compact flash drive of the associated background label ECU. During the operation the Laserscanner measures the distance to all objects in its view. These distance measurement values are compared with the background map's occupancy values for each individual cell. If a distance measurement lies within a cell with an occupancy probability of

$$p_{cell\ x, cell\ y} \geq 0.5 \tag{1}$$

the measurement will be marked as background. Otherwise it is labeled as valid foreground. In Fig. 5 the scan of one Laserscanner from Fig. 3 can be seen. The scan data is labeled using the background map from Fig. 4. All measurements considered to be background are drawn in a yellow color. All remaining measurements are drawn in a red color, marking these as valid foreground and therefore being road users.

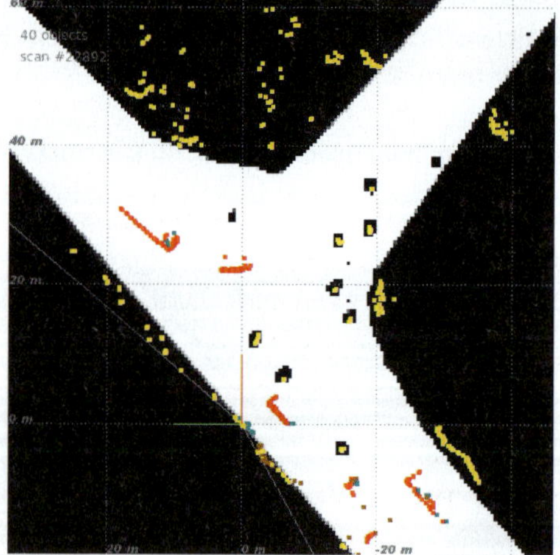

Fig. 5.   Background labelled scan from one Laserscannerat intersection

The background labeled scan data from all infrastructure sensors on the intersection is transferred to the fusion ECU which processes the scan data in four consecutive steps:

<u>Scan fusion</u>: The scan data from each infrastructure Laserscanner is transformed to a common coordinate system and added to a fused scan data frame. The background label information remains untouched resulting in a common scan data representation not distinguishable from a regular Laserscanner

except a higher measurement density and less shadowing effects. In order to avoid a time offset between measurements from different Laserscanners, the sensors need to be precisely synchronized. This is done by syncing each Laserscanner using certain time slots within a NTP-synchronized real time clock.

<u>Object tracking</u>: The tracking algorithm identifies objects from the distance measurements of the Laserscanners. Since only measurements will be used that are marked as valid foreground, the processing power needed is much lower, even though the total amount of data has been multiplied by the usage of several Laserscanners at once. The tracking algorithm has been adapted and optimized to comply with object shapes not common for a single Laserscanner tracking. Cars usually appearing as an "L"-shape in a single Laserscanner scan, might now appear with its full rectangular shape as it is seen from different angles.

<u>Object classification</u>: Once objects are identified by the tracking algorithm, the classes of the objects need to be identified using the classification algorithm. This is achieved by a well trained artificial neural network as proposed in [2]. The following road user classifications will be determined:
- ▶ passenger car
- ▶ truck or bus
- ▶ bike
- ▶ pedestrian
- ▶ unknown object

As with the tracking algorithm, the processing power required is reduced by focusing on road user objects while improving classification reliability and robustness.

<u>UDP object output</u>: In the remaining step, the classified object data is transmitted via Ethernet using an UDP protocol specified in the SAFESPOT project [1].

5   Conclusion

In this paper a method has been shown, which eliminates background measurements of a Laserscanner based infrastructure sensor using a semi-automatic generated background map. Removing these measurements which are of no interest reduces the necessary processing power of a tracking and classification algorithm and therefore reduces the cost, size and power consumption of the fusion and tracking ECU as well as improving robustness and reliability of the tracking and classification algorithms.

6   Acknowledgement

SAFESPOT [1] is part of the 6th Framework Program, funded by the European Commission. The partners of SAFEPROBE and INFRASENS thank the European Commission for supporting the work of this project. The authors would like to thank in particular TeleAtlas for supplying the map data used in this paper, and also all partners within SAFESPOT especially from SAFEPROBE, INFRASENS and SINTECH for their cooperation and fruitful teamwork on data fusion and data processing.

References

[1]   SAFESPOT – Co-operative Systems for Road Safety 'Smart Vehicles on Smart Roads'; online at: www.safespot-eu.org

[2]   F. Ahlers, Ch. Stimming, Laserscanner Based Cooperative Pre-Data-Fusion, In: Advanced Microsystems for Automotive Applications, Berlin, Germany, 2008.

[3]   Weiss, T., Wender, S., Fuerstenberg, K. Dietmayer, Online Mapping for the Robust Detection of Driving Corridors, PReVENT ProFusion e-Journal, Vol.2, 2007.

Marc Sowen, Florian Ahlers
Ibeo Automobile Sensor GmbH
Research Department
Merkurring 20
22143 Hamburg
Germany
marc.sowen@ibeo-as.com
florian.ahlers@ibeo-as.com

Keywords: laser scanner, cooperative safety, infrastructure, background detection, data fusion, environment perception

User Needs for Intersection Safety Systems

M. Wimmershoff, A. Benmimoun, RWTH Aachen University

Abstract

Nowadays, most so called 'accident black spots' have been eliminated from the road networks. However, intersections can still be regarded as accident black spots. Depending on the region and country, from 30% to 60% of all injury accidents and up to one third of the fatalities occur at intersections. Thus, the European R&D project INTERSAFE-2 aims to develop and demonstrate a co-operative intersection safety system, in order to significantly reduce injury and fatal accidents at intersections. In line with this project, the user needs for intersection safety systems and the users' impressions respectively attitudes towards such systems were determined with the aid of a questionnaire which has been executed in six European countries. The results of the Europe-wide questioning are presented within this paper.

1   Introduction

In 2008 5300 accidents with seriously injured people or fatalities happened at intersections and good visible T-junctions out of town in Germany. Each fifth bad accident met this constellation [1]. Thus, intersections still count to the accident black spots. According to [1] the main reason for these accidents is human failure. In order to significantly reduce injury and fatal accidents at intersections, the European Commission co-funds a R&D project called INTERSAFE-2, which aims to develop and demonstrate a co-operative intersection safety system. INTERSAFE-2 is the successor of the project INTERSAFE, which was a subproject of PReVENT and ran from 2004 to 2007. INTERSAFE-2 goes beyond the scope of INTERSAFE in that it integrates warning and intervening systems, while INTERSAFE only provided warning systems. Furthermore, INTERSAFE-2 takes special requirements for trucks into account.

In order to determine the user needs and user expectations for passenger car and truck drivers with regard to intersection assistance a questionnaire has been executed in six European countries. In addition the users' impressions and attitudes towards potential intersection assistance should be evaluated within the questionnaire. Therefore the questionnaire was divided into three parts.

The first part consists of questions about ADAS in general in order to gain an overview, if the subjects are already familiar with advanced driver assistant systems (ADAS) and how they think about it. The second part comprises intersection assistant relevant questions, whereat the intersection assistant was introduced as a system that detects critical intersection situations (e.g. miscalculation of speeds and distances, or failure to see traffic signs/ lights by the driver) and supports the driver with additional information, warnings and/ or interventions. As a third part of the questionnaire the demographical data of the subjects were collected. The demographical data was intentionally placed at the end of the questionnaire due to the assumption that a subject starts with a high motivation to fill in the questionnaire that decreases with time.

Within this paper the evaluation of the survey is described in detail.

2   Demographical Data

The questionnaire has been executed in six European countries:
> Finland (FI),
> France (FR)
> Germany (DE)
> Great Britain (UK)
> Romania (RO)
> Sweden (SE)

The overall number of subjects amounts for passenger cars to N = 106, whereby 52 subjects were questioned in Germany, 19 in France, 15 in Great Britain, 9 in Romania and 11 in Finland. The total amount of interviewed truck driver is N = 44 (20 subjects in Germany, 5 in Finland, 10 in Great Britain and 9 in Sweden). This sample size might not be sufficient to be representative but quite suitable to derive the user demands.

65% of the passenger car subjects are male and 34% are female, while the truck interviewees are 92,5% male and 2,5% female. The age distribution for both vehicle types is shown in Fig. 1. 88% of the passenger car interviewees own a vehicle and have most experience with a mid-size car (57%). The truck drivers have of course most experience with trucks (74%).

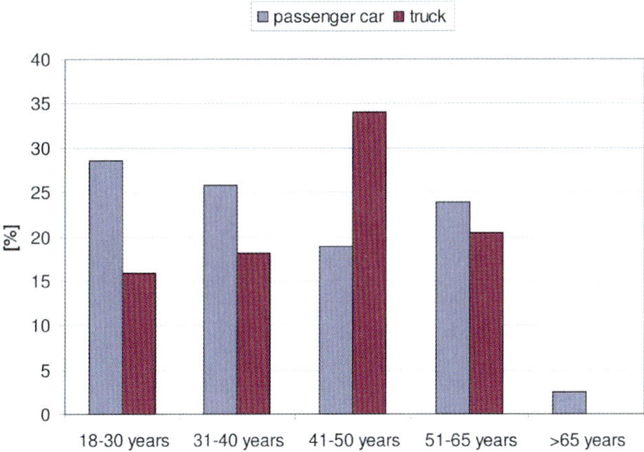

Fig. 1.   Age distribution of subjects

The majority of the passenger car subjects drive 10.000 km per year to 20.000 km per year and thus are quite representative. 40% drive mostly on interurban roads, while 34% drive their car mostly in the inner city. A fifth part of the interviewees drive in the majority of cases on highways. The main purposes of vehicle usage are rides to work (49%) and private use/holiday (36%).

Two third of the truck interviewees drive more than 100.000 km per year and 62% drive mostly on highways. About 10% mentioned that they drive mostly in the inner-city. These results are reflected in the answers for the question for which purpose the trucker mainly uses his vehicle (67% long-haul transport, 15% medium distance transport and 18% local distribution/city traffic).

3   ADAS in general

The questionnaire revealed that more than 80% of the interviewed passenger car drivers are in general willing to use ADAS (sum of answer "1" and "2" on a scale of "-2: not at all" and "2: absolutely"; truck 73%). Beyond, 60% of passenger car drivers already use ADAS, whereby as the most frequent ADAS the navigation system (36%) is mentioned. Remarkable is that 100% of the truck drivers already use ADAS, whereby 95% use cruise control. The whole evaluation of the question "Do you already use driver assistance systems" is shown in Fig. 2.

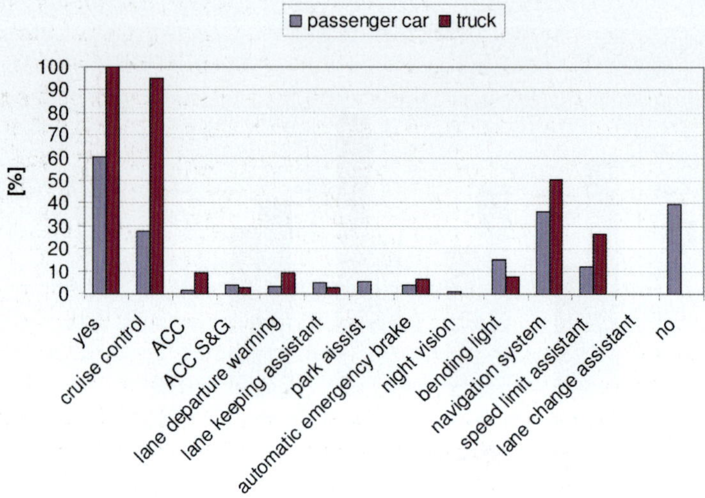

Fig. 2.   Evaluation of the question: „Do you already use ADAS?"

The question which ADAS the subjects would like to have in their car or truck shows that there is high demand for driving support provided by technical systems. Five out of fourteen suggested systems were mentioned by at least 50% of the questioned passenger car subjects. These systems are navigation system (77%), night vision (59%), ACC (54%), park assist (54%), and speed limit assistant (52%).

The intersection assistant is mentioned by more than 30% of all subjects, which is remarkable since the subjects do not have any information about the system and its function at this stage of the questionnaire.

Truck drivers are more reserved than passenger car drivers with regard to the ADAS question. They marked three systems with a frequency higher than 40%: lane change assistant (52%), ACC (51%) and lane departure warning (41%).

When the subjects are asked on which trips they want to be assisted, the evaluation shows that almost one third of the passenger car drivers (28%) want to be assisted at intersections (truck: 13%), which matches very well with the previous questions. The most frequent mentioned trips on which the passenger car subjects want to be assisted are highways (80% passenger car, 68% truck), interurban (67% passenger car, 40% truck) and inner-city roads (59% passenger car, 42% truck).

An important outcome of the questionnaire arises by evaluating the questions if the subjects accept an information/ warning, an intervention and a takeover of the vehicle guidance by an ADAS. As an example of a takeover of the vehicle guidance, the automatic acceleration or deceleration as it is done by ACC was given. The analysis clearly points out that an information/ warning by an ADAS is accepted by passenger car and truck drivers (mean value is 1,52 for passenger car and 1,26 for trucks on a scale from "-2: not at all" to "2: absolutely"), while intervention (0,42 for passenger car and 0,19 for trucks) respectively completely take over (0,47 for passenger car and 0,28 for trucks) are accepted only with a bare majority and a high standard deviation. Especially the bare majority for a takeover is remarkable because more than 50% of both interviewed groups mentioned that they would like to have ACC. One reason for this is found by examining the question, which addresses principal reservations of the subjects against ADAS. The evaluation of this question shows that 35% of the passenger car interviewees have principle reservations against ADAS (truck 19%).

4   Intersection Assistant

A remarkable result of the questionnaire is that more than 50% of the passenger car drivers and almost 40% of the truck drivers were already involved in critical intersection situations. The most frequent cause for these critical situations is for passenger car interviewees the oversight of a right of way traffic sign (answer "5" in upper diagram of Fig. 3, 18%) and oversight of crossing traffic because of view occlusion or glare (answer "8" in upper diagram of Fig. 3, 14%). Truck drivers answered most frequently besides miscellaneous reasons (answer "9" in lower diagram of Fig. 3, 15%) that the oversight of cyclists while turning right (answer "1b" in lower diagram of Fig. 3, 10%) and the traffic light violation (answer "6" in lower diagram of Fig. 3, 7%) leads to a critical intersection situation. The detailed results with regard to this question and the index for the given answers are displayed in Fig. 3.

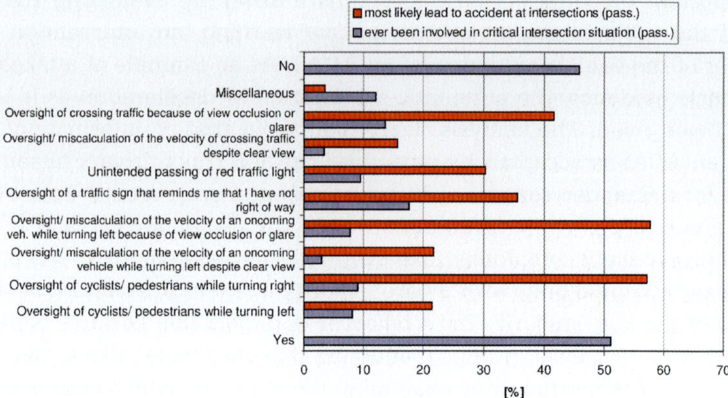

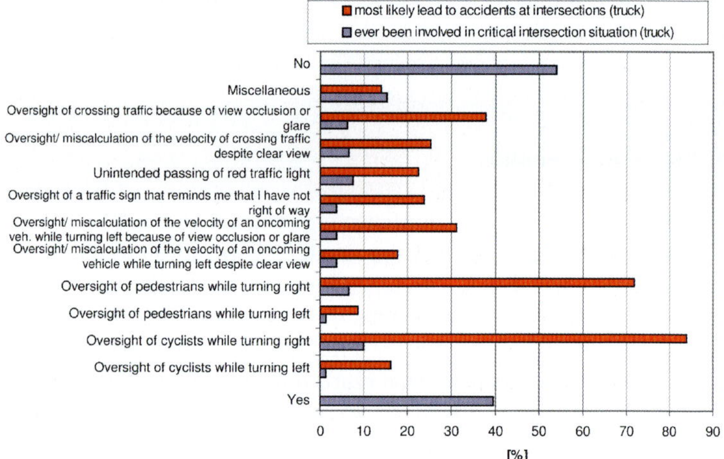

Fig. 3.   Results of the question: „Have you ever been in a critical intersection situation?" and "What do you think, which one of the following situations could most likely lead to an accident at intersections with regard to you?" (top: passenger car, bottom: truck)

The subjects of the passenger cars think that oversight/ miscalculation of the velocity of an oncoming vehicle while turning left because of view occlusion or glare (58%) and the oversight of cyclists/ pedestrians while turning right (57%) lead most likely to an intersection accident. The evaluation of the answers of the truck drivers shows a clear result. 84% think that the oversight of cyclists while turning right respectively oversight of pedestrian while turning right (72%) could most likely lead to an accident with truck involvement at intersections.

From these two questions five scenarios can be derived, which are mentioned most frequently by the subjects. These scenarios are consistent with the relevant accidents scenarios for an intersection safety system which are identified out of the analysis of accident statistics in [2]. The scenario oversight of crossing traffic because of view occlusion or glare corresponds to turn into/straight crossing path which can be found in [3]. Traffic light violation is a subset of turn into/straight crossing scenarios [3]. The oversight of oncoming traffic/ miscalculation of velocity while turning left corresponds to the scenario left turn across path [3]. VRUs are addressed within right turn across path; parallel or oncoming opponent VRUs which matches with the answer oversight of a pedestrian/ cyclist while turning right. In addition, especially the relevant scenarios for trucks derived from accidentology [2] addresses these scenarios, which were very frequently mentioned by truck drivers. The oversight of a right of way traffic sign corresponds to right of way conflicts.

Furthermore, the survey revealed that 72% of the passenger car subjects, who were already involved in a critical traffic situation, mentioned that an intersection assistant could have supported them in these specific situations (truck: 66%) and 91% of the passenger car drivers think that an intersection assistant could assist them at intersections (truck: 76%; sum of subjects marked "1" or "2" on a scale from "-2: not at all" to "2: absolutely"). This statements reflects the meaning of the subjects about an intersection assistant and shows a high user acceptance. This is underlined by the outcome, that 73% of passenger car drivers think that considerably less accidents would happen, if all drivers would have an intersection assistant (truck: 62%).

More than 50% of the passenger car drivers want to be assisted all the time by the intersection assistant and not only in special environmental situations (for example darkness, rain, fog or sun glare). Most truck drivers also want to be assisted all the time (37%), but in addition one third of them answered further on that they want to be assisted in poor environmental conditions.

An intersection assistant that works only at intersections which are especially equipped (e. g. car-to-infrastructure communication) or an intersection assistant which indicates only the right of way regulation are accepted by the most drivers. This shows a positive attitude towards an partial introduction of an intersection assistant at accident black spots.

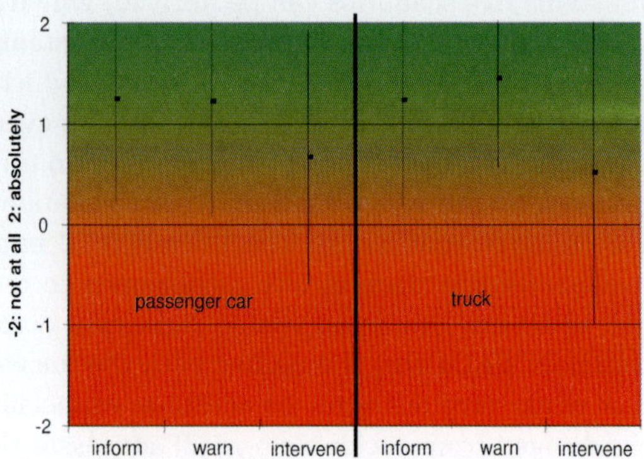

Fig. 4.   Evaluation of the questions: "The intersection assistant should inform me about the vehicle having right of way"; "The intersection assistant should warn me against an imminent accident"; "The intersection assistant should intervene if there is an imminent accident"

A traffic lights assistant, which indicates when the lights will change, is wanted by more than half of the interviewees (63% passenger car, 62% truck), but shows a high standard deviation (1,22 for passenger cars and 1,66 for trucks), i.e. the opinions of the subjects are not consistent.

In addition, it is very important that the intersection assistant offers the possibility to switch it off because this is mentioned by more than 90% of passenger car and 79% of truck drivers.

Fig. 4 shows the evaluation of the questions, if the subject thinks that the intersection assistant should inform, warn and/or intervene if there is an imminent accident. The outcome shows that the meaning of passenger car and truck drivers is consistent. An information or a warning is more accepted than an intervention of the intersection assistant. The reasons for that are on the one hand, as described in chapter 3, the principle reservations against ADAS and on the other hand the subjects mentioned that they do not want to be patronised by an ADAS because they want to preserve the control of the vehicle themselves.

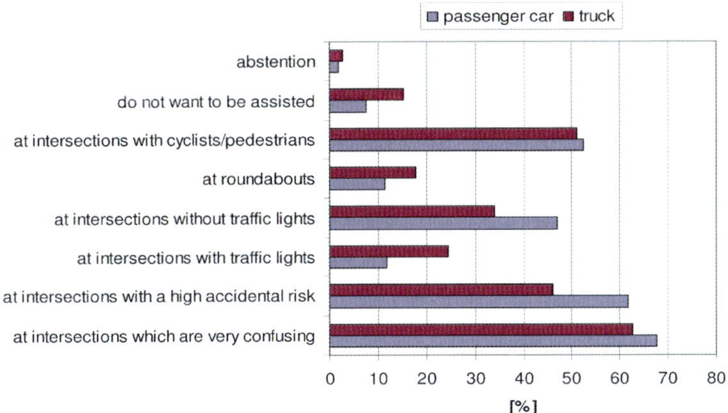

Fig. 5   Evaluation of the question: "At which intersections do you want to be assisted by an intersection assistant?"

Fig. 5 shows that there is only a small minority of the subjects who do not want to be assisted by an intersection assistant: less than 10% for passenger car and 15% for truck. Also remarkable is that only a minority of the passenger car subjects wants to be assisted at roundabouts or intersections with traffic lights (compare Fig. 5). Especially the minority for intersections with traffic lights is notable and might be evoked by the meaning that an intersection with a traffic light has to be safe due to the traffic light itself.

5   Conclusion

The survey reveals that most of the interviewees have experience with ADAS (60% of questioned passenger car and 100% of questioned truck drivers already use ADAS) and that ADAS in general are very positive regarded by the subjects.

The fact that intersections are accident hot spots is confirmed by the survey. More than 50% of passenger car drivers mentioned that they have been involved in critical intersection situations. This shows on the one hand the high potential of an intersection assistant and on the other hand that the benefits of an intersection assistant are easy to communicate to the potential users as they have already experienced the addressed scenarios.

The scenarios identified by the subjects as most relevant are confirmed by the accident analysis. But while accident research shows that often inattention and distraction are important accident causes, the subjects see the reasons to be more related to external factors, e. g. occlusion.

The intersection assistant in general is rated very well and the drivers have very positive attitude with regard to it. This conclusion is underlined by the outcome that 91% of the passenger car and 76% of the truck drivers think that an intersection assistant could support them at intersections. Even more, there is only a minority of subjects who mentioned that they do not want to be assisted at any intersection by an intersection assistant.

A downsized version of an intersection assistant, for example an intersection assistant which works only at intersections which are especially equipped or an intersection assistant which indicates only the right of way regulations, is generally accepted by the subjects, but the opinions are more spread. The opinion about a traffic lights assistant is in accordance.

The subjects have no reservations against an intersection assistance system which interacts with the driver in form of an information or warning. Different to this, the acceptance of an intervention by an intersection assistant is not as high. This has to be taken into account when designing such a system (e. g. HMI layout) and its market introduction. It can be expected that the reservations of the drivers against interventions will diminish after they have the opportunity to experience the function and its benefits.

References

[1] N.N., ADAC Unfallforschung 2008, Kreuzungsunfälle, www.adac.de, article available at http://www.adac.de/Verkehr/sicher_unterwegs/Unfallforschung/ Kreuzungen/default.asp

[2] Meinecke, M. et. Al, INTERSAFE-2, Deliverable D3.1, User Needs and Operational Requirements for the Intersection Safety Assistance System, Chapter 2: Accident Review and Relevant Scenarios

[3] German In-Depth Accident Study, accident database 07/1999-08/2007 (accidents of 2007 skipped), Dresden

Dipl.-Ing. M. Wimmershoff
Institut für Kraftfahrzeuge Aachen, RWTH Aachen, Driver Assistance Department
Steinbachstr. 7
D-52074 Aachen
Germany
wimmershoff@ika.rwth-aachen.de

Dipl.-Ing. A. Benmimoun
Steinbachstr. 7
D-52074 Aachen
Germany
benmimoun@ika.rwth-aachen.de

Keywords: intersection, ADAS, user needs

Driver Assistance

Evaluation of ADAS with a supported-Driver Model for desired Allocation of Tasks between Human and Technology Performance

A. P. van den Beukel, M. C. van der Voort, University of Twente

Abstract

Partly automated driving is relevant for solving mobility problems, but also cause concerns with respect to driver's reliability in task performance. The presented supported driver model is therefore intended to answer in which circumstances, what type of support enhances the driver's ability to control the vehicle. It became apparent that prerequisites for performing tasks differ per driving task's type and require different support. The possible support for each driving task's type has been combined with support-types to reduce the error causations from each different performance level (i.e. knowledge-based, rule-based and skill-based performance). The allocation of support in relation to performance level and driving task's type resulted in a supported driver model and this model relates the requested circumstances to appropriate support types. Among three tested ADAS systems, semi-automated parking showed best allocation of support; converting the demanding parallel parking task into a rather routine-like operation.

1   Introduction

The expectations of technical solutions for solving mobility problems are very high. The reasoning behind the expected advantages is generally that assistance systems are in comparison to human drivers superior with respect to precision of operation and ability to operate under severe circumstances. Being more precise and having faster reaction times, automated cars are presumed to cause less accidents and therefore reduce congestion [1]. Demonstrations with prototype vehicles – for example: Darpa Urban Challenge – show already the technical capabilities for completely automated driving. Due to reasons of practicality, liability and user preferences [2] automation of the complete driving task is, however, not likely to become reality in the mid-term future. For this reason there will exist transitions between human (driver) operation and system (vehicle) operation when specific driving tasks are automated.

Human Factors experts argue that partly automated driving also has drawbacks. If a driving task is supported by an assistance system, this could cause for example mental underload and loss of skills, resulting in a decrease in driver reliability [3]. Endsley & Kiris [4] provided insight that automation of a task can reduce situation awareness by placing the operator out-of-the-loop and gave evidence that improving single aspects of the driving task (like reaction time) does not necessarily mean that the driving task as a whole improves. However, several studies show positive effects of automated driving on road efficiency [5]. In view of both the relevance automated driving has for solving mobility problems, and the expressed concerns, it is important to answer the question: In what driving circumstances is automation beneficial? Therefore this paper presents a supported-driver model which helps to specify these circumstances and relate them to appropriate support types. Therewith this model assists the development of appropriate applications for partly automated driving.

Within the presented model, circumstances are being defined by both the type of subtasks a driving task consist of, as well as by the performance levels required to execute these tasks. Development of the model is therefore based on existing models for human task performance. The overall goal of assistance should be to enhance the ability of the joint driver-vehicle system to remain in control throughout the travel [6]. The model therefore refers to the abilities to remain in control for defining when automation is beneficial. Three Advanced Driver Assistance Systems (ADAS) will be evaluated to what respect they are in line with the supported-driver model. The results will also be used to appraise the supported driver model as a possible means for further developing driver assistance, like facilitating the transitions between driver and vehicle operation for partly automated driving.

2 Models of the driving Tasks

To compile a model for desired allocation of support among driving tasks, this section captures existing models for human task performance. It is important to acknowledge first, that the purposes for which these models have been generated differ. Some models are intended to evaluated (new) ADAS with respect to driving safety, other models reflect typical driving behaviour to be applied within traffic simulations in order to assess traffic efficiency. Consequently, some models are more normative and others rather descriptive. As the aspired model is intended to help the development of driver interfaces which might not yet be existing, our model will be descriptive too.

2.1 Driving as Series of Control Loops

Several models exist which describe driving tasks as sets of control loops, combining input with feedforward and feedback control. Within these models specific names are defined for the steps within a control loop. McRuer [7], for example, considers driving as a sequence of operations, characterised by: "visual search", then "decision-making" and subsequently "effecting the desired actuation". Other authors use different names, but in fact these models all consider driving as a set of control loops, being performed within one or several sequences of input, throughput and output. Dirken [8] uses the same distinction in his Human-machine-interface (HMI) model to explain how human and machine cooperate and interact to execute a task. The HMI-model defines a sequence of input, throughput and output both at the human and machine-side that will be cycled through to achieve a task, possibly during multiple phases. See Fig. 1 for a representation of the model. Within this model, human input considers perception of information. This information is applied within the step throughput to evaluate and make decisions. The result of these decisions, human's reactions, are represented as output. Human output is then considered to be input for a machine. It contains information which will be processed within the machine by means of an input, throughput and output phase as well.

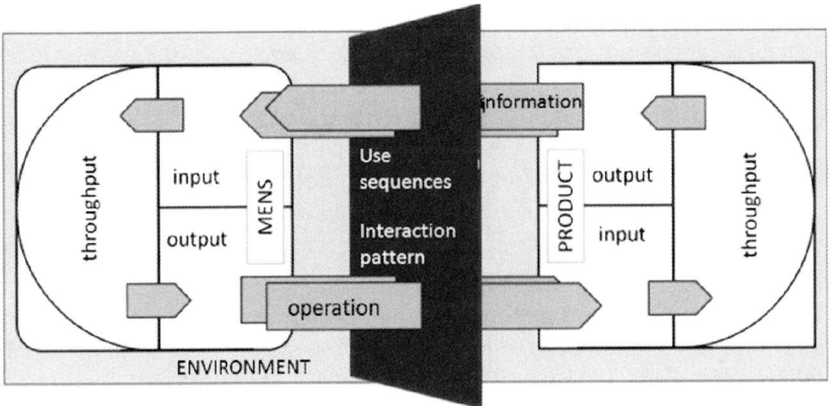

Fig. 1.   Human-Machine-Interface Model (Dirken, 1997) visualizes Task Performance as a Series of Control Loops.

2.2 Rasmussen's Model for Task Performance

Rasmussen [9] makes a general distinction for performing tasks, based on the degree of experience and distinguishes three levels: the knowledge-based level, the rule-based level and the skill-based level. Between these levels Rasmussen also differentiates whether the control-loop is being phased through in a closed-loop or in an open-loop fashion. At the highest knowledge-based level, human behaviour is goal-controlled and depends upon feedback-correction. Knowledge-based tasks are therefore performed within a closed-loop process. This is the level at which people develop new ways of problem solving. It requires attention and effort. Steps involved in this control loop are demanding, especially the decision making. At the rule-based level, tasks are involved that are more familiar. The results of (previous) actions are available and have become like rules to the person that is performing. Once a rule is chosen the actions are carried out in a rather automatically fashion. The control process is mostly a closed-loop one, but the steps within this loop are less demanding. At the lowest skill-based level, highly practiced routines are carried out. Actions are completely automated. The control process is an open-loop one, i.e. there is no continuous feedback mechanism. Only when something goes wrong in this open-loop mode, this will trigger task performance to be carried out at a higher level. With this distinction in levels, Rasmussen considers the amount of mental effort needed for executing a task and therewith addresses a dependency on individual differences in task performance.

2.3 Driving Task as a hierarchical Structure of Control Levels

Michon [10] proposed that the driving task could be structured at three levels. This model resulted in a fundamental hierarchy of the different levels within the driving task as a whole. These levels are: the strategic, tactical and operational level. At the strategic level drivers prepare their journey; this concerns general trip planning, route, transportation mode, time, etc. At the tactical level drivers exercise manoeuvring control, allowing to negotiate the directly prevailing traffic circumstances, like crossing an intersection, obstacle avoidance, etc. Here drivers are mostly concerned with interacting with other traffic and the road system. The operational level involves the elementary tasks that have to be performed to be able to manoeuvre the vehicle, mostly performed automatically and unconsciously (like: steering, using pedals or changing gears). Executing a tactical or operational task always facilitates achieving the goals of a task on a superordinated level. Although this model is intended to be an objective descriptive model, the allocation of subtasks might differ per superordinated driving task. For example; regulating speed is considered an operational task in most general circumstances. However, being a subtask of reverse parking, regulating speed might be considered a tactical task.

Michon's hierarchical structure of driving tasks relates to the general HMI-model for executing a task, because each of the three levels within Michon's model can be explained whilst using the HMI-model. The differences between driving task levels can be derived from the duration and amount of recurrences in which the HMI-model is being 'walked through'. The input-throughput-output-loop for a strategic task usually is of long duration; because evaluation if a strategic task has succeeded (e.g. if a goal has been achieved) can only be made upon arrival. Within a strategic task loop, several tactical or operational task loops can be performed. Michon's hierarchical structure of the driving task and Dirken's HMI-model are valuable in helping explaining how a driving task is being executed. When mental models and situation awareness are known that drivers have during execution of a specific driving task, these models also help to make predictions how the driver could react. Nonetheless, these models do not directly supply insight in desired task allocation for the investigated ADAS. A next step is therefore to relate Michon's descriptive model of the driving task's hierarchy to the general model for task performance from Rasmussen.

2.4 Relation between Driving Task's Hierarchy and Performance Level

The relation between the driving tasks' hierarchy and the general model for task performance from Rasmussen is visualised in Fig. 2.

Levels of task performance (Rasmussen)		Hierarchy of driving tasks (Michon)		
		strategic tasks	tactical tasks	operational tasks
	knowledge-based	navigating in unfamiliar area	controlling skidding car	novice on first lesson
	rule-based	choice between familiar routes	passing other vehicles	driving unfamiliar vehicle
	skill-based	route used for daily commuting	dealing with familiar intersections	vehicle handling in familiar circumstances

Fig. 2.   Relation between driving task's hierarchy and performance level. Adopted from Hale et al. [12]

In general, each type of driving task from the hierarchical ladder can be performed at either knowledge-based, rule-based or skill-based level. The examples in Fig. 2 show that the performance level which a driver uses to execute a task, is dependent on driving experience and contextual knowledge. Experienced drivers for example could need knowledge-based performance when passing an unfamiliar crossing. For experienced drivers most driving tasks cluster in the three cells on the diagonal from upper-left to lower right. Knowledge-based behaviour is involved at the strategic level, rule-based behaviour at the tactical level and skill-based behaviour at the operational level. As the examples show, exceptions reflect differences between skilled and novice performance, and between familiar and unfamiliar situations. Therefore, it becomes clear that task allocation can only be applied in relation to driving experience and contextual knowledge.

Error causes and the probabilities of causes, differ per performance levels [11]. As skill-based tasks are comprised of very familiar actions, which humans perform almost automatically, inattention is the main cause of errors for tasks performed at this level. Rule-based tasks are familiar to the performer. Upon correct recognition of a situation or condition, the performer can apply a stored rule to steer towards a known end goal. Subsequently, the denominating error cause is misinterpretation. Successful performance of the knowledge-based task depends heavily upon the performer's fundamental knowledge, diagnosis, and analysis skills. Unlike rule-based tasks, the performer is not per se able to steer towards an end goal. In general, drivers operate more homogeneously and predictable at skill-based and rule-based levels than at knowledge-based level. The probability for an error to occur is therefore lower for tasks performed at these levels. Altogether, it is recommendable that a support system supports driving tasks to be performed at rule-based or skill-based level rather than at knowledge-based level [12]. This is important insight in order to specify those subtasks for which support is most appropriate.

3   Support Levels

To develop the required supported driver model, we now focus on possible support levels. Six different levels can be distinguished and are indicated in Fig. 3. In a sequence from passively supporting drivers, by assisting, to actively supporting drivers, by taking over tasks, these levels are: Augmenting, Advising, Warning, Mitigation, Intervention and Automation. The differences between these support levels can be explained by identifying what step within the HMI-model is being supported. These steps can be any of the input, throughput or output steps at either the human's or machine's side. The first level, augmenting, supports the human's input by emphasizing relevant and important information, like offering a camera view with highlighted pedestrians during night. Advising and warning especially supports the driver's throughput in decision making and evaluation of input. It might also support in drawing attention to specific input. Advises might be given upon and without human's request.

Warnings are independent from driver's request. The fourth level, mitigation, aims at tempering criticality of a dangerous situation, for example: pretensioning brakes when the vehicle approaches an obstacle with high relative velocity. Mitigation is a support level that only comes into operation when the driver gives the corresponding output, e.g. braking. Automation and intervention are both support levels in which the machine takes over human's part in the control loop for a specific (sub)task. The difference is that intervention does take over suddenly, in critical situations without prior request, whereas automation is intended to be activated voluntarily for a period of time. Often intervention is intended as safety enhancement, whereas automation is rather regarded to increase comfort.

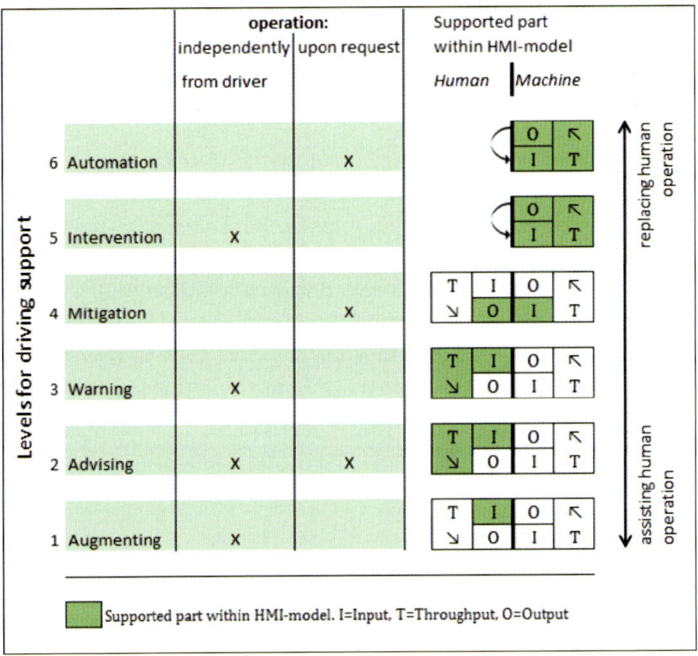

Fig. 3.   Classification of support levels

Fig. 3 visualizes the differences between support levels by identifying what step within the HMI-model is being supported. The supported steps are either the input, throughput or output step at the human's side, or the control loop is shortened by automating the human's steps within a subtask.

Related to the classification of support levels is the division in responsibility. Although it is difficult to define a precise separation in responsibilities the fourth level, mitigation, is considered to mark a change from support levels (i.e. levels 1, 2 and 3) within which the human remains responsible for executing a (sub)task and those support levels whereat the machine is responsible for executing the involved (sub)task. For the support levels automation and intervention, it is clear that the machine is fully responsible as the human is placed out of the control loop. When support of a (sub)task takes place only within the human's input or throughput step, as applies to augmenting, advising and warning, the output remains unsupported and the driver can theoretically operate without being affected by the previous supported control steps. Though assisted by a means of support, the driver is considered to remain responsible for tasks applied with those support types. (Nonetheless, everyone will understand that advises and warnings are intended to positively influence human behaviour. Therefore developers have the responsibility to provide correct information. Hence, false warnings or wrong advises might cause deteriorated driving behaviour). Mitigation is a support level with divided responsibility. The machine's control loop is being supported to raise its effectiveness, this will however only be put into effect when the related human output is performed.

A driving task can be composed of several subtasks and when support is applied, the support type might differ per subtasks. Subsequently, responsibility for performing a driving task can be divided over the driver and the machine per subtask. The overall goal of automotive design should be to enhance the ability of the joint driver-vehicle system to remain in control throughout the travel [6]. The identification of responsibility per support type is therefore aimed at ensuring that the allocation of support among the subtasks enables this joint driver-vehicle system to remain as good as possible in control. The allocation of support to subtasks under the condition that responsibility within a subtask is non-ambiguously be either at the driver or at the machine, seems therefore a plausible approach.

4   Introducing a Supported Driver Model

In paragraph 2.4 the relationship between the driving tasks' hierarchy and performance levels has been explained and being visualized in a driving task vs. performance matrix. It became apparent that the level of performance for executing a type of driving task depends on the level of (driving) experience and the contextual knowledge (i.e. familiarity with the location). To continue developing the required supported-driver model, we now implement the preferred support types within the matrix.

4.1 A supported Driver Model dependent on Driving Tasks' Type and Performance Level

The allocation of support is based upon selecting which support types help best to avoid causation of errors and these causes differ per level a tasks is being performed with. For the driving task types, it has also been distinguished what the main prerequisites are for proper execution of those tasks. Support types have been selected based upon their contribution in gaining these prerequisites. The resulting scheme is visualized in Fig. 4.

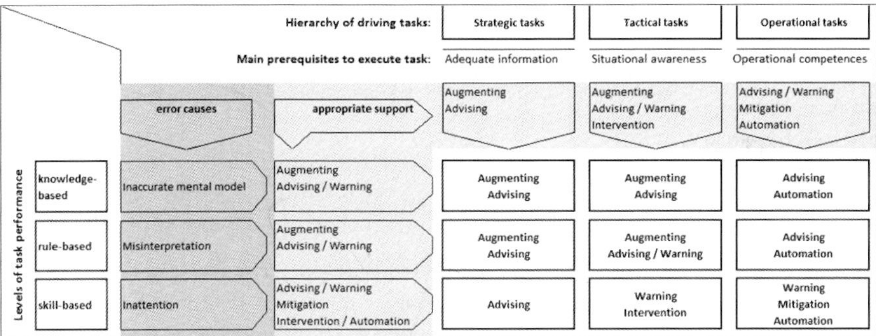

Fig. 4.   Preferred Support Types dependent on Driving Task and Performance Level.

As the strategic tasks involve preparation of a journey, adequate information is an important prerequisite to perform these tasks. In general, augmenting sensoric input and advising with regard to the interpretation of this input are adequate support types. However, the error causation on skill-based level is inattention, therefore the preferred support for a strategic task performed at this level is only advising. At the tactical level drivers are mostly concerned with interacting with other traffic and the road system. Compared to the other driving task types, the consequences from errors made at this level are most dangerous. Intervention is therefore an appropriate support to avoid errors caused by inattention. Also warning generally applies to tactical tasks. For tactical tasks performed at knowledge- or rule-based level, misinterpretation is the dominating cause of error, therefore advising or warning is appropriate for this task as well. Perception and interpretation are especially important to gain good situational awareness. Augmenting is therefore a support type that applies especially to tactical tasks, performed at knowledge- and rule-based level. Advising and warning are similar types of assistance, both offering informational support. The difference is whether this information is requested or not. The same difference is with automation and intervention: Automation is voluntarily, but intervention is not. Besides, intervention is normally intended for a short instance. The consideration whether this support should be voluntarily or not is dependent on the situation. If an operational task is operated at knowledge-based level for example, the support type advising is especially useful to assist learning and to develop appropriate competences.

Automation of an operational task performed at knowledge-based level avoids mistakes made by novice drivers due to lack of knowledge and competences. Therewith this support type can be an important benefit for road safety. Hence, automation of operational tasks reduces the need to develop operational competences. Consequently, this takes away the risk that the drivers make operational mistakes. However, it also takes away the ability for the driver to develop these competences. Therefore automation of an operational task should only be applied if it is ensured that this support will be executed under all circumstances.

4.2 Advantages of the supported Driver Model

In comparison to existing models for the driving task, this assisted driver model helps to specify circumstances for adequate support. These circumstances are differentiated with regard to type of driving task and performance level. This allows to evaluate whether support types applied to existing assistance systems enhance the ability of the joint driver-vehicle system to remain in control under relevant circumstances. Moreover, the enhancement can be compared with the related unassisted driving task as well. Since, unassisted subtasks can be placed within the driving task vs. performance matrix as well, these can be evaluated with regard to their relation to error occurrence and error severity. Likewise, section 6 will evaluate three existing ADAS systems. Moreover, the model can be used to examine existing driving tasks and to explore what kind of new support is most adequate for which subtask. Therewith the model can be used for developing future applications for assited driving, intended to help reducing mobility problems.

4.3 Considerations when applying the assisted Driver Model

The following aspects are important to consider when applying the model:
> ▶ A driving task consists of one or more subordinated tasks, called subtasks;
> ▶ With respect to the performance levels of the involved driving tasks, it is desirable that support enables driving tasks to be as much as possible executed on rule-based and skill-based level, because drivers then behave more homogeneously and predictable;
> ▶ As performance levels are dependent on driving experience, the model

needs to be verified for a chosen group with defined experience;

The considerations to allocate support types, as explained in the first paragraph of this section, give reason for the following general recommendations for the development of ADAS:

- ▶ Enable single subtasks to be performed at preferably skill-based or rule-based level;
- ▶ When possible reduce the amount of subtasks the human is involved in, but avoid that subtasks are being replaced by subtasks which require a higher performance level;
- ▶ Automation of operational tasks should be applied to reveal the driver and therewith easing the driver in executing tactical tasks;
- ▶ Within a single subtask, avoid support types without a clear separation of responsibility between driver and technology;
- ▶ Preferably, automation of a subtask should be ensured under all applicable circumstances, because loss of skills only then remains without harm;
- ▶ Besides, the transitions between human and machine operation involved in non-permanently automated subtasks most often relate to the existence of system boundaries. These system boundaries are often reached in critical circumstances. Therefore applying non-permanently automated subtasks is potentially dangerous;
- ▶ When support is temporary, provide clarity with regard to the moment of termination of a supported subtask. Provide also clarity upon who is responsible for terminating the support.

5 Advanced Driver Assistance Systems

This section introduces three assistance systems, which will be evaluated with the assisted driver model in the next section. The evaluation focuses on the appropriateness of the applied support to enhance the ability to remain in control within the related driving circumstances. The evaluated systems are intended to support reverse parallel parking, cruising and changing lanes and are described in the next paragraphs.

5.1 Semi-Automated Parallel Parking (SAPP)

With this assistance system the vehicle parks in parallel parking slots semi-automatically. One of the first suppliers who developed this system and introduced it on the market is Valeo.

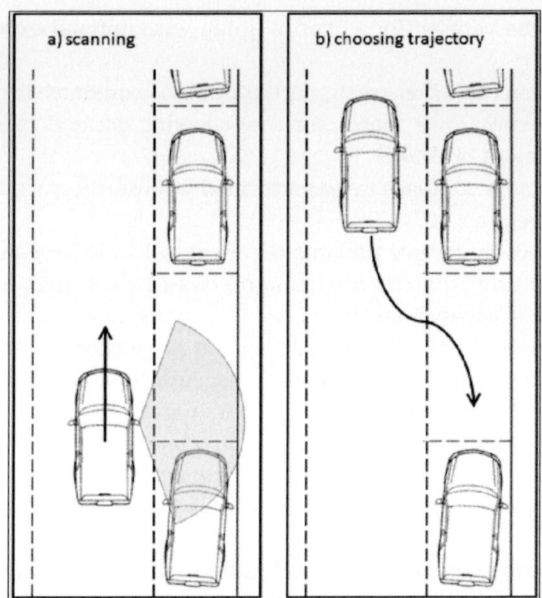

Fig. 5.   Process steps for using SAPP.

The Valeo-system is taken as a basis for this case. The system functions as follows, see Fig. 5. After activation, the system scans for available parking places. For scanning and assessment of the possibilities, the driver needs to pass the complete parking slot. A display indicates when a parking place has sufficient size. As soon as the driver selects reverse driving, the vehicle steers automatically and obtains the correct path for reverse parking. During this reverse parking manoeuvre the driver operates backing up-speed with the gas pedal, brake and probably clutch. Therewith the driver remains responsible for the moment and speed of backing up and can make intermediate stops of the manoeuvre when traffic circumstances require to do so. The most important technologies this system consists of are: parking place measurement, based on radar or a camera; electric powered steering, trajectory calculation and rear vision.

5.2 Adaptive Cruise Control (ACC)

Cruise control is a system to automatically remain speed. Adaptive Cruise Control (ACC) systems use either a radar or laser setup to monitor vehicles in front of the host vehicle and use this information to slow down when approaching another vehicle and to accelerate again to the preset speed if traffic allows to do so. With other words: ACC automates keeping distance and remaining speed. The driver preselects a desired speed; this is displayed in the instrument cluster. Furthermore, the driver chooses between three predefined headway settings. These settings involve fixed time based distance stages, therefore the distance to the vehicle in front changes with speed. For the first commercially available ACC systems, the following system boundaries are typically [13]:

- ACC operates only from 30 km/h onwards;
- ACC cannot be used in stop-and-go traffic. When approaching stationary traffic, the driver must take control of the vehicle by braking in good time;
- The time gap to a vehicle ahead is usually not less than 1 s. When driving straight ahead, the response of the ACC to a vehicle cutting in close may be delayed. The vehicle that cuts in is not detected before it is positioned on the same lane as the ACC vehicle. As a result the time gap can be below 1 s for a brief period;
- The system has limited deceleration capabilities. A typically maximum deceleration of 2,0 m/s^2 is provided. Large speed differences, e.g. quickly approaching a heavy goods vehicle, can therefore not be regulated;
- Detection of the leading vehicle may be lost when cornering due to the system's limited side field of vision. When cornering, the ACC vehicle is not accelerated to the requested speed for an instance in order not to drive up too close to the leading vehicle.

When the system has reached its functional limits, the driver is requested to assume control by releasing a warning signal. Furthermore, ACC is marketed as a convenience system and driver interventions always have higher priority than the ACC control. Because of automating de- and accelerating, ACC is widely regarded as a key component of any future generations of smart cars.

5.3 Lane Change Assist (LCA)

The Lane Change Assist system is intended to support changing lanes by warning if a vehicle is within the driver's blind spot area or if an approaching vehicle does not allow a safe lane change. See Fig. 6. from driver A's perspective, vehicle C is within the blind spot area and vehicle B approaches with a higher relative velocity.

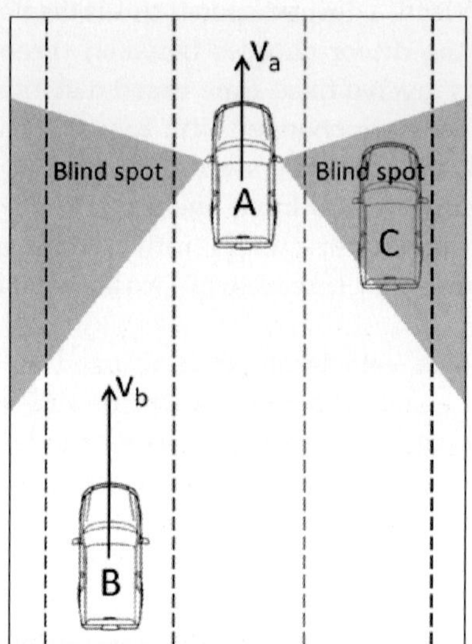

Fig. 6.   Critical Factors involved in changing Lanes

Monitoring the vehicle's blind spot area is especially a problem for lorries and delivery vans. Hence, freight vehicles are available with Blind Spot Detection (BSD). Monitoring relative velocity of approaching vehicles from the rear has been introduced to the market by MobilEye and this system is taken as a basis for this case study [14]. The system uses camera-technology to measure relative speed. The system does not support overtaking within situations with oncoming traffic. The main functions are: (a) tracking of dynamic targets moving towards the vehicle, (b) detection of obstacles at close range, (c) detection of vehicles in the blind spot and (d) detection of critical and non-critical objects in the lateral and rear areas. Apart from monitoring the host vehicle's blind spot area, it also detects approaching vehicles and motorcycles from a distance of 60m. This data is geared towards indicating to the driver whether it is safe to change lane or not. With MobilEye's system the visible alert whether or not it is safe to change lanes, is continuously available at the side mirrors.

6 Evaluation ADAS with supported Driver Model

6.1 Assessment of SAPP with Respect to preferred Support Types

As explained before, SAPP supports the task of reverse parallel parking. To assess this assistance function with respect to the preferred support types, it is important to first review how reverse parallel parking is in general executed by a driver with average experience. This is schematically visualized in the task-level matrix, see Fig. 7. The task is being executed in five subtasks. First a parking spot has to be defined which is likely to have the appropriate size (1). This selection is a strategic task, executed at knowledge-based level. Next, the appropriate trajectory needs to be defined to manoeuvre the vehicle along (2). Then, the operation needs to be appropriately timed to avoid hindrance of other road users (3). This is a tactical task, performed at rule-based level. In addition to defining the trajectory, speed needs to be regulated to move the vehicle along the trajectory (4). Finally, the appropriate trajectory needs to be evaluated and possibly be refined (5). When the vehicle is being parked between obstacles –which is often a reason for reverse parallel parking– defining and refining the trajectory (steering the wheel) is a delicate task requiring a sequence of closed-loop control cycles.

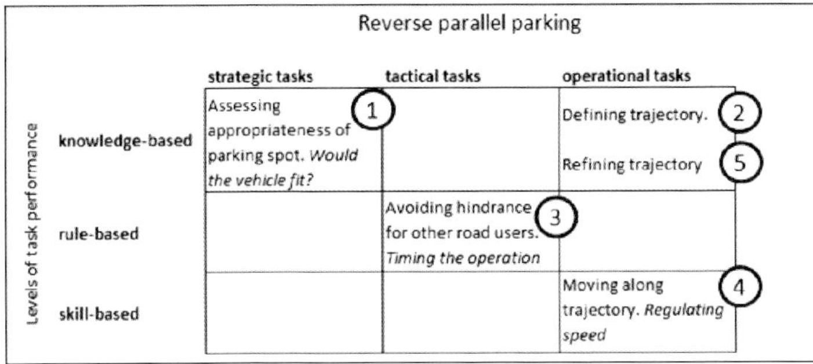

Fig. 7. Classification of Subtasks for Reverse Parallel Parking with Respect to Driving Task-Hierarchy and Performance Level.

Next, Fig. 8 visualises a comparison of the SAPP system with the preferred supported driver model. The SAPP system automatically scans for parking slots with an appropriate size. This information is available for the user upon request. In this manner the SAPP system changes the attention requiring knowledge-based task of recognising an appropriate parking spot into a skill-based task of requesting information. Furthermore, the 'knowledge' part, which involves estimating and continuously assessing the ideal trajectory, is taken over by the SAPP-system. Automation of this subtask is in line with the preferred support model, because choosing the appropriate trajectory is a demanding (knowledge-based) operational task.

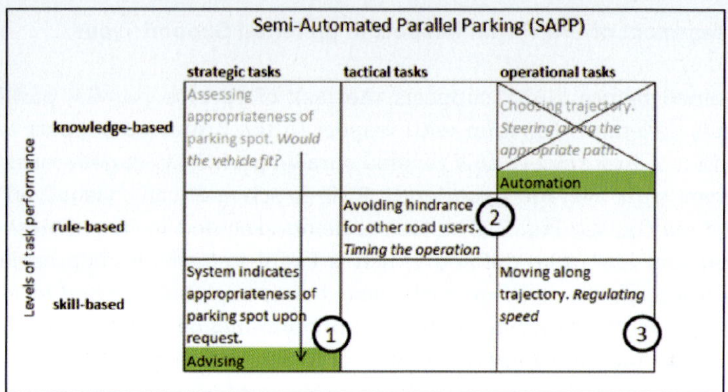

Fig. 8.   Supported Subtasks for Reverse Parallel Parking when applying SAPP.

In general, SAPP helps to change reverse parallel parking from a rather knowledge-based driving task to rule- and skill-based subtasks. The human's subtask involves supervision which he can perform at a rule-based level, the rule applies not to hinder other traffic and by breaking the driver can temporarily stop the manoeuvring, allowing vehicles to pass, before continuing again. Furthermore, automation of the trajectory reduces the amount of parallel subtask the driver is involved in. Therefore the driver can be more concentrated on the subtasks he is responsible for. One might put forth that automation of choosing trajectory reduces parking skills. However, as long as this subtask is being automated under all circumstances, this is of no matter.

6.2 Assessment of Adaptive Cruise Control (ACC) with Respect to preferred Support Types

ACC supports two subtasks: remaining speed and keeping distance. To assess this assistance function with respect to the preferred support types, it is again important to review how both subtasks relate to the task-level table assuming that these tasks are executed by a person with average driving experience. See Fig. 9.

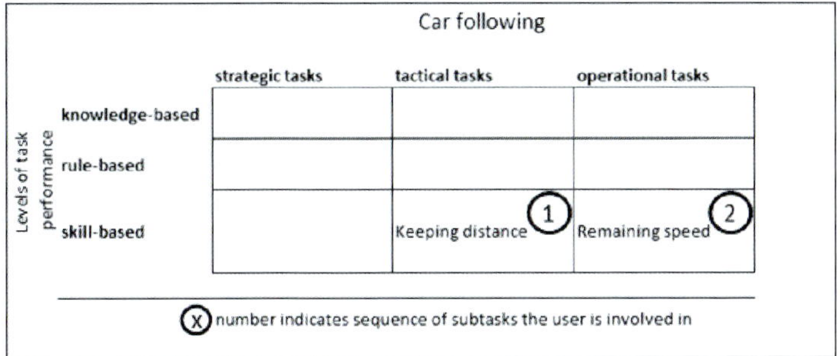

Fig. 9. Classification of Subtasks for Car following with Respect to Driving Task Hierarchy and Performance Level.

For an average driver, the subtask to remain speed is an operational task executed at skill-based level. The classification of 'keeping distance' within the model is less obvious. It depends on traffic circumstances whether keeping distance is rewarded a tactical or operational task. Under calm traffic conditions and within familiar areas, keeping distance is an operational task. However under circumstances with heavy traffic, keeping distance is a tactical driving task. Also experienced drivers need to consider keeping distances which are long enough to avoid collision, but short enough to avoid other drivers continuously merging in front. It is within such densed traffic however for which Adaptive Cruise Control offers an advantage over normal cruise control. Therefore, keeping distance will be considered a tactical task within this research. Experienced drivers execute this task with skill-based performance.

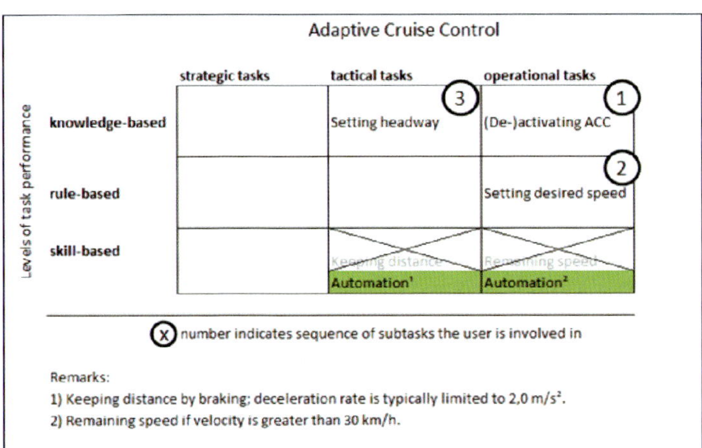

Fig. 10. Supported Subtasks for Cruising when applying ACC.

Fig. 10 visualises a comparison of the ACC system as described in section 5.2 with the preferred supported driver model. From this comparison we conclude that the ongoing subtasks 'keeping distance' and 'remaining speed' are being supported by means of automation. For the driver new subtasks are being introduced to set speed and set headway. Besides, the driver first need to activate the system to start operating. Activation and deactivation of ACC can be done manually by pressing the activation switch. Also when the driver uses the brakes, the system deactivates. When the driver brakes during a reflex-like response, deactivation is sometimes not immediately noticed. In general, a driver needs to consider beforehand if traffic circumstances are appropriate to apply adaptive cruising, therefore the (de-)activation task (1) is considered an operational task on knowledge-based level. As described before, keeping distance is considered a tactical task. Consequently, setting headway (3) is also a tactical task. However it will be performed at knowledge-based level. The reason is that the headway-settings are represented by a graphic within the instrument cluster and relate to three fixed time based distance stages. This means that the distance to the vehicle in front changes with speed. Evaluating and refining the made settings therefore require adjustments over time.

When comparing ACC with the supported driver model and considering the frequency for operating subtasks, it is concluded that the new subtasks are performed less frequent, but require higher performance levels. This is not in line with the proposed supported driver model and might explain why drivers adapt their behaviour when using ACC [15]. Furthermore, it has to be noticed that keeping distance is not being automated under all circumstances. The driver is required to intervene by braking when a deceleration rate greater than 2,0 m/s^2 is necessary. Because the driver is then suddenly placed in this tactical task, this could be a possible treat. Therefore, ACC reduces under system-favourable circumstances the driver's subtasks, but suddenly increases driver's workload under critical circumstances.

6.3 Assessment of Lane Change Assist (LCA) with respect to preferred Support Types

Again the corresponding driving task is divided into subtasks. For changing lane the involved subtasks are indicated in Fig. 11 as follows: (1) Verifying the possibility for lane access, answering the question; "Is the lane to change to available?". If the lane is available the subtask for actually changing lane (2) and regulating speed (3) follow. As the first subtask involves judgement about the behaviour of other traffic, this is a tactical task which experienced drivers will operate at rule-based level under normal circumstances. One might argue if changing lane (2) and regulating speed (3) are tactical, respectively operational tasks. Obviously, steering itself and operating the pedals are in common driving situations operational tasks. However, in situations of heavy traffic applying these operations might need cleverly timing, hence tactfulness. Therefore, changing lane is considered a tactical task as well.

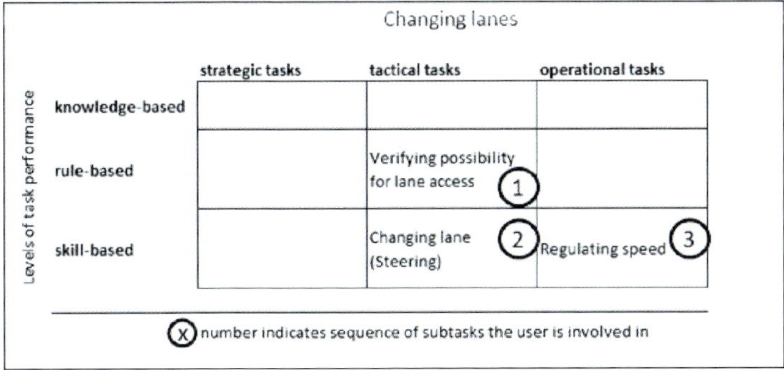

Fig. 11.   Classification of Subtasks for changing Lanes with Respect to Driving Task Hierarchy and Performance Level.

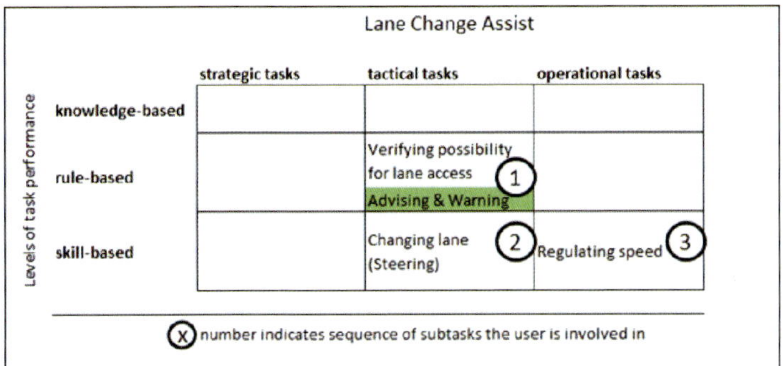

Fig. 12.   Supported Subtasks for changing Lanes when applying LCA.

When comparing the supported subtasks as offered by the Lane Change Assist system with the supported driver model for classification of preferred support levels, it can be concluded that LCA is in line with this model. The assistance system does not reduce the amount of involved subtasks, but among the subtasks it facilitates the most demanding one. Hence, it complies with the advise, as formulated in section 4.3, to allocate support so that it enables driving tasks to be as much as possible being executed on rule-based and skill-based level.

7   Conclusions

As mentioned in the beginning of this paper, when solutions for partly automated driving are being developed, it is important to answer in which circumstances, what type of support enhances the driver's ability to control the vehicle. Existing models how humans perform (driving) tasks are useful to gain insight in what subtasks can be distinguished. However, these models do not provide insight in distinguishing which subtask is in need for what kind of support. It turned out to be necessary to identify which errors can be avoided by what type of support. The performance levels distinguished within Rasmussen's model relate to typical error causes per level. This allows to recommend in general that a support system should support driving tasks to be performed at skill-based or rule-based level rather than at knowledge-based level, because drivers then operate more homogeneously and predictable.

It became clear that prerequisites for performing tasks differ per driving task's type and those prerequisites require different support. Therefore, support-types have been selected based upon their contribution in providing these prerequisites and have been combined with support-types which reduce the error causations from each different performance level (i.e. knowledge-based, rule-based and skill-based performance). This combination of allocating support in relation to performance level and driving task's type resulted in the supported-driver model and therewith relates the requested circumstances to appropriate support types. In this respect the model successfully answers the paper's initial question. It has to be noticed however, that the allocation of subtasks with respect to performance level and driver task type might sometimes be an arbitrary choice. During normal driving conditions, for example, it depends on your point of view whether lane changing is a tactical task or operational task. This ambiguity thus influences the recommended support-types, generated with the model. It needs to be noticed as well, that the model does not yet relate subtasks to timing (i.e. duration and recurrence of a task). Timing will have influence on the workload though and should therefore be related to the choice of support types too. Preferably, this aspect will be taken in consideration during further development of the model.

Nevertheless, the model has successfully been used to evaluate three existing ADAS systems with respect to the ability of the joint driver-vehicle system to remain in control under relevant circumstances. According to the model, the semi-automated parallel parking (SAPP) system, showed best allocation of support by automating the knowledge-based subtask (choosing trajectory) and adapting the remaining subtasks for the driver to rule- and skill-based level. Therewith the demanding reverse parallel parking task becomes a rather routine like operation. In this respect the SAPP system is a good example of appropriate support. As the model is successful in examining existing driving (sub)tasks with respect to possible and appropriate support types, the model is expected to be beneficial for the development of future applications of partly automated driving too.

References

[1] European Commission, Communication on the Intelligent Car Initiative "Raising awareness of ICT for smarter, safer and cleaner vehicles", COM(2006) 59-final, European Commission, Brussels, 2002.

[2] Stevens W., Evolution to an Automated Highway System. In: P.A. Ioannou (Ed.), Automated Highway Systems, Plenum Press, New York, pp. 109-124, 1997.

[3] Rothengatter, T., et. al., The driver. In: J.A. Michon (Ed.), Generic Intelligent Driver Support, A comprehensive report on GIDS, Taylor & Francis, pp. 33-52, 1993.

[4] Endsley, M., Kiris, E., The Out-of-the-loop performance problem and level of control in automation, Human Factors, 37, no. 2, pp. 381-394, 1995.

[5] Van Arem, B., et. al., The impact of Cooperative Adaptive Cruise Control on traffic-flow characteristics, IEEE Transactions on Intelligent Transportation Systems, 7, no. 4, pp. 429-436, 2006.

[6] Hollnagel, E., A function-centered approach to joint driver-vehicle system design. Cognition, Technology & Work, 8, no. 3, pp. 169-173, 2006.

[7] McRuer, D.T., et. al., New results in driving steering control. Human Factors 19, pp. 381-397, 1977.

[8] Dirken, J.M., Product-ergonomie. Ontwerpen voor gebruikers, Delft University Press, Delft, 1997.

[9] Rasmussen, J., Skills, Rules and Knowledge: Signals, Signs and other Distinctions in Human Performance Models, IEEE Transactions on Systems, Man & Cybernatics, Vol. 13, pp. 257-266, 1983.

[10] Michon, J., A critical view of driver behavior models: what do we know, what should we do? In: Evans, L., Schwing, R.C., Human behavior and traffic safety, Plenum Press, New York, pp. 485-524, 1985.

[11] Rasmussen, J., A taxonomy for describing human malfunction in industrial installations. In: Journal of occupational accidents, 4, pp. 311-333, 1982

[12] Hale, A., et. al., Human error models as predictors of accident scenarios for designers in road transport systems, Ergonomics, 33, pp.1377-1388, 1990.

[13] Leveson, N. (2005). Adaptive Cruise Control – System Overview, Massachusetts institute of technology, Anaheim USA, 2005.

[14] Gat, I., et. al., Monocular Vision Advance Warning System for the Automotive Aftermarket, SAE World Congress & Exhibition 2005, Detroit USA, 2005.

[15] Hoedemaker D.M., Behavioral reactions to advanced cruise control: results of a driving simulator experiment. In: Van der Heijden, R.E.C.M., Automation of Car Driving. Exploring societal impacts and conditions, Delft University Press, Delft, pp. 101-121, 2000.

Arie Paul van den Beukel, Mascha C. van der Voort
University of Twente
Laboratory Design, Production and Management
PO Box 217
7500 AE Enschede
The Netherlands
a.p.vandenbeukel@ctw.utwente.nl
m.c.vandervoort@ctw.utwente.nl

Keywords: ACC, ADAS, automation, driver assistance, driver performance, driving task's hierarchy, LCA, partly automated driving, semi-automated parallel parking (SAPP), supported driver model, task allocation

Lane Recognition Using a High Resolution Camera System

R. Schubert, G. Wanielik, Technische Universität Chemnitz
M. Hopfeld, K. Schulze, IAV GmbH

Abstract

Vision-based driver assistance systems are contributing to traffic safety and comfort in a variety of applications. However, due to the limited resolution provided by state-of-the-art cameras, the observable area is very limited. Thus, high resolution image sensors have been developed recently for the automotive area. In this paper, the potential of such camera systems is analyzed on the example of a lane recognition system. In particular, the possible augmentation of the observable area is discussed. However, it is also shown which algorithmic adaptations are necessary in order to benefit from such sensors. For that, a new lane model is proposed which consists of several continuous segments and, thus, is also suitable for higher detection ranges. Finally, first results from a practical implementation of this model are presented and discussed.

1 Introduction

Vision-based driver assistance systems have found their way from technical development into serial production. The application areas for such systems are very diverse and include the recognition of lanes [1], vehicles [2], or pedestrians [3]. Among those applications, lane recognition systems have received particular attention recently, as they enable important functions such as Lane Departure Warning (LDW) or Lane Keeping [4].

However, due to the limited resolution of most automotive camera sensors (VGA), the observable area of vision-based systems is very limited. For instance, a reliable detection of lane markings is considered hardly possible beyond 50 meters. This limitation is particularly important for collision avoidance systems which require a much higher detection range. Thus, one common approach to overcome the described limitation is to combine additional sensors with a camera (for instance a Radar system [2]). Another consequence of the limited resolution is the fact that it is very difficult to use one camera for different applications. If the focal length of the vision system is for instance chosen for lane recognition purposes – that is, for a long range – the aperture angle will be very limited, which prevents the use of such a system for pedestrian recognition.

Recently, high resolution cameras for automotive applications have become available. These sensors yield a notable potential for vision-based driver assistance systems as they promise a significantly augmented observation area. However, they also imply new challenges for perception algorithms. In this paper, the potential benefits of a high resolution camera system are analyzed both theoretically (section 2) and on the practical example of a lane recognition system (section 3). As state-of-the-art algorithms for estimating the course of a lane cannot be considered suitable for the enhanced detection range, a new approach is proposed. This method is based on a lane model consisting of several continuous, circular segments and, thus, promises a higher flexibility without loss of stability. The basic idea of this model is presented in section 4.2. The parameters of the model can be estimated from the detected lane markings in the image using a Bayesian estimator. In this paper, the Unscented Kalman Filter (UKF) is used for this purpose. Its fundamentals are briefly explained in section 4.1. Finally, first results from practical tests of the proposed system are presented and discussed in section 5.

2 Benefit Analysis of High Resolution Cameras for Driver Assistance Systems

In the following section, the benefits of a high resolution camera compared to VGA resolution are analyzed. In particular, equations for both the range and angle gain are derived. For sake for comparability, the sensors' dimensions are assumed to be equal for both cameras. They only differ in the number of picture elements (pixel) per line and row, respectively.

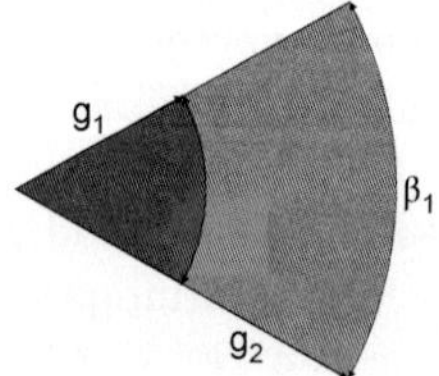

Fig. 1.   Range gain of a high resolution camera. The darker area illustrates the limited detection range of a standard VGA camera. By replacing the VGA sensor with a higher resolution sensor, a significant gain of range can be achieved as shown by the light grey circle segment.

Range Gain

A camera can be modelled as a convex lens. The following two equations describe the projection of an object G which has a distance g to the lens' principal axis and an image B which is generated in the distance b from the principal axis, turned upside down and scaled. The projection also depends on the lens' focal length f:

$$\frac{B}{G} = \frac{b}{g} \tag{1}$$

$$\frac{1}{f} = \frac{1}{b} + \frac{1}{g} \tag{2}$$

One gain of a higher resolution camera could be an increase of range (compare Fig. 1). At first, an equation to compare the ranges g_1 and g_2 of both cameras is derived. As a starting point, an equation for the range of one camera g with respect to focal length f, object dimension G, and image dimension B is developed from equation 2:

$$b = \frac{fg}{g - f} \tag{3}$$

and inserted into equation 1. Thus, the formula for the distance g becomes:

$$g = \left(\frac{G}{B} + 1 \right) f \tag{4}$$

To compare the two camera ranges, the quotient g_1/g_2 can be written as

$$\frac{g_1}{g_2} = \frac{\left(\dfrac{G_1}{B_1} + 1 \right) f_1}{\left(\dfrac{G_2}{B_2} + 1 \right) f_2} \tag{5}$$

Since both cameras create an image of the same object, it is safe to define $G = G_1 = G_2$. Additionally, both cameras are using identical lenses and, thus, $f = f_1 = f_2$ can also be defined. With those definitions, equation 5 becomes:

$$\frac{g_1}{g_2} = \frac{B_2}{B_1} \cdot \frac{B_1 + G}{B_2 + G} \tag{6}$$

Typical objects will be in the size of meters which is several magnitudes more than the pixel dimension, as the latter is (currently) measured in 10^{-6} m. Thus, if $B_1 << G$ and $B_2 << G$ and is true, equation 6 becomes

$$\frac{g_1}{g_2} = \frac{B_2}{B_1} \tag{7}$$

With p describing the number of pixels per line and w describing the sensor width for both sensors, B becomes

$$B_1 = \frac{w_1}{p_1} \qquad\qquad B_2 = \frac{w_2}{p_2} \tag{8}$$

In the beginning of this section, it was assumed that both cameras' imaging sensors are equal in size and, thus, it follows that $w = w_1 = w_2$ which gives as a result the maximum distance g an object G can be away from the camera is directly proportional to the sensor's pixels per line

$$\frac{g_1}{g_2} = \frac{p_1}{p_2} \tag{9}$$

In case camera 2 has a sensor with VGA resolution (640x480) and the reference object can be seen in the picture when its distance g_2 is not higher than 50 m and camera 1 has a sensor with SXGA resolution (1280x1024), the object can be seen in camera 1's picture at distance $g_1 = \dfrac{1280}{640} \cdot 50\,m = 100\,m$.

Angle gain

Another advantage of a high resolution camera may be an increased aperture angle and, thus, an extended lateral field of view (cp. Fig. 2). The aperture angle β depends on the sensor's diagonal length e and focal length f [5]

$$\beta = 2 \arctan \frac{e}{2f} \tag{10}$$

As the diagonal length is assumed to be identical for the two compared camera sensors, the aperture angle β only depends on the focal length f. From equation 2, f can be expressed as

$$f = \frac{bg}{b+g} \tag{11}$$

Since g represents the distance between object and camera and shall be constant in this case for both cameras, the image distance b needs to be replaced by another expression. From equation 1, one obtains

$$b = g\frac{B}{G} \tag{12}$$

Inserting equation 12 in equation 11 gives the following formula for the focal length f

$$f = \frac{g \cdot \dfrac{B}{G}}{1 + \dfrac{B}{G}} \tag{13}$$

In equation 13, f depends on the distance g of an object G to the camera and its picture size B. In order to compare the focal lengths of those two cameras, a quotient can be written as

$$\frac{f_1}{f_2} = \frac{\left(g_1 \cdot \dfrac{B_1}{G_1}\right)}{\left(g_2 \cdot \dfrac{B_2}{G_2}\right)} \cdot \frac{1 + \dfrac{B_2}{G_2}}{1 + \dfrac{B_1}{G_1}} \tag{14}$$

Assuming that the object is the same for both cameras results in $G = G_1 = G_2$. With the additional assumption that it has the same distance from both cameras, $g = g_1 = g_2$ is obtained. Considering these conditions simplifies equation 14 to

$$\frac{f_1}{f_2} = \frac{B_1}{B_2} \cdot \frac{G+B_2}{G+B_1}$$

(15)

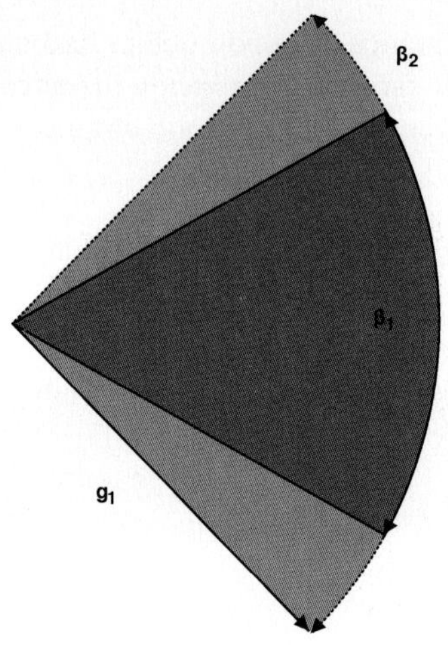

Fig. 2.   Angle gain of a high resolution camera. Although both sensors' detection ranges are equal, the higher resolution sensor's (lighter) aperture angle is increased compared to the VGA sensor (darker). Both sensors use objectives with identical focal lengths.

The object G's dimension (in meters) is several magnitudes higher than the pixel dimensions B_1 and B_2 and (currently 10^{-6} m). With the conditions $G \gg B_1$ and $G \gg B_2$, equation 15 can be simplified to

$$\frac{f_1}{f_2} = \frac{B_1}{B_2}$$

(16)

Using this relation between the focal length of both cameras in equation 10 for calculating the aperture angle β gives

$$\beta = 2\arctan\left(\dfrac{e}{2\dfrac{B_1}{B_2}f_2}\right) \tag{17}$$

For a hypothetical comparison between two cameras – where camera 2 has double the pixel count of camera 1, that is, ratio $\frac{B_1}{B_2}=2$, a focal length f_2 = 15 mm – and both camera sensors have a diagonal length of $e=\frac{2}{3}$ the original aperture angle is $\beta_1=59°$, while the aperture angle with a higher resolution camera calculates to $\beta_2 \approx 97°$.

Summary

In this section, the expected gains of a high resolution camera compared to a VGA camera have been described from a theoretical perspective. From the geometrical relations, two corner cases have been derived. Obviously, arbitrary combinations of both cases are also possible.

Furthermore, it should be noted that in practice, the detection range of a camera does not increase proportionally with the resolution. Given a constant sensor size, the signal to noise ratio (SNR) of the sensor is decreasing with growing resolution. That is, increasing only the resolution does not automatically lead to a higher image quality.

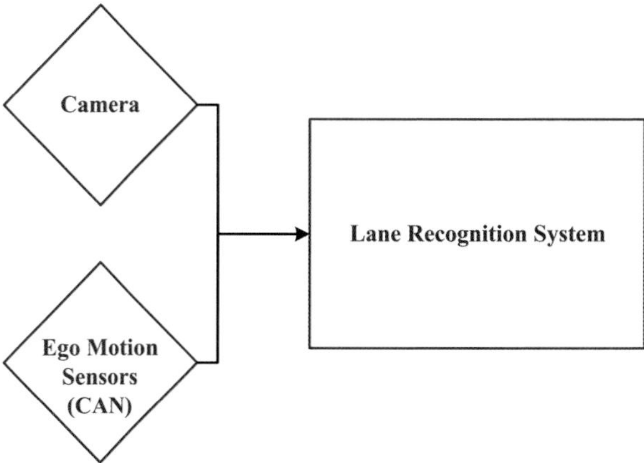

Fig. 3.   Block diagram of the implemented lane recognition system

3   System Architecture

For the system described in this paper, a KWD LOGLUX i5 camera [6] has been used which is using a resolution of 1280 x 1024 pixels. However, for estimating the course of a lane, camera images are not the only useful information. In fact, data from the vehicle's internal CAN bus about the ego motion (in particular about velocity, acceleration, and yaw rate) can significantly improve the stability of the estimation. Thus, camera images and ego motion data are processed within the system. Fig. 3 illustrates this basic structure.

4   Lane Estimation

4.1   Filter Architecture

In this section, the statistical framework for estimating the course of a lane will be presented. In particular, the *Unscented Kalman Filter* (UKF) will be explained as a recent development of estimation theory. The content of this section is based on [7] and will be summarized here for convenience.

The UKF is an unbiased, minimum-mean squared error estimator of a dynamic system with the state vector $\vec{x}$ and the covariance matrix P [7]. In general, the structure of the UKF is similar to the well-known Extended Kalman Filter (EKF), that is, it predicts the state of the system using a (typically nonlinear) state transition equation $f(\vec{x})$ and corrects this prediction by incorporating the observations which may come from different sensors.

The particularity of the UKF derives from the fact that for predicting the state vector of the system, not only one single value, but a whole probability distribution has to be transformed. While this is an easy task in the linear case, it causes significant problems if the state transition equation is nonlinear. The EKF's approach to that issue is a linearization of $f(\vec{x})$ around the current value of $\vec{x}$. The covariance matrix can then be predicted using the Jacobian of $f(\vec{x})$.

However, this approach has the unpleasing property of underestimating the uncertainty in many cases [8]. Although that property can partly be compensated by adding more process noise, it generally deteriorates the performance of the filter and in the worst case causes it to divert from the true value.

Another approach for transforming probability distributions in a nonlinear manner is to sample the distribution and transform the samples separately. If the samples are drawn randomly, this approach is called particle filter. While this filter yields good tracking results, it requires a large amount of computational resources due to the large number of particles which are necessary to adequately represent the distribution. Other approaches are thus using deterministically chosen samples to transform the probability distribution. One special case of those so called Sigma Point Kalman Filters is the Unscented Kalman Filter.

The UKF performs the so called Unscented Transformation in order to calculate the sigma points (i. e. the samples) from the probability distribution. In particular, the sigma points are obtained by applying the following equation:

$$\vec{\chi}(k) = \left[\vec{x}(k) \cdot \vec{x}(k) + \sqrt{(\lambda + L)P(k)} \cdot \vec{x}(k) - \sqrt{(\lambda + L)P(k)} \right] \tag{18}$$

where L is the size of the state vector and λ is one of the three scaling parameters of the UKF which heavily depend on the type of the distribution [9].

After applying the state transition equation to the sigma points, that is $\vec{\chi}(k+1) = f(\vec{\chi}(k))$, the predicted mean and covariance of the state vector can be calculated as follows:

$$\vec{x}^-(k+1) = \sum_{i=0}^{2L} w_i^m \cdot \vec{\chi}(k+1)$$

$$P^-(k+1) = \sum_{i=0}^{2L} w_i^c \cdot \left[\vec{\chi}(k+1) - \vec{x}^-(k+1)\right] \cdot \left[\vec{\chi}(k+1) - \vec{x}^-(k+1)\right]^T \tag{19}$$

In those equations, are weight factors which are calculated from the scaling parameters , and :

$$w_0^m = \frac{\lambda}{L+\lambda}$$

$$w_0^c = \frac{\lambda}{L+\lambda} + \left(1 - \alpha^2 + \beta\right) \tag{20}$$

$$w_i^c = w_i^m \frac{\lambda}{2(L+\lambda)}$$

With the update step being complete, the state is now corrected. For that, the state is at first transformed into the measurement space using the (in general also nonlinear) function h in order to obtain the expected observations y^- and their covariance U^-.

$$\vec{Y}^-(k+1) = h\big(\vec{\chi}(k)\big)$$

$$\vec{y}^-(k+1) = \sum_{i=0}^{2L} w_i^m \vec{Y}^-(k+1) \tag{21}$$

$$U^-(k+1) = \sum_{i=0}^{2L} w_i^c \cdot \big[\vec{Y}^-(k+1) - \vec{y}^-(k+1)\big] \cdot \big[\vec{Y}^-(k+1) - \vec{y}^-(k+1)\big]^T$$

Finally, the correction step combines the predicted state and the observations $\vec{y}$ as follows:

$$P_{xy}(k+1) = \sum_{i=0}^{2L} w_i^c \cdot \big[\vec{\chi}(k+1) - \vec{x}^-(k+1)\big] \cdot \big[\vec{Y}^-(k+1) - \vec{y}^-(k+1)\big]^T$$

$$K(k+1) = P_{\vec{y}}(k+1) U^-(k+1)^{-1} \tag{22}$$

$$x(k+1) = \vec{x}^-(k+1) K(k+1) \cdot \big[\vec{y}(k+1) - \vec{y}^-(k+1)\big]$$

$$P(k+1) = P^-(k+1) - K(k+1) U^-(k+1) K(k+1)^T$$

In contrast to the EKF, the UKF calculates the covariance of the system state correctly to the third order of the Taylor series expansion (for Gaussian distributions). Furthermore, it does not require an analytical linearization of the model equations, which would be hardly possible for the model described in the next section.

4.1 Geometrical Lane Model

State of the Art

As shown in section 2, a high resolution sensor can be used to significantly increase the detection range of a camera. However, with the extended detection range, some new problems arise when modelling the course of a lane. Existing models describe the course of the lane for a limited area in front of the vehicle. In this limited area, the curvature change from one scene to the other is rather limited.

Among the existing descriptions for the course of a lane are the *clothoid model* [10] and the simpler *circular model* [11]. Both allow the estimation of an accurate model in the vicinity of the vehicle. While the circular model is able to describe the lane with one parameter less than the clothoid model and, thus, leads to a more stable estimation, the clothoid model is able to cope with (constant) changes in curvature.

Extended Circular Model

Basically, the Extended Circular Model is using all parameters, properties and assumptions that apply also to the circular model. Additionally the Extended Circular Model introduces n circle segments, where $n = 0,1,2,...$ is the segment number, starting with $n = 0$ for the first segment in front of the vehicle.

The centre of the first segment's circle is assumed to be located at the y-axis of the segment coordinate frame. For positive curvatures $(c_n>0)$, the centre's position is at radius r, while for negative curvatures it is at $-r$. Also the segment coordinate system K_s is assumed to be rotated to the world coordinate system K_w by y_s. The segments have a fixed length l_n and each segment has only one parameter: the curvature c_n. The lane width ω is assumed to be equal in all segments. Thus, the state vector X for the lane becomes

$$X = \begin{pmatrix} Y_s \\ a_0 \\ w \\ c_0 \\ \vdots \\ c_n \end{pmatrix} = \begin{pmatrix} distortion\_angle \\ offset\_to\_the\_left\_lane\_marking \\ lane\_width \\ curvature\_in\_segment\_0 \\ \vdots \\ curvature\_in\_segment\_n \end{pmatrix} \tag{23}$$

Modelling a steady lane with concatenated circle segments is possible if both segments' circle centres and the intersection at the segment border are on the same line (compare Fig. 4).

Creating the model from the parameters in the state vector X is an iterative process. Starting with the first segment $n = 0$ in front of the vehicle, every segment's parameters have to be calculated. The parameters include the segment's circle centre position M_n, the intersection point S_{n+1} with the next segment $n + 1$, and a description of the straight line g_{n+1} passing through the points M_n, S_{n+1} and M_{n+1}. This first segment $n = 0$ is a special case, since there is no segment $n = -1$ and, thus, there can be no description for the straight line g. Thus, the first step is to calculate the coordinates for the circle centre in the first segment from the parameters a_0, y_s, and c_0 which is very similar to the calculations of the circular model in [11].

The circular model can describe a turn to the right, a turn to the left and a straight line with only one parameter: the curvature c_n, where

$$
\begin{aligned}
C_n &< 0: \text{turn to the right} \\
C_n &> 0: \text{turn to the left} \\
C_n &= 0: \text{straight line}
\end{aligned}
\tag{24}
$$

Additionally, the radius R_n of the circle can be calculated by:

$$
R_n = \frac{1}{c_n} \quad c_n \neq 0
\tag{25}
$$

In case the first segment's curvature c_0 describes a turn, the position of the circle centre M_0 can be calculated from the vehicle's rotation angle y_s, the offset a_0 and the radius R_0 by

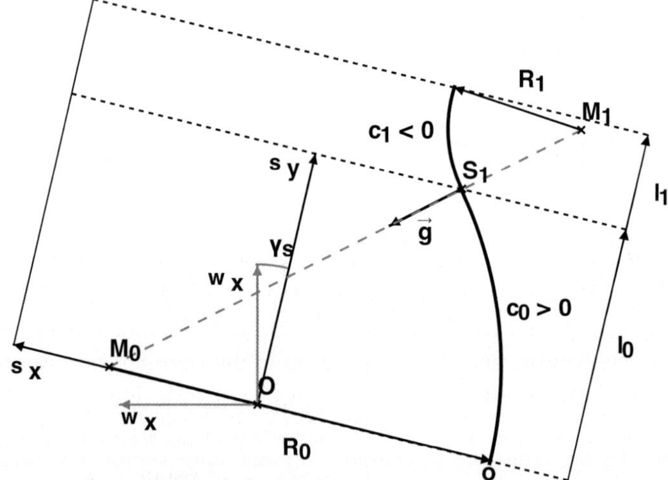

Fig. 4.   The concatenation of two circle segments results in a steady lane in the case that both circle centres and the intersection at the segment border are on the same line.

$$M_0 = \begin{pmatrix} x_0 \\ y_0 \end{pmatrix} = \begin{pmatrix} -R_0 \sin(y_v) \\ R_0 \cos(y_v) + o \end{pmatrix} \left(c_n \neq 0 \right) \tag{26}$$

With the centre's location M_0 and the radius R_0, the position of any point on the circumcircle can be calculated with either a given x or y. The segment length l_0 will be the given x and, thus, starting from

$$R_n^2 = (x - M_{x0})^2 + (y - M_{y0})^2 \tag{27}$$

A solution for y can be derived from this equation as follows:

$$y = \sqrt{(R_n)^2 - (x - M_{x0})^2} \quad (c < 0)$$
$$y = \sqrt{(R_n)^2 + (x - M_{x0})^2} \quad (c > 0) \tag{28}$$

The intersection point S_1 at the segment border l_0 can be calculated by

$$S_1 = \begin{pmatrix} S_{x1} \\ S_{y1} \end{pmatrix} = \begin{pmatrix} l_0 \\ \dfrac{}{\sqrt{(R_0)^2 \mp (x - M_{x0})^2}} \end{pmatrix} \tag{29}$$
$$\left(c \neq 0 \right)$$

This point is essential to know, since at this point the lane described by the first segment ends and the second segment begins. This way, there will be no discontinuity – which is required in order to obtain a smooth lane.

As a last step the circle centre M_0 and the intersection point S_1 will be used to generate a vector $\vec{g}$. This vector is the directional vector of a straight line through the points M_0 and S_1 and will be normalized by its absolute value.

$$\vec{g}_1 = \begin{pmatrix} g_{x1} \\ g_{y1} \end{pmatrix} = \frac{1}{|M_0 - S_1|} \cdot \overline{M_0 - S_1} \tag{30}$$

In combination with the point S_1, the vector $\vec{g}$ will be used to calculate the next segment's circle centre point M_1. The requirement to place the points M_0, S_1 and M_1 on the same line will be fulfilled this way. It is important that every segment's vector $\vec{g}$ always points in the same direction g_{yn}. may be defined to be always positive, so the following correction can be applied

$$g_{y1} < 0 : g_1 = \begin{pmatrix} -g_{x1} \\ -g_{y1} \end{pmatrix} \tag{31}$$

At this point, all parameters for the first circle segment (M_0, S_1) are calculated and the vector $\vec{g}_1$ is known. Essential knowledge is represented by the intersection point S_1 and the vector $\vec{g}_1$. These parameters define where the reference line begins in the second segment and where the next circle centre will be located.

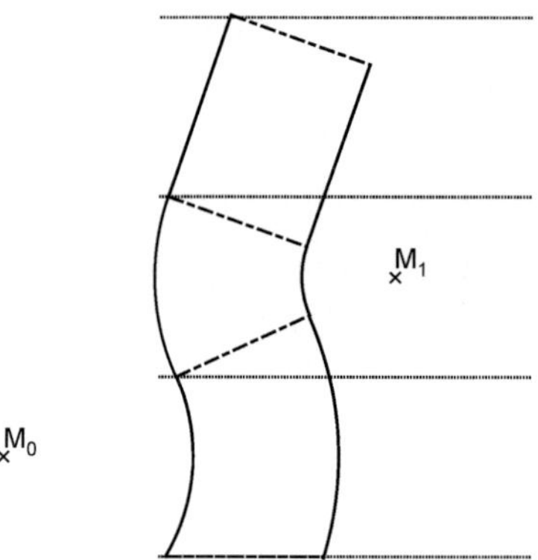

Fig. 5. The Extended Circle model can concatenate any number of circle segments. The segments' individual curvatures decide whether there will be a left/right turn or a straight line. The reference lane (left) intersects the segment borders (dotted line). Since the lane width is constant, every other lane's intersection point may not be located at the segment border (right).

Those essential parameters need also be calculated in case the first segment's curvature is equal zero and thus is a straight line. From the known parameters v and o the intersection point with the second segment is given as follows

$$S_1 = \begin{pmatrix} S_{x1} \\ S_{y1} \end{pmatrix} = \begin{pmatrix} l_0 \\ \tan(y_s)l_0 + o \end{pmatrix} \tag{32}$$

The other essential parameter is the vector $\vec{g}_1$. This vector is the straight line's normal vector and can be calculated from the point S_1 at the end of the segment and the given offset o as

$$\vec{g}_1 = \frac{1}{\sqrt{(s_{y1} - o)^2 + (s_{x1})^2}} \cdot \begin{pmatrix} -(S_{y1} - o) \\ S_{x1} \end{pmatrix} \tag{33}$$

Again, this vector will be normalized by its absolute value and the orientation may be corrected as in eqn. 31. In both cases ($c_0 = 0$; $c_0 \neq 0$) the essential parameters S_1, $\vec{g}_1$ are derived from the first segment. Together with the curvature c_n the n-th parameters allow calculation of the $n + 1$-st parameters. In case the curvature c_n describes a turn, the circle centre M_n can be calculated as

$$M_n = S_n + R_n \cdot \vec{g}_n \tag{34}$$

using S_n as the starting point and going R_n times the direction $\vec{g}_n$ of . This leads to the segment's circle centre position M_n. With the centre's coordinates and the segment end $\sum_{i=0}^{n} l_i$ the intersection with the $(n + 1)$ -th segment becomes

$$S_{n+1} = \begin{pmatrix} S_{x_{n+1}} \\ S_{y_{n+1}} \end{pmatrix} = \left(\frac{\sum_{i=0}^{n} l_i}{\sqrt{(R_n)^2 \mp (\sum_{i=0}^{n} l_i - M_{x_0})^2}} \right) \tag{35}$$

Using the correct solution is analogue to eqn. 26. The next step is to calculate $\vec{g}_{n+1}$ from the circle's centre M_n and the intersection with the next segment S_{n+1} analogue to equation 29:

$$\vec{g}_{n+1} = \begin{pmatrix} g_{x_{n+1}} \\ g_{y_{n+1}} \end{pmatrix} = \frac{1}{|M_n - S_{n+1}|} \cdot \overrightarrow{M_n - S_{n+1}} \tag{36}$$

The resulting vector is normalized to unity but may need a correction as shown in equation 31. All essential parameters have now been calculated for the case of a segment with a turn.

The other case is a segment with a curvature c_n which is equal to zero and, thus, is describing a straight line. A line can be expressed as an analytical function

$$y = m \cdot x + n \tag{37}$$

Using the vector $\vec{g}_n$ and S_y the parameters rise m and offset n can be calculated as

$$m = -\frac{g_{x_n}}{g_{y_n}} \tag{38}$$

$$n = S_{y_n} - m \cdot S_{x_n}$$

By inserting the segment end coordinate $\sum_{i=0}^{n} l_i$ into eqn. 31 the intersection with the next segment S_{n1} becomes

$$S_{n+1} = \begin{pmatrix} S_{x_{n+1}} \\ S_{y_{n+1}} \end{pmatrix} = \begin{pmatrix} \sum_{i=0}^{n} l_i \\ m \cdot \sum_{i=0}^{n} l_i + n \end{pmatrix} \tag{39}$$

and the vector $\overrightarrow{g_{n+1}}$ is copied from the previous segment's vector

$$\overrightarrow{g_{n+1}} = \overrightarrow{g_n} \tag{40}$$

Summary

In this section, the Extended Circular Model has been described. Starting from the state vector X, the segment parameters have been developed for the reference lane. The first segment was treated as a special case which is identical to the well known Circular Model. All other segments were steadily connected at their segment borders. This generated a smooth reference lane with n segments, where every segment is described only by the curvature parameter c_n.

5 Results & Conclusions

The lane model proposed in this paper has been implemented and tested using the concept vehicle Carai from Chemnitz University of Technology. For evaluation purposes, the output of a lane recognition system using the circular model with a VGA camera has been compared to one using the Extended circular model and the high resolution camera described in section 3. In both cases, a detection range of 90 meters has been defined.

Fig. 6 shows two sample images of the evaluation. It turned out that the circular models estimates the correct course of the lane within a distance of approximately 50 meters. Beyond that range, the single curvature parameter is not able to represent the lane correctly. In comparison, the Extended circular model is able to correctly estimate the lane's curvature over the whole detection range. Furthermore, the image shows that the estimated lane is continuous and yet able to model different curvatures at different ranges.

While the extension of the models state space provides a way to improve the estimation accuracy significantly for large detection ranges, the evaluation also shows that the stability of the tracking is slightly better using the circular model approach. Though it is well-known that a filter's stability decreases with higher state space dimensions, future work will deal with ways to further improve this aspect. In particular, models which explicitly take into account the correlations between the segments' curvatures will be analyzed.

Fig. 6.   Comparison of the circular model (top) using a VGA camera and the Extended circular model (bottom) using a XVGA camera. In both cases, a detection range of 90 meters has been defined.

References

[1] Cramer, H. et. al, Modelle zur multisensoriellen Umfeld- und Situationserfassung, 3. Workshop Fahrerassistenzsysteme, In German, 2005.

[2] Richter, E. et. al, Radar and Vision based Data Fusion - Advanced Filtering Techniques for a Multi Object Vehicle Tracking System, Proceedings of the IEEE Intelligent Vehicles Symposium, pp. 120-125, 2008.

[3] Fardi, B. et. al, A Fusion Concept of Video and Communication Data for VRU Recognition, Proceedings of the 11th International Conference on Information Fusion, 2008.

[4] Rohlfs, M. et. al., Gemeinschaftliche Entwicklung des Volkswagen "Lane Assist", 24. VDI/VW 2008 Gemeinschaftstagung Integrierte Sicherheit und Fahrerassistenzsysteme, pp. 15-34, In German, 2008.

[5] Bässmann, H., Kreyss, J., Bilderverarbeitung Ad Oculos, Springer, In German 2004.

[6] Kamera Werk Dresden, Datenblatt Loglux i5, http://www.kamera-werk-dresden.de/, retrieved on 23.1.2009.

[7] Wan, E., van der Merwe, R., The Unscented Kalman Filter, in: Kalman Filtering and Neural Networks, John Wiley & Sons, Inc., 2001.

[8] Julier, S., Uhlmann, J., A General Method for Approximating Nonlinear Transformations of Probability Distributions, Technical Report, Dept. of Engineering Science, University of Oxford, 2001.

[9] Julier, S., Uhlmann, J., Unscented filtering and nonlinear estimation, Proceedings of the IEEE, 92, pp. 401-422, 2004

[10] Gern, A., Multisensorielle Spurerkennung für Fahrerassistenzsysteme, PhD thesis, University of Stuttgart, In German 2005.

[11] Cramer, H. Modelle zur multisensoriellen Erfassung des Fahrzeugumfelds mit Hilfe von Schätzverfahren, PhD thesis, Chemnitz University of Technology, In German, 2004.

Robin Schubert, Gerd Wanielik
Chemnitz University of Technology
Professorship of Communications Engineering
Reichenhainer Straße 70
09126 Chemnitz
Germany
robin.schubert@etit.tu-chemnitz.de
gerd.wanielik@etit.tu-chemnitz.de

Martin Hopfeld, Karsten Schulze
IAV GmbH
Department of Safety Electronics / Surrounding Field Sensors
Kauffahrtei 25
09120 Chemnitz
Germany
martin.hopfeld@iav.de
karsten.schulze@iav.de

Keywords: lane recognition, ADAS

Automatic Generation of High Precision Feature Maps of Road Construction Sites

A. Wimmer, T. Weiss, F. Flögel, K. Dietmayer, University of Ulm

Abstract

Road construction sites on highways are a demanding environment for drivers as the lane width is reduced. Especially for trucks, lateral control is a challenge. If the driver slightly drives over the lane markings, other vehicles cannot use the neighbouring lane. This leads to a reduction of the road capacity, thus causing traffic jams. A driver assistance system which supports the driver in the task of lateral control highly benefits from the use of accurate infrastructure (feature) maps. This paper presents an approach for the automatic detection, classification, and mapping of specific elements of road works like guard rails, safety barriers, traffic pylons, and beacons with laser scanners.

1   Introduction

Over the past years, the freight volume on high- and freeways has constantly increased [1]. This demands a constant maintenance and repair of the road surface and infrastructure such as bridges. If a road construction site is set up, the width of the lanes usually is reduced. This makes lane keeping a difficult task; especially for heavy duty vehicles. If a truck slightly rides over the lane markings, other car drivers often do not dare to overtake (see Fig. 1). As a result, not all lanes are fully utilisable, which may lead to traffic jams. In order to keep up a high throughput rate even in demanding surroundings like road construction sites, a "Road Work Assist" is proposed. This driver assistance system should support the driver in lateral control.

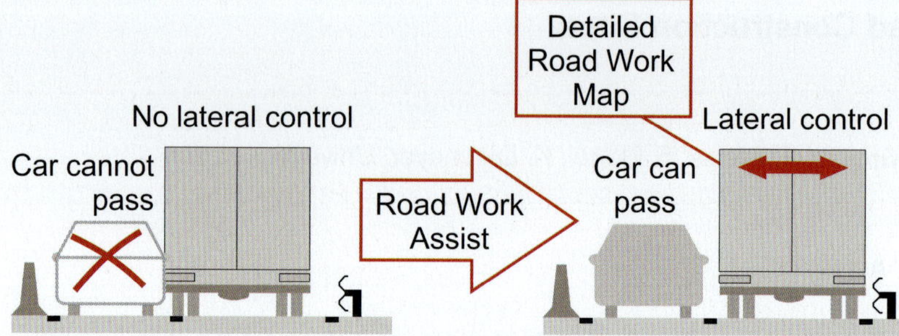

Fig. 1.   If trucks drive over their lane markings in road construction sites, car drivers may not be able to pass. A "Road Work Assist" is proposed for controlling the truck laterally to make the neighbouring lane fully utilisable. It uses the data of the detailed "Road Work Map" described in this paper.

There are different concepts of lane keeping support for heavy duty vehicles. One idea is to supply the road with magnetometers which can be detected by onboard sensors of the vehicle [2]. Two main drawbacks arise from this concept. On the one hand, there are additional costs for the auxiliary hardware in the pathway. On the other hand, there is a lack of flexibility. Once the magnetometers are installed, a change in traffic routing – as it is done in road construction sites – cannot be realised without extra effort. Another concept for lateral guidance is to detect lane markings with a video camera based system [3]. At road works this poses additional requirements on the detection algorithm as there are different kinds of lane markings. Depending on national regulations, one type of line may overrule the original one, e.g. yellow lane markings replace the white lines. This extra complexity may lead to false detection of a lane recognition algorithm. Therefore, a third concept is proposed which uses the information of accurate feature maps for an assistance system for lateral control. Once the detailed map of the road construction site has been generated, it can be send to approaching vehicles via wireless communication techniques. The cars and trucks equipped with a "Road Work Assist" use the data of their onboard sensors together with the information of the map. The vehicles localise themselves within the feature map by using a front facing laser scanner which detects traffic pylons or beacons as well as the road delimiting elements such as guard rails and safety barriers. With the additional information of the map not only the current position within the lane is known, but also a look-ahead capability can be realised. This ability is especially useful for the lateral control of long and big vehicles like trucks. This paper presents an approach for the automatic detection and mapping of elements used at road construction sites.

2   Sensor Setup

The mapping vehicle is equipped with several sensors. The highly accurate positioning system ADMA (Automotive Dynamic Motion Analyzer of Genesis Elektronik GmbH) provides a global position estimate within centimetre accuracy. It fuses precise yaw rate and acceleration sensors with RTK-GPS (RTK-GPS: Real-Time-Kinematic Global-Positioning-System). A four layer laser scanner ALASCA XT of Ibeo Automobile Sensor GmbH is mounted on the vehicle's rear and acquires a vertical distance profile of the surrounding area for detecting the carriageway's delimiting elements such as crash barriers and steel or concrete (Jersey) barriers (see Fig. 2). The angular resolution of the scanner is up to 0.125 degrees. A front facing laser scanner can be used for the detection of traffic pylons and beacons. For the recognition of lane markings, a video camera is suitable.

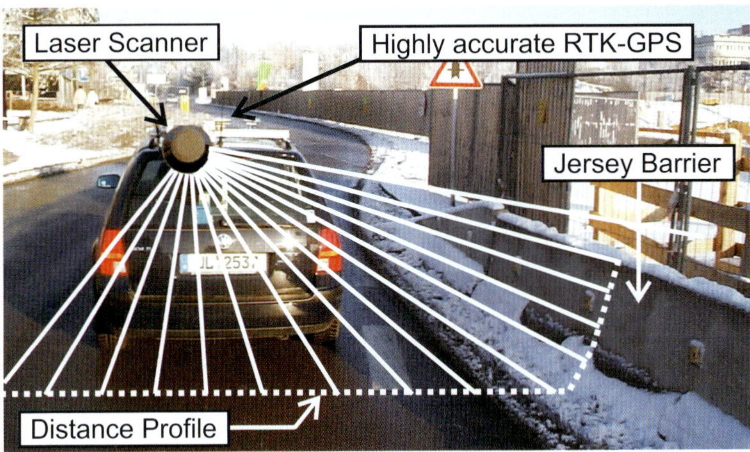

Fig. 2.   Mapping vehicle with a vertically mounted laser scanner. Typical elements of road construction sites like concrete (Jersey) barriers can be detected.

3   Detection and Mapping of Road Construction Sites

There are various regulations [4] which specify the use of safety and guiding elements of road construction sites. Different types of barriers are frequently used. On the one hand, they protect construction workers from the traffic passing by. On the other hand, they separate lanes with opposing driving direction. These safety barriers have to fulfil certain specifications, depending e.g. on the velocity level. One attribute among others is the ability to absorb energy in a potential crash. Additionally, they have to be strong and high enough to prevent trucks from entering the opposing traffic. However, these regulations do not specify a certain form, but mainly the physical characteristics. For that reason, several different types of safety barriers have been developed [5].

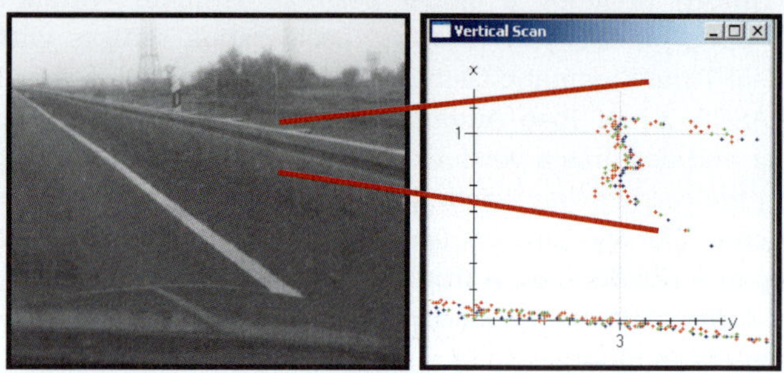

Fig. 3.   Image of a standard guard rail and the corresponding measurement of the four layer laser scanner.

Extensive data has been collected on various road construction sites in southern Germany. One result was that there are some distinctive groups of barrier types. A guard rail is standard not exclusively in road construction sites (Fig. 3). Concrete or Jersey barriers (Fig. 4) exist in three different heights and are frequently used at road works. Additionally, there are some other, usually smaller steel safety barriers (Fig. 5). The figures also show the corresponding data recorded by the four layer laser scanner.

Fig. 4.   Image of a concrete (Jersey) barrier and the corresponding measurement of the four layer laser scanner.

The key task for detection and mapping of safety barriers will be to distinguish between these elements and background. Furthermore, a separation of the different types of safety barriers is necessary, if they have to be represented differently in a precise infrastructure map. For that purpose, some pre-processing is done on the raw laser scanner data. After that, feature extraction algorithms are applied to compute significant characteristics which are handed over to a classifier.

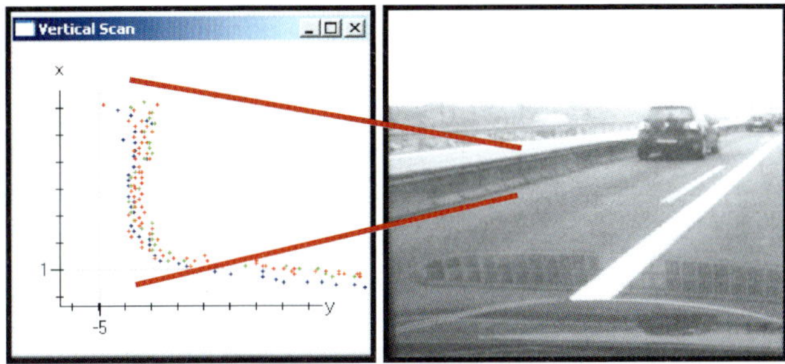

Fig. 5.   Image of one of the various steel safety barriers in road construction sites and the corresponding measurement of the four layer laser scanner.

3.1 Sensor Pre-Processing

Some pre-processing of the raw laser scanner data is done based on previous work [6]. For easier detection of distinctive structures, the contour of the distance profile is sharpened by applying a third order Gaussian filter. As the mapping vehicle might slightly roll during measurement, the data is rotated to street plane for compensation. The next step is to find regions which are candidates for the position of possible safety barriers. For that purpose the scan data is grouped into slots. For each slot, the sum of all measurements is calculated and weighted by a distance criterion based on the effects of the scanner's angular resolution. Fig. 6 depicts an example, where the candidates show the correct position of a Jersey barrier and a guard rail. However, one candidate has been found in the background. For separating background from safety barriers, the use of a classifier is proposed.

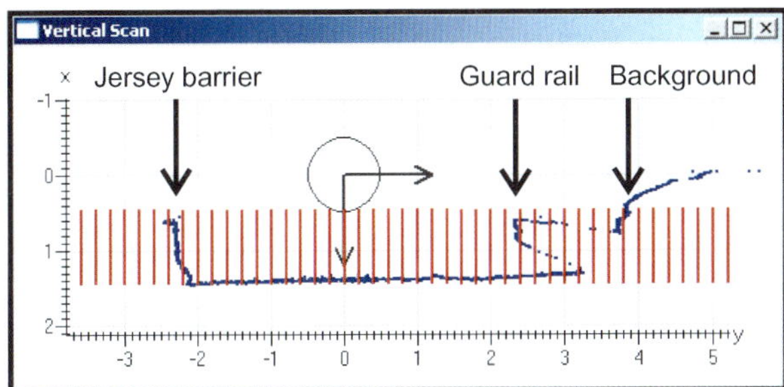

Fig. 6.   The vertical scan data is grouped into slots for locating possible candidates of safety barriers.

3.2 Feature Extraction

In order to make use of a classifier, characteristic features of every structure have to be extracted. The various types of safety barriers usually have a typical outline that can be measured by the laser scanner. This outline is best represented by a hull curve applied on the scan data. Those measurement points which are closest to the vehicle – as this is the border that should be incorporated into the map – are connected, if they do not exceed a certain maximum distance. The result is a hull polygon of the whole structure, which is the basis for some of the following features.

All guard rails have a typical and distinctive form with a grooving in the middle of their profile. This grooving can be extracted by a geometric algorithm. Therefore, the first feature can be calculated as the height, and the second feature as the depth of this grooving (Fig. 7). For Jersey barriers the value of feature 2 (depth) will be much smaller than for guard rails. This means that this is a characteristic suitable for distinguishing between these types of safety barriers.

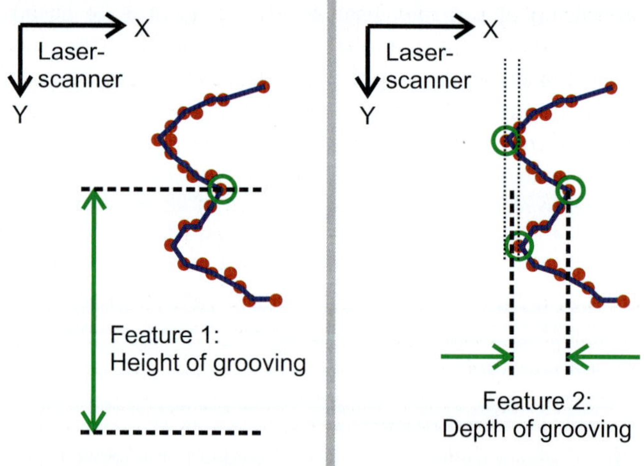

Fig. 7.   Extraction of feature 1 (height of grooving, left) and feature 2 (depth of grooving, right).

There are two characteristics which can be directly extracted from the hull polygon. The total length of the contour is used as feature 3. By only summing up the vertical parts – up to a certain angle – of the contour, feature 4 is calculated (see Fig. 8).

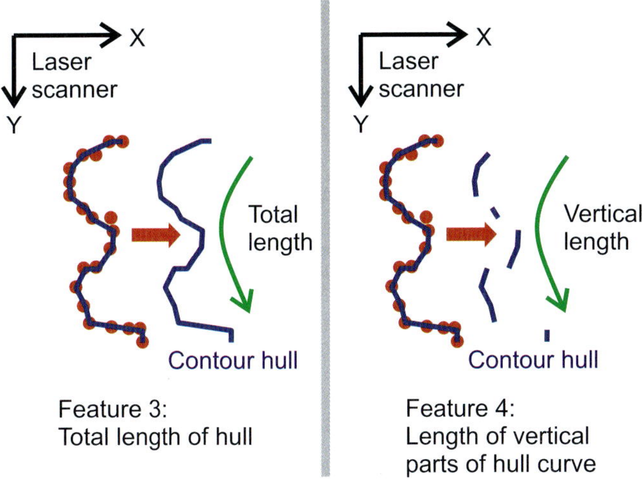

Fig. 8.   Extraction of feature 3 (total length of hull, left) and feature 4 (length of vertical parts of hull curve, right).

Feature 5 gives the total height over ground of the structure. In general, that is a strong characteristic for most of the different classes. Finally, feature 6 is calculated as a measure for the "smoothness" of the contour. Artificial structures like barriers have a rather smooth surface compared with measurements on the background vegetation. For that, a regression line is fit into the scan points and the corresponding variance is computed.

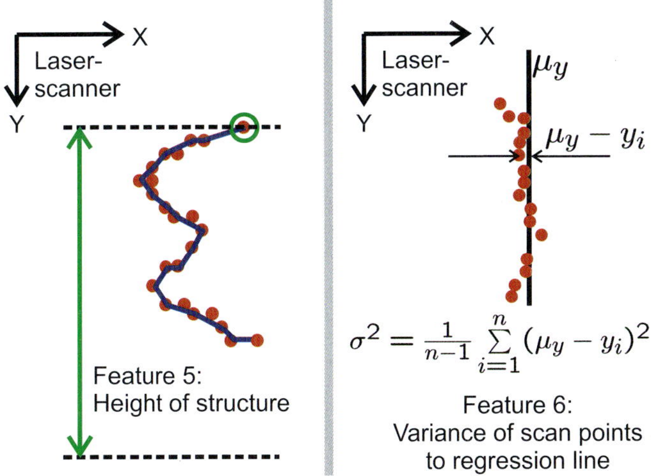

Fig. 9.   Extraction of feature 5 (height of structure, left) and feature 6 ("smoothness" of structure, right).

These features are calculated to have measures which are characteristic for each of the classes of road work elements. For evaluation, a subset of real data is labelled for every type of barrier. After that, the distribution of the feature values can be derived for every class and each of the six features. Fig. 10 exemplarily shows the result of the vertical length of the hull curve (feature 4) for the standard Jersey barrier.

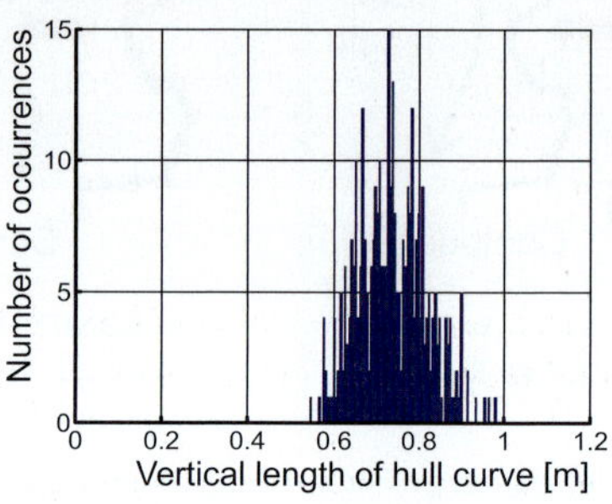

Fig. 10.   Number of occurrences of the values of feature 4 (length of vertical parts of hull curve) for class 4 (Jersey barrier).

As can be seen in Fig. 10, the number of occurrences are Gaussian distributed, which is basically true for all classes and features. For that, mean and standard deviation are calculated to represent each distribution for every class and feature. Further evaluation has shown that some types of barriers have similar mean and an analogue variance, which means that they can be grouped to one single class. As a consequence, five classes have been derived to represent all kinds of different barriers used at common road construction sites. Class 1 represents the standard guard rail; class 2 is a group of small steel and concrete safety barriers. Class 3 and 4 represent two types of Jersey barriers, which basically vary in their height. Finally, class 5 is used for another steel safety barrier. Fig. 11 shows exemplarily the class-conditional probability density function of all classes for the hull curve's length (feature 3).

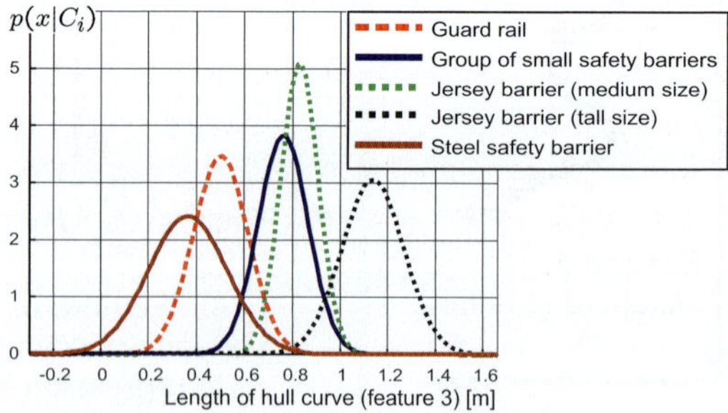

Fig. 11.   Class-conditional probability density functions of feature 3 (length of the hull curve) for all classes C_i.

3.3 Classification

The Gaussian distribution of the feature values motivates the use of a Bayesian Classifier. By using labelled data, the class-conditional probability density functions $p(x/C_i)$ for feature x and class C_i have been derived. With that knowledge, the probability that class C_i is true for a given feature value x can be calculated:

$$P(C_i \mid x) = \frac{p(x \mid C_i)P(C_i)}{\sum_{j=1}^{N} p(x \mid C_j)P(C_j)} \tag{1}$$

where $p(C_i)$ is the prior probability of class C_i. This formula is extended for the use of the six dimensional feature vector x.

A Bayesian classifier has been chosen for its flexibility in adding new classes. Extensive data has been collected on various road construction sites. However, it might occur that another type of barrier cannot be represented with one of the above mentioned 5 classes. In that case, this type has to be added to the classifier as additional class. For learning classifiers this is a time consuming process due to the recursive structure of the learning algorithm. When using a Bayesian classifier, only the class-conditional probability density function has to be derived from labelled data by calculating mean and standard deviation for the new class. Additionally, tests have shown that in this application the classification result mainly depends on the performance of the feature extraction algorithm.

For evaluation of the classifier's performance, over 3300 barriers on various construction sites have been labelled manually and compared to the automatic classification result. The total detection rate evaluates to approximately 97% and indicates the ability of the classifier to distinguish between background and actual safety barriers. If the different types of safety barriers have to be separated, e.g. distinguishing between Jersey barrier and steel safety barrier, the rate of correctly classified objects is slightly decreased to about 95%. False class allocation does not necessarily have a negative impact on the map building process, as the extraction algorithm for characteristic points is the same for most types of barriers (see next section). The false detection rate, meaning that background is erroneously mistaken as some type of safety barrier, is below 2%.

3.4 Automatic Generation of Maps

For the map building process, three characteristic points which best represent the structure are derived for every class. For guard rails, these points are
directly taken from the feature extraction algorithm. Upper and lower points,
as well as the position of the grooving are incorporated into the map. Other
safety barriers are also represented by three characteristic points. In this case
the middle point is calculated to describe the position of the typical bend of
concrete barriers (Fig. 12).

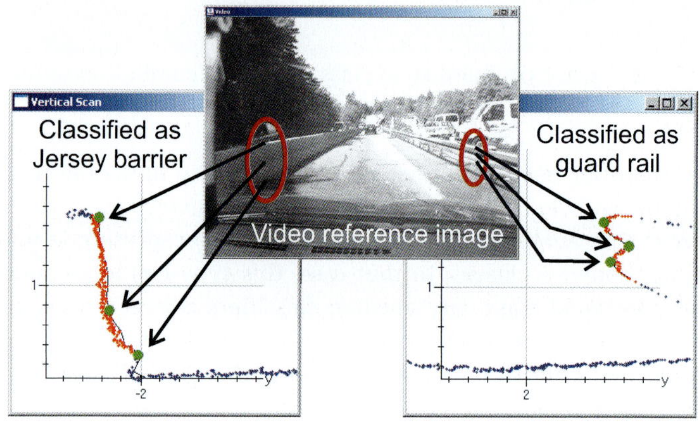

Fig. 12.   Real data example of a detected guard rail (right) and a Jersey barrier (left). The automatically extracted points that best represent
the structure are incorporated into the map (green dots).

For the generation of a precise infrastructure map these points now have to
be transformed into a global reference coordinate system. A highly accurate
position estimate of the vehicle is obtained from the ADMA platform. As the
relative location of the laser scanner to the vehicle coordinate system is known,
a global position of the three characteristic points can be calculated and is
incorporated into the map.

Additional robustness can be achieved by applying post-processing algorithms
on the data. An outlier detection based on a distance criterion automatically
removes false detections in the background. For the interpolation between
single nodes and for the reduction of measurement noise, a smoothing spline
is calculated. Fig. 13 is an example for a "Road Work Map".

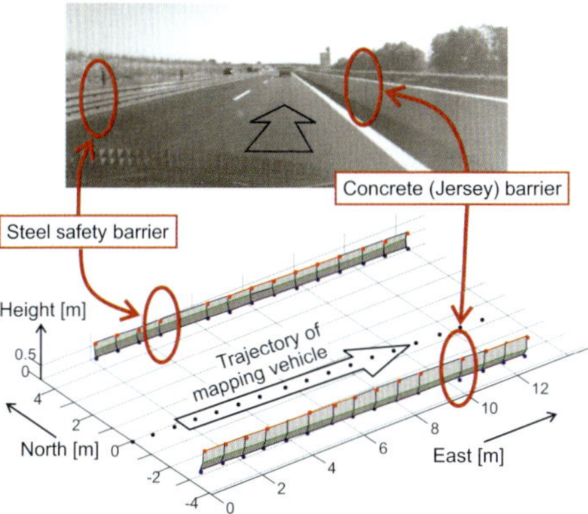

Fig. 13.   Detail of a 3D Road Work Map and a video image of the scenery for visualisation.

The concept of a "Road Work Assist" implies the transfer of this road work map to the approaching vehicles via infrastructure to car communication. For that reason the amount of data has to be as small as possible. An algorithm is proposed to reduce the number of nodes without loosing much accuracy. First, a smoothing spline is fit into the data. If some nodes do not contribute to additional position information, they are neglected. The maximum loss of accuracy is a parameter. If a maximum tolerance of 3 cm is applied, the number of nodes can be reduced by nearly 80%.

3.5   Analysis of Absolute Accuracy of the Map

The proposed concept for detection and mapping of road construction sites is evaluated with respect to absolute accuracy. For that purpose a Jersey barrier and a guard rail – which are easily accessible – have been chosen for manual position measurement. Highly accurate static GPS measurements are acquired and with data of the land surveying office a global estimate within σ_{static} = 3 mm accuracy can be achieved. The position estimate of the onboard GPS system is in the range of $\sigma_{dynamic}$ = 1 – 2 cm. In total that gives an accuracy range of the global position measurements of about +/- 4 to 7 cm (3σ). In the following plots the global position has been transformed into a local Cartesian coordinate system (tangent plane) for better visibility.

Fig. 14 shows the detailed results of the measurements on the Jersey barrier. The lines indicate the reference position, derived from three static GPS measurements. The circles indicate the positions of the automatically detected barrier as described in the previous sections. For that plot, post-processing and especially smoothing algorithms have not yet been applied.

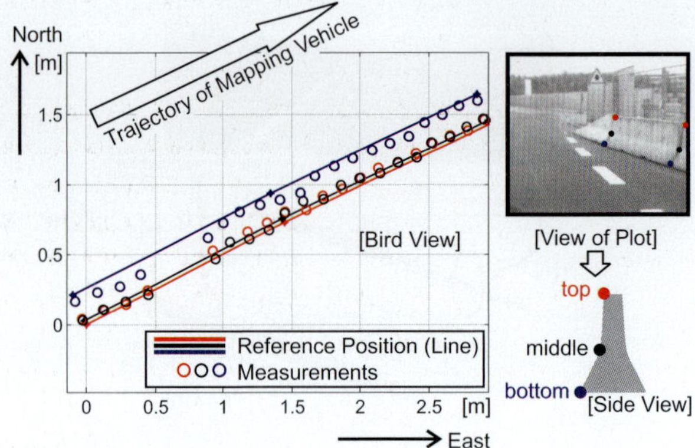

Fig. 14. Evaluation of the absolute accuracy of the detection and mapping of a Jersey barrier.

The performance of mapping guard rails is evaluated similarly. The reference measurements of the barrier are indicated by circles and connected with a red line (Fig. 15). This time the guard rail has been detected and mapped twice by two drive-bys, including a lane change manoeuvre for proving the repeatability of the concept. The blue and black stars represent the vehicle's position for every time step. On the detailed extract it is clearly visible that the detected guard rail position for both measurements is well within the total accuracy that is achievable with such an experimental setup for referencing the algorithms.

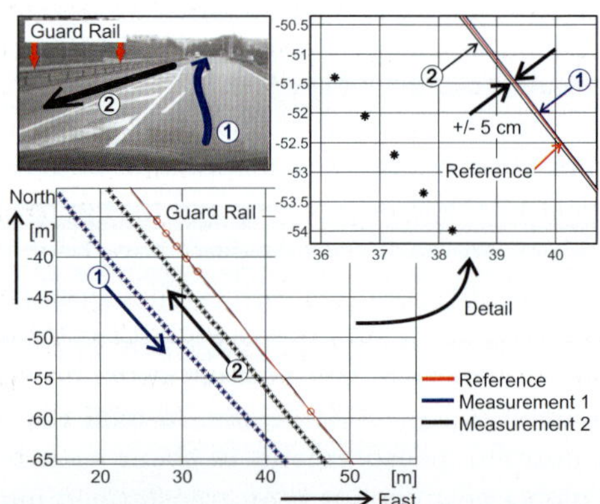

Fig. 15. Evaluation of the absolute accuracy by automatically mapping a guard rail. The barrier is measured twice by two drive-bys (blue and black dots).

4 Detection of Traffic Pylons and Beacons

The above mentioned safety barriers are important elements for the installation of road construction sites. For temporal road works with only a short duration, it is also possible to use traffic pylons [4]. In the case of a change in traffic routing, especially at the beginning and the end of a road construction site, but also when one lane is routed to e.g. the emergency lane, guiding reflector posts (beacons) are set up, sometimes in addition to Jersey barriers. These beacons and pylons are also important elements which have to be incorporated into the "Road Work Map".

For that objective the data of the front facing laser scanner is used. First, suitable candidates have to be extracted from the scan data. In previous works [7], the generation of an online occupancy grid has been developed. Every time step a measurement grid is calculated from the scan data. For this, the surrounding area is divided into grid cells. The cell's value is a measure for the likelihood of occupancy. In principle, every scan point within the grid cell increases its value whereas the area between a reflection and the sensor is assumed to be free and thus the value is decreased. These ideas provide the basis for a more sophisticated method for the generation of the measurement grid based on a detailed sensor model.

This grid is used for a contour extraction algorithm. As the size of beacons and traffic pylons for use in road construction sites is prescribed [4], suitable candidates can be found based on the contour dimensions. Fig. 16 illustrates the contours (blue) and the corresponding candidates (yellow). This candidate list is further reduced by exploiting the fact that beacons and traffic pylons have to be fully reflective. The laser scanner is able to measure the energy of the object's echo, thus detecting reflecting surfaces. Those contours which contain scan points that are classified as reflectors will undergo further investigation (Fig. 16, red). Finally, a tracking algorithm verifies the stationary candidates and removes false detections in the background. Fig. 17 shows a road construction site with several beacons; all detected elements are marked by a red circle.

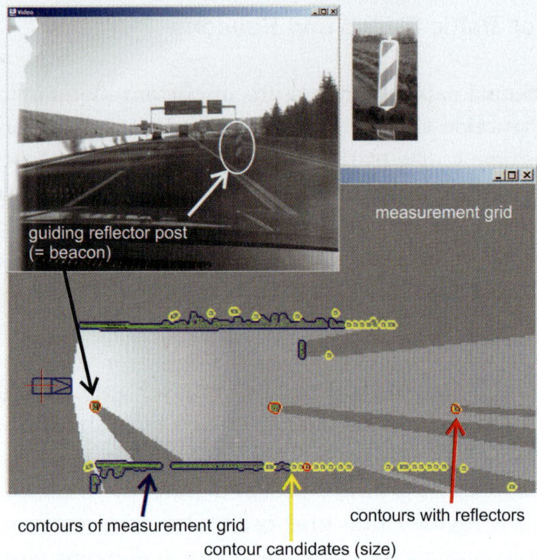

Fig. 16. Extraction of possible candidates for beacons. The results of the contour extraction on the measurement grid are indicated by a blue polygon. All contours with appropriate size are marked in yellow colour. Contours which contain scan points classified as reflector are shown in red.

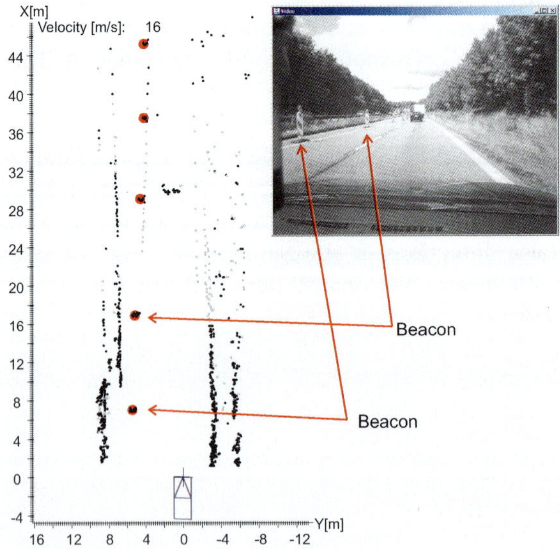

Fig. 17. Bird view of detected and tracked beacons (red) in a road construction site.

5 Conclusions and Future Work

This paper has proposed a new approach for the automatic detection and mapping of safety and guiding elements in road construction sites. A vertically scanning laser scanner acquires a distance profile which is used for the detection of safety barriers. After feature extraction and classification characteristic points, which best represent the structure, can be incorporated into a highly accurate infrastructure map. Outlier detection and interpolation algorithms make manual post-processing unnecessary. Traffic pylons and beacons can be detected from the data of a front facing laser scanner by using a contour extraction and exploiting the element's reflectivity.

Future work will include the detection and mapping of lane markings in road construction sites.

6 Acknowledgements

The authors would like to thank Ibeo Automobile Sensor GmbH for their support and all the invaluable suggestions, hints and tips.

References

[1] European Commission (Ed.), Panorama of Transport, Office for Official Publications of the European Communities, Luxembourg, 2007.

[2] Wang, J.-Y., Tomizuka, M., Robust H_∞ Lateral Control of Heavy-Duty Vehicles in Automated Highway System, Proceedings of the American Control Conference, San Diego, California, 1999.

[3] Martini, S., Murdocco, V., Lateral Control of Tractor-Trailer Vehicles, Proceedings of 2003 IEEE Conference on Control Applications, CCA 2003.

[4] Bundesministerium für Verkehr, Bau- und Wohnungswesen (BMVBS), Abteilung Straßenbau (Ed.), Richtlinien für die Sicherung von Arbeitsstellen an Straßen (Safety Instructions for Road Construction Sites), BMVBS (German Ministry of Transport), 1995.

[5] Knoll, E. (Ed.), Der Elsner (Handbook for Traffic and Transportation), Otto Elsner Verlagsgesellschaft mbH & Co. KG, Dieburg, Germany, 2004.

[6] Weiss, T., Dietmayer, K., Automatic Detection of Traffic Infrastructure Objects fort he Rapid Generation of Detailed Digital Maps using Laser Scanners, Proceedings of the 2007 IEEE Intelligent Vehicles Symposium, Istanbul, Turkey, 2007.

[7]   Weiss, T., et. al., Robust Driving Path Detection in Urban and Highway Scenarios Using a Laser Scanner and Online Occupancy Grids, Proceedings of the 2007 IEEE Intelligent Vehicles Symposium, Istanbul, Turkey, 2007.

[8]   Duda, R., et. al, Pattern Classification, John Wiley and Sons, New York, USA, 2000.

[9]   Fritz, H., Longitudinal and Lateral Control of Heavy Duty Trucks for Automated Vehicle Following in Mixed Traffic: Experimental result from the CHAUFFEUR project, Proceedings of the 1999 IEEE International Conference on Control Applications, Kohala Coast-Island of Hawaii, Hawaii, USA, 1999.

Andreas Wimmer, Thorsten Weiss, Francesco Flögel, Klaus Dietmayer
University of Ulm,
Institute of Measurement, Control, and Microtechnology
Albert-Einstein-Allee 41
89081 Ulm
Germany
andreas.wimmer@uni-ulm.de
thorsten.weiss@uni-ulm.de

Keywords: driver assistance, road construction site, infrastructure map, laser scanner, lateral control

Laserscanner Based ADAS for Motorcycle Safety

B. Rössler, K. Kauvo, IBEO Automobile Sensor GmbH

Abstract

Motorcyclists and moped drivers are road users with a particularly high accident risk since motorcycle accidents are severe in nature, due to the relative minor protection of motorcyclists. Furthermore, today the field of driver assistance systems is mostly dedicated to the passenger vehicle and heavy goods vehicle sector. Driver assistance systems for Powered Two Wheelers (PTWs), which rely on onboard vehicle sensors, have not been not considered so far. In this paper we present first results of the European funded Marie Curie research project MYMOSA and especially the Integrated Safety work within its work plan. The objective of the Integrated Safety work package is the development of an integrated safety system capable to detect impending dangerous situations and accident scenarios. As a consequence of the proposed activities in these projects a long-term reduction of at least 20% of injuries and fatalities of motorcyclists is foreseen.

1   Objectives

Motorcyclists and moped drivers are road users with a particularly high accident risk. Motorcycle accidents are severe in nature, due to the relative minor protection of the riders. Furthermore, given the young age of many victims, these accidents often result in a high loss of life expectancy for casualties and high social-economic costs for severely injured motorcyclists.

The general objective of the MYMOSA project is to improve Powered Two Wheeler (PTW) safety and lead to a significant reduction of injuries and fatalities of motorcyclists. This objective will be gained by co-operation of researchers from various universities, research institutes and companies and the collaborative generation of multidisciplinary know-how (accidentology, accident dynamics, and biomechanics).

One work package of MYMOSA is the Integrated Safety, for which the objective is the development of an integrated safety system, capable to detect impending dangerous situations and accident scenarios. Furthermore, the rider should be informed of the foreseen risks or the PTW behaviour should be directly influenced, with the purpose of reducing the injury risk. As a consequence of the proposed activities in this project a long-term reduction of at least 20% of injuries and fatalities of motorcyclists is foreseen.

There has been very little research how to implement ADAS systems and applications specifically for motorcycles. Honda and Yamaha have equipped some demonstrator motorcycles with extensive amount of ITS technologies within the Advanced Safety Vehicle (ASV) Project. [1] These projects rely partly on wireless inter-vehicle communication or camera technologies.

Until these days onboard sensors for hazard object detection and classification are not utilized in PTWs. In this part of the MYMOSA project Laserscanner technology is used to track and classify objects and road users for collision avoidance or mitigation purposes.

2   System Description

The principle idea of the MYMOSA integrated safety approach is sketched in Fig. 1. It is a common opinion that an integrated approach to PTW safety is the best way to improve the overall safety level of motorcycle [2], since it could drastically reduce fatalities and injuries to approximately zero. Despite its importance for the safety improvement of riders, up to now this sector has not been investigated systematically. The proposed approach considers the rider, the motorcycle, and the surrounding environment as a unique system where information is continuously exchanged through sensors and communication devices with the objective of an increased safety level. More precisely, the collected information will be used to warn (and inform) the rider about dangerous situations and to influence (and stimulate) actions or otherwise to take actions in place of the rider itself. On the other hand the sensory system of the PTW should continuously assess potentially dangerous driving situation and detect the potential risk based on the stored information about the most relevant accident scenarios.

The principle of an integrated safety system can be implemented at different detail levels but the following three items must be considered to develop a comprehensive multidisciplinary approach:
   ▶ Sensors,
   ▶ actuators and Human Machine Interfaces (HMI),
   ▶ data elaboration and decision logic.

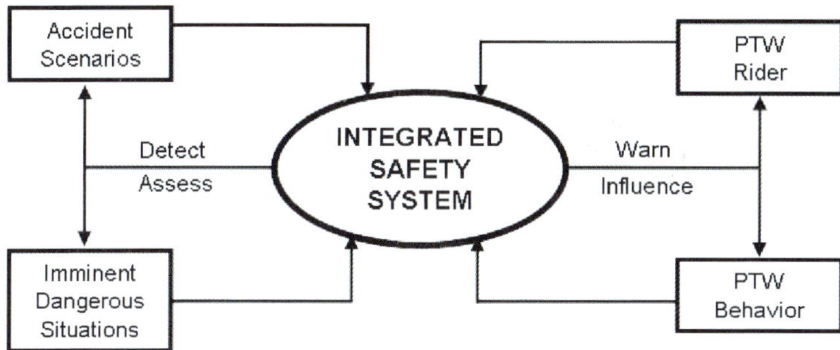

Fig. 1.   Sketch of MYMOSA integrated safety

3   Hardware Setup

For the development and testing of novel algorithms for sensor data processing
for the MYMOSA Integrated Safety System, the PTW platform of the parallel running European research project PISa was used. For this demonstrator
setup an Ibeo LUX Laserscanner was integrated in order to perceive the area
around the front of the motorcycle to detect potential hazardous situations for
the rider.

Fig. 2.   Ibeo LUX Laserscanner on the demonstrator

Fig. 2 shows the integration of the current version of the Ibeo LUX Laserscanner
into the demonstrator between the headlights. The used Laserscanner provides
a horizontal scan area of up to 110° and a resolution of up to 0.125°. The
sensor scans simultaneously in four scan planes, which are named as yellow,
green, blue and red from top to bottom. This is presented in Fig. 3. The vertical
resolution between the layers is 0.8°.

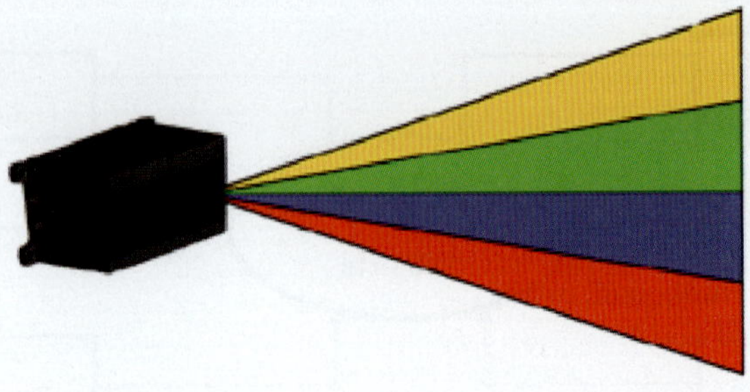

Fig. 3.   Four individual scan planes of LUX

The height of the Laserscanner from the ground in the demonstrator installation is 88.0 cm, which is significantly more than what is used in a typical car installation. Therefore for the motorcycle installation the Laserscanner is tilted slightly. The selected tilting (pitch angle) is 2.2°, which means that the lowest layer (red) hits the ground at the distance of 15.5 m, see Fig. 4.

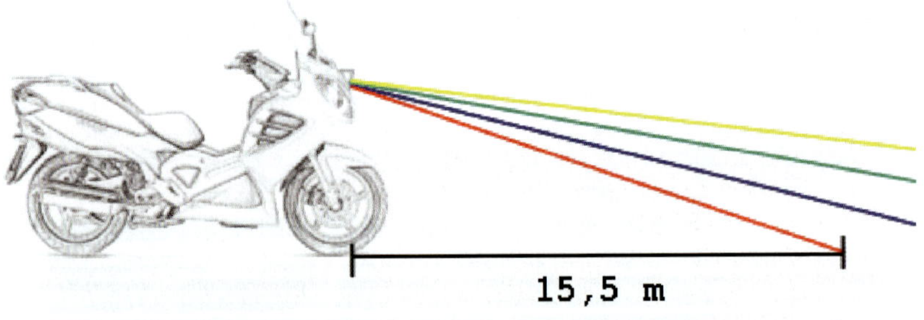

Fig. 4.   Pitch angle of the Laserscanner installation

The roll angle of the PTW is measured continuously with an inertial measurement unit (IMU). As an IMU an MTi-G from Xsens is used. The MTi-G is an integrated GPS and inertial measurement unit with a navigation and altitude and heading reference system processor. It is based on MEMS inertial sensors and a miniature GPS receiver and it also includes additional aiding sensors: a 3D magnetometer and a static pressure sensor.

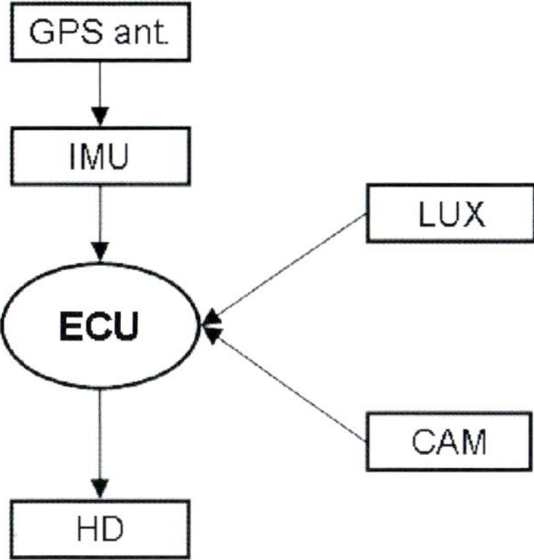

Fig. 5.   Hardware components of the demonstrator

Fig. 5 presents additional hardware components installed in the demonstrator. IMU and Ibeo LUX Laserscanner are the main measurement devices. A camera (CAM) is needed as a reference during the software development phase. Ibeo´s electronic control unit (ECU) provides the needed processing power for all algorithms. Results are stored in a hard drive (HD) for evaluation and offline data analysis. Fig. 6 shows the installations in the under-seat storage of the demonstrator. The orange device in the middle is the IMU and bigger grey case on the right is the ECU.

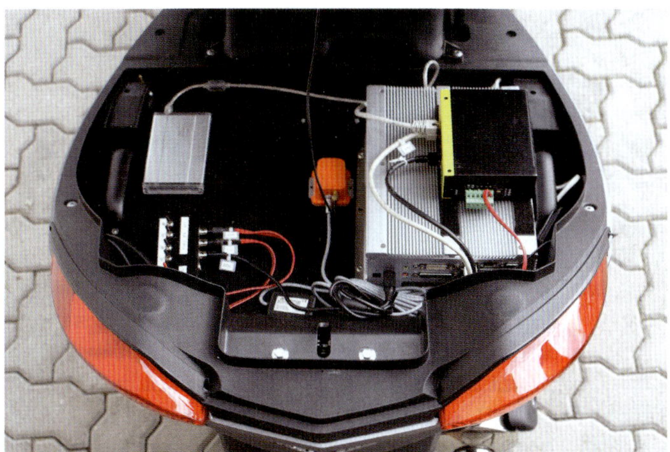

Fig. 6.   Under-seat storage in demonstrator

In order to build a Laserscanner-based integrated assistance system for PTWs it is necessary to do data preprocessing on the raw scan data. One essential task of such preprocessing is the ground detection. For a typical mounting position on a vehicle, a part of the scan data comes from measurements on the ground, or objects that are very close to the ground, such as grass, speed bumps, or lane markings. The task of the ground detection module is to separate the ground objects and near-ground objects from the relevant objects in the scan data.

By definition, ground objects stick out of the ground no more than 20 cm. All ground objects are not deleted from the scan data, but marked as ground. The following processing steps can then decide if and how to use this data. The algorithm uses a heuristic to determine if a scan point comes from a ground object or not. In general, *objects* have scan points from several scan planes in close proximity, while data is much more smeared on *ground objects*. Also, ground typically appears first in the lower scan planes. [3]

Fig. 7 presents the main scan data processing steps. At the bottom is the raw, unprocessed, scan data received from an Ibeo LUX Laserscanner. The two following steps *dirt & rain*, and *ground detection* are included into the data preprocessing phase. The upper three steps are referred to as the high level scan processing.

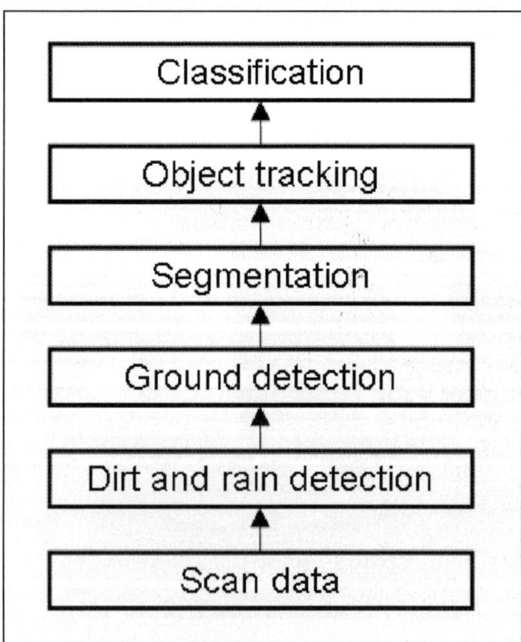

Fig. 7.   Scan data processing steps

The topmost step "classification" is needed to determine the road user´s type like passenger car, lorry, bike, or pedestrian. This information is needed when developing a driver assistance system and applications for PTWs in the future. This depicted simplified procedure is similar for passenger cars and PTWs.

At the *ground detection* step scan points are labelled as ground points if they are echoed from the ground. That step in preprocessing needs specific attention, when implemented into PTWs where the situation is more challenging compared to passenger vehicles. The main challenge for PTWs is the additional degree of freedom while driving, i.e. the rolling of the motorcycle especially in curves.

With the current installation tests indicated that at roll angle values over 10° a large portion of scan points comes from measurements on the ground objects. Fig. 8 shows a snapshot of the ground detection step, where some data is labelled as ground. At that situation the rider is doing a left curve with a roll angle of -9,1°. The grey straight lines indicate the total scanning area from -55° up to 55°, brown points are ground objects and red points are echoed from relevant objects. The lorry is detected between the ground points on the left, two parked passenger cars and light pole on the right. Additionally smeared echoes from the bushes at the far right can be made out. The object tracking and classification is not applied yet. All the brown points cannot reliably be detected as ground objects with the algorithms and parameters developed for passenger cars so far. For PTWs scan data should be processed differently.

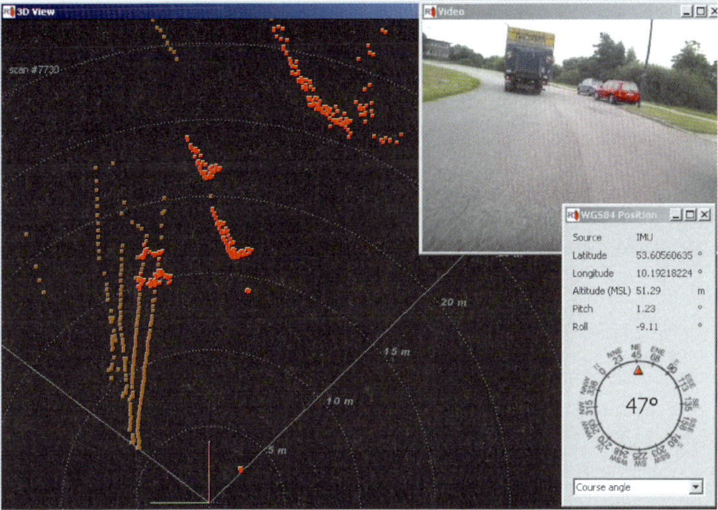

Fig. 8.   Ground point labelling depending on the roll angle of the PTW

Therefore, a so called 2-sided labelling was developed. Also calculation was significantly improved by adding several speed improvements to the current ground detection algorithms. This was needed because the used scanning frequency of the Ibeo LUX Laserscanner is 12,5 Hz, which means that there is a 80 ms time window to make all the needed calculations. That 2-sided labelling needs more computing time of the ECU, so speed improvements were essential.

Generally ground point labelling for PTWs is done in the following way: if the rider is making a left turn some amount of points are labelled as ground points on the left hand side and on the right hand side some amount of points are labelled as object points. To determine the exact amount of points, two angles α and β, are defined as follows:

$$\left.\begin{array}{l} \alpha = 0, \Phi < -29,4° \\[2mm] \alpha = \dfrac{379}{15950} * \Phi + 0,70, \ \Phi >= -29,4° \\[2mm] \beta = \left|\dfrac{379}{15950} * \Phi\right| - 0,50, \ \Phi <= -21,0° \\[2mm] \beta = 0, \Phi > -21,0° \end{array}\right\}, -55° < \Phi <= -4° \tag{1}$$

$$\left.\begin{array}{l} \alpha = 0, \Phi < -29,4° \\[2mm] \alpha = \dfrac{379}{15950} * \Phi + 0,70, \ \Phi >= -29,4° \\[2mm] \beta = \left|\dfrac{379}{15950} * \Phi\right| - 0,50, \ \Phi <= -21,0° \\[2mm] \beta = 0, \Phi > -21,0° \end{array}\right\}, -55° < \Phi <= -4° \tag{2}$$

,where Φ = roll angle

As an example if the roll angle is -10° i.e. left turn: α = 26,4° β = -15,0°

Fig. 9 shows the labelling sectors and angles α and β when rolling. In that case it is assumed that the motorcycle is making a left turn i.e. a rolling angle is negative. Inside the green area points are labelled as ground and inside the red area points are labelled as objects. New values for the angles α and β are calculated at every scan if the absolute value of the roll angle is more than 4°. Otherwise the general ground detection algorithms are applied.

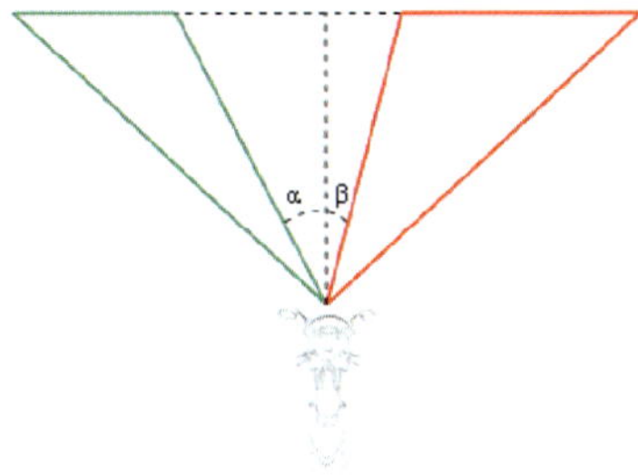

Fig. 9.   Labelling sectors when rolling

4   Conclusions

This paper we present some first results of integrating a Laserscanner into a demonstrator. This is the very first step to make it possible to develop a special driver assistance system for motorcycles. These first results show that it is successfully possible to use the proven Laserscanner technology also in motorcycles. The presumable diminishing of the size of electronics and mechanics will make it even easier to integrate a Laserscanner into a motorcycle and develop an integrated safety system for PTWs.

References

[1]   Honda Completes Development of ASV-3 Advanced Safety Vehicles, Honda Motor Co., Ltd, Japan, 2005.

[2]   APSN - Roadmap of future Automotive passive safety technology development, European Vehicle Passive Safety Network, 2004.

[3]   Dittmer M, Kibbel J, Salow H, Willhoeft V, Team-LUX, DARPA Urban Challenge 2007, Technical Paper, IBEO Automobile Sensor GmbH, 2007.

Bernd Rössler, Kimmo Kauvo
IBEO Automobile Sensor GmbH
Merkurring 20
22143 Hamburg
Germany
bernd.roessler@ibeo-as.com
kimmo.kauvo@ibeo-as.com

Keywords: powered two wheelers, motorcycle, ADAS, laser scanner

Graceful Degradation for Driver Assistance Systems

S. Holzknecht, E. Biebl, Technische Universität München
H. U. Michel, BMW Group Research and Technology

Abstract

In a future car-IT architecture consisting of a centralized grid-cluster of good-performance electronic control units connected via a switched high-bandwidth communication network, new possibilities for driver assistance systems come up. Working on raw data in combination with strong central processing units allows advanced signal processing. In case of breakdown which results in reduced available communication bandwidth and/or reduced available calculating power, graceful degradation will help to keep the system running. That means either the data acquisition can work at lower performance (lower angular resolution, lower update rate etc.) to reduce the calculation power demand as well as the network data rate, or a part of the signal chain can dynamically be shifted to smart-sensors and make them work similar to current sensors like Bosch's Adaptive Cruise Control [1]. In this case, only a list of objects instead of raw data is transmitted to the electronic control unit where a (brake- or accelerate-) decision could be made with low calculation power while accepting a lower overall system performance.

1 Introduction

This work is based on results of an interdisciplinary research project, organized in a collaboration of Technische Universität München and BMW Group Research and Technology. This project is part of the CAR@TUM initiative, where the decades-worth cooperation between the Technische Universität München and the BMW Group is intensified and restructured. The field *"IT-Architecture"* is one of six initiated high tech-projects and has its scope on new approaches and design rules for an automotive Electrics and Electronics (E/E) Architecture.

1.1 Novel Car IT Architecture

In today's luxury class car-IT architecture there are more than 100 electronic control units, and most of them need to be cross-linked via various bus systems. Because of the risen safety and comfort needs, this trend even ranges to the subcompact class. This leads to high complexity in developing processes and more possibilities of error. An approach to solve the issues coming up with such architecture is breaking up the fixed association of software and hardware. A centralized grid cluster of a few good-performance electronic control units (ECU), connected among themselves and to sensors and actuators with one homogeneous high-bandwidth network [2] represents a novel IT architecture. To enable different applications to work on one ECU, virtualization comes into operation. Based on this future-proof car IT architecture several driver assistance systems get new possibilities in signal processing and failure management.

1.2 Advantages in Signal Processing

Due to the availability of a high bandwidth network, we can work on raw data from compatible sensors and process it in one of the central ECUs. There, computing power compared to embedded solutions is relatively cheap. Thus, advanced algorithms with a high demand for CPU usage can be used. The system-wide existence of raw data in principle permits early as well as late data fusion. All customer applications can access data of the surroundings so that the number of sensors can be reduced and costs can be saved, e.g. adaptive cruise control and park distance control could be done with the same laser scanner devices. Furthermore, because of the availability of raw data additional information like reflectivity, processing algorithms can be adapted to each desired customer application in order to work at its best performance.

1.3 Advantages in Failure Management

In case of an ECU breakdown the local soft real-time and non real-time software which has been running will be relocated by an appropriate software-management system to a remaining ECU. After that, the available calculation power for all applications is less. For that reason graceful degradation coming along with a lower need for processing power helps to keep the applications running, even though at lower overall performance.

In a failure situation resulting in a reduced available communication bandwidth, maybe caused by a broken network edge followed by an adaptive rerouting, graceful degradation accompanied with a lower need for transmission capacity also helps to keep the applications running.

2 Concept of Graceful Degradation

In current car IT architectures both a breakdown of ECUs and a breakdown of network components produces a malfunction of the affected application(s). With our novel architecture only a part of the assigned computing and network capacity is lost. The better way is now to *degrade* the application instead of failing. To guarantee that the application will do its work well enough in degraded mode, a trade-off to work *gracefully* has to be found. In our case a laser scanner system representing driver assistance systems will be analyzed for its possibilities to reduce the demand for the mentioned capacities in case of a failure.

2.1 Sensor Hardware

Our utilized sensor type is a one-layer laser scanner (LIDAR) named LD-OEM 1000, made by SICK, Inc (Fig. 1). A rotating scan head which is transmitting laser pulses gets data of the surroundings by measuring the round trip time of the reflections. This smart sensor outputs measured values from the surroundings as raw data via a 10Base-T Ethernet data interface. Thanks to the integrated digital signal processor (DSP), some processing of data can be done within this sensor.

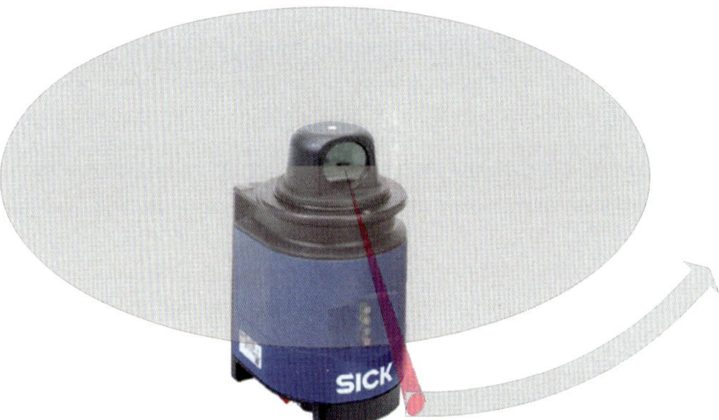

Fig. 1.   Laser Scanner LD-OEM 1000

The raw output data rate and consequently the network load, depends on following parameters, each with LD-OEM technical features:

- ▶ Acquisition angle / 0° - 360°
- ▶ Angular resolution / 0.125° - 1°
- ▶ Scanning frequency / 5 Hz - 20 Hz
- ▶ Need of echo amplitudes / yes - no

The acquisition angle can be configured for the specific purpose which is needed by the customer application, e.g. a 180° configuration with a mid-bumper mounted sensor is required for park distance control, whereas a 16° - 80° acquisition angle - depending on the speed range - is required for an adaptive cruise control [e.g. in 3,4], see Fig. 2.

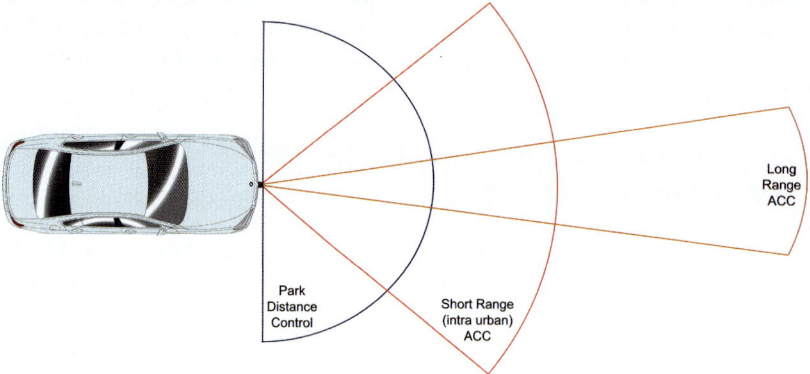

Fig. 2.   Acquisition Angle Depending on Application

Depending on the dimensions of the objects to be detected, different angular resolutions can be used. For pedestrian detection, higher resolution is necessary in contrast to car detection, where less resolution suffices, both at the same ranges.

The minimum space s with an angular resolution α between two reflecting measure points at the same object distance d, shown in Fig. 3, derives from the law of cosine:

$$s = \sqrt{2d^2(1-\cos\alpha)} \tag{1}$$

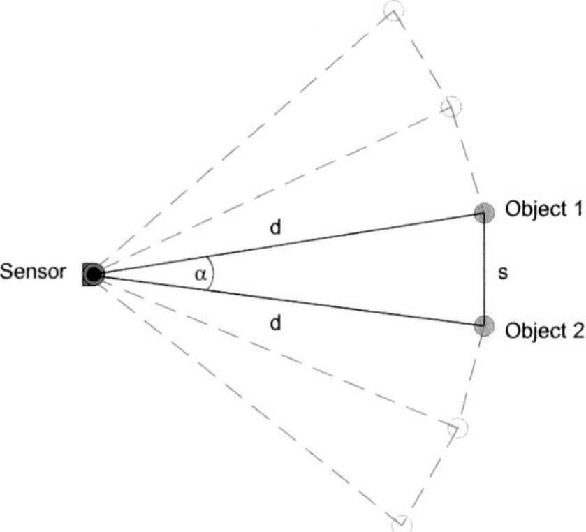

Fig. 3. Correlation of Angular Resolution and Minimum Space between Objects

According to equation 1, the requirement for angular resolution can be calculated. In our pedestrian detection algorithm, two possibilities of recognition are used: leg pair pattern (case 1) and simple width checking (case 2), each case with an attached likelihood of 45% and accordingly 20% for the visualization. To detect a leg pair pattern, we assume legs have a width between 10 and 20 cm and a pair of them have a gap not exceeding 60 cm. If there is no gap, the width checking comes into operation where all objects whose width is between 30 and 80 cm will be recognized. In case 1, assumed we want to detect a 5 cm gap. For autonomous braking in urban areas (speed 50 kph, dry road, stopping distance approx. 20 m) 0.14° angular resolution would be required. In case 2 where 30 cm need to be detected for the same scenario 0.86° suffices.

Scanning frequency is equivalent to pictures of the surroundings per second which means at higher speeds a higher update rate is required. At 50 kph and 15 Hz scanning frequency every 67 ms and accordingly 0.93 meters the surroundings are scanned. This should be enough for an autonomous brake system in urban areas while in our park distance control scenario 5 Hz is sufficient.

In some cases, available echo amplitudes can be interesting for data processing algorithms. As an example, a scanned segment with a width of 2 meters and a high amplitude area of 50 cm in the middle (license plate) is most likely a bumper. A similar reflectivity-homogeneous segment could also be a wall. Further, identifying faulty measurements by monitoring the same values over a dedicated period to detect potential outliers is also possible.

2.2 Degrading Options

The before mentioned possibilities of configuring the laser scanners can reduce data rates and computing power to a certain degree. Using the internal DSP enables further decreasing: parts of the signal chain can be shifted into the DSP under acceptance that no raw data will be available. However, the DSP is not able to replace the whole signal chain. This also would not make sense because of needless redundancy.

2.3 Data Rate Calculation

For raw data the *TCP* Payload rate TPR_{raw} of a LD-OEM 1000 can be calculated as follows:

$$TPR_{\mathrm{raw}} = \left(\frac{\text{Acquisition Angle}}{\text{Angular Resolution}} \cdot (2\,\text{Byte}_{\text{Distance}} + 2\,\text{Byte}_{\text{Echo}}) + 3\,\text{Byte}_{\text{Overhead}} \right) \cdot \text{Scanning Frequency} \qquad (2)$$

With the help of DSP the *TCP* payload rate TPR_{dsp} is much less but it depends on the software and the surroundings scene.

The demand for computing power correlates with the data input amount in the ECU and the desired output, e.g. raw data input and filtered output. These values will be measured or calculated by means of our demonstrator.

3   Demonstrator

Our measurement setup consists of two LD-OEM 1000. Our aim is detecting and tracking pedestrians with the aid of data fusion. In figures 4 and 5 our demonstrator and the relevant detail of the construction schematic are shown.

Fig. 4.   LIDAR Scanners Mounted on Demonstrator

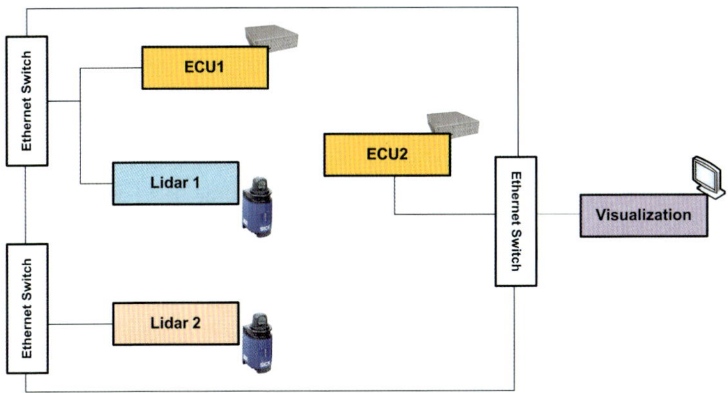

Fig. 5.   Construction Schematic

The functioning of the two different working modes (normal operation and graceful degradation using DSP support) is described in Fig. 6. The thickness of the arrows corresponds qualitatively to the data rate.

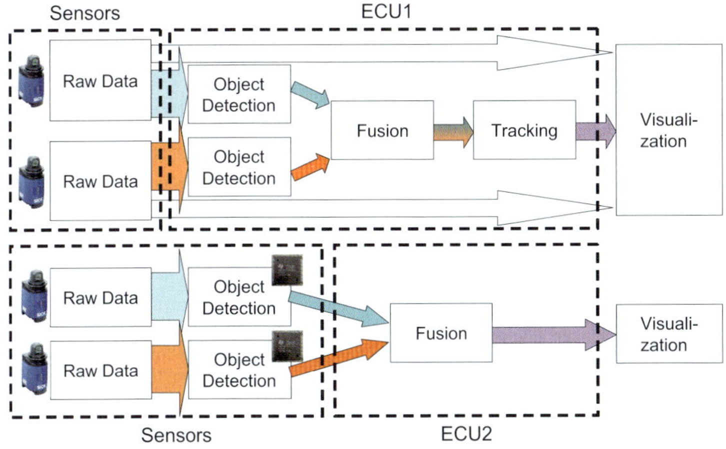

Fig. 6.   Normal and Degraded Mode

The upper part specifies normal operation mode, where raw data from the sensors is processed in ECU 1 to display the detected and tracked objects as well as the raw data on a screen. This chart is also valid for degraded modes that use reconFig.d laser scanners without applying the internal DSP. These modes are set up manually to be integrated in our results.

The lower part specifies graceful degradation mode, where processed data (detected persons) is being sent to ECU 2 and passed on to the visualization.

ECU 1 fails and causes via an appropriate software-management system a relocation of the LIDAR application to ECU 2.

In Fig. 7, the sensor alignment showing its acquisition area near the vehicle is displayed. Fig. 8 represents our demo application. It graphs raw data of the surroundings where the color represents the reflectivity. Furthermore the detected pedestrians appear with percentage information which informs about the likelihood that the identified object is a pedestrian.

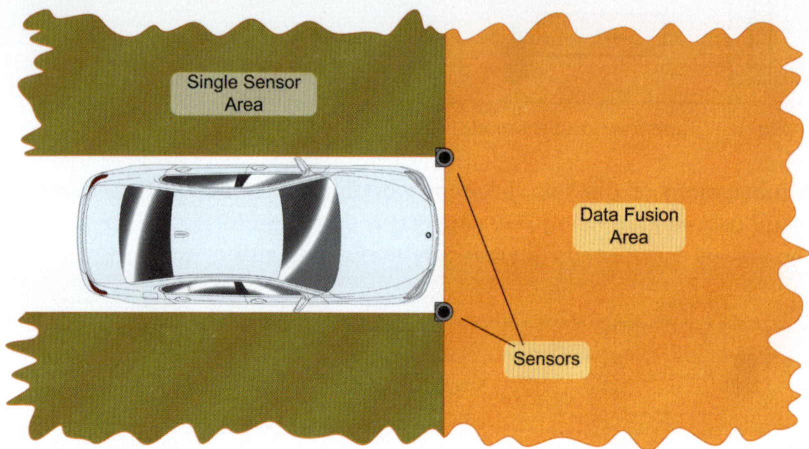

Fig. 7.   Sensor Alignment

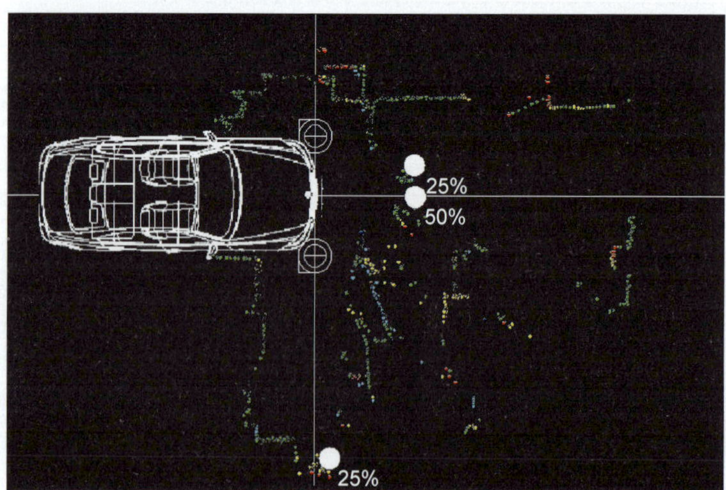

Fig. 8.   Visualization of Raw Data and Detected Pedestrians

4   Results

The following table shows a comparison of normal operation, two levels of graceful degradation by reconfiguring the sensor (mode 1 and 2) and graceful degradation by using the sensor's DSP (mode 3). In normal operation (see Fig. 6 upper part) to ensure being as close as possible to realistic driver assistance systems not only pedestrian detection but also tracking via Kalman filtering is used. In degraded mode 1 just the scanning frequency is lowered, in degraded mode 2 the angular resolution and the transmission of echo amplitudes is changed, in degraded mode 3 a combination of reducing raw data rate as well as DSP data processing is used. Analyzing the DSP data (see Fig. 6 lower part) takes about 1 second, which means we have a picture update rate of 1 Hz. Afterwards the ECU transforms coordinates, fuses objects and passes data created by this procedure to visualization which does not need too much processing power.

Normal Operation		*Degraded Operation*		
		Mode 1	*Mode 2*	*Mode 3*
Scanning frequency	*10 Hz*	*5 Hz*	*5 Hz*	*5 Hz*
Angular resolution	*0.25 °*	*0.25 °*	*0.125°*	*0.25 °*
Aquisition angle	*270 °*	*270 °*	*270 °*	*270 °*
Echo Amplitudes	*yes*	*yes*	*no*	*no*
TCP payload at each sensor	*43430 Byte / s*	*21715 Bytes /s*	*21715 Bytes /s*	*41 Bytes / s averaged*
Normalized computing power in ECU	*1*	*0.450*	*0.303*	*0.032*
Update rate	*10 pictures / s*	*5 pictures / s*	*5 pictures / s*	*~1 picture / s*

To evaluate the used computing power a profiler is used. The chosen tool is OProfile [5], which is a system-wide profiler for Linux systems, capable of profiling all running code at low overhead. Each measured CPU usage is normalized to the maximum which performs at normal operation mode. The measurements were done in the space of half an hour under similar circumstances (same surroundings, similar pedestrian volume).

As expected, reducing the update rate will reduce the used computing power analogical because the data has to be processed less often (normal mode vs. degraded mode 1). The influence of using echo amplitudes just for visualization on CPU usage is relatively high, also the data rate is affected (degraded mode 1 vs. 2). In this case, because of the low distance from sensor to scene, the same amount of pedestrians is detected, so the Kalman Filter is used in the same way. The ECU as well as the communication network is used hardly at mode 3, which shows that a simple driver assistance system that uses late data fusion can work also at adverse circumstances. The *TCP* payload is in that case 4 bytes per detected person plus 23 bytes Header, both per second.

Values in Degraded Mode 3 can easily be changed, e.g. if we want to have 5 Hz update rate we can reduce the acquisition angle to 108 ° (whereas this angle needs not to be a continuous piece of the surroundings) and increase the angular resolution to 0.5 ° at the same DSP processing capacity.

5   Future Prospects

Graceful Degradation can be widened to all parts of the signal chain. In case of processing raw data in a centralized ECU, Graceful Degradation can be applied e.g. to the tracking filter. Algorithms can be changed to work better adapted to its momentary requirements, for instance for simple motion smoothing, low-pass filtering can replace Kalman filtering at lower demand for computing power.

In case of using computing power of built-in processors in smart sensors, some front parts of the signal chain (instead of the whole algorithm) can be shifted out of the ECU, e.g. performing segmentation on raw input data helps reducing the demand for communication bandwidth as well as appropriate processing power.

Analyzing the particular demand of data of the surroundings at all points (angular resolution, scanning frequency, not essential data redundancy by fusion ...) for different scenarios of graceful degradation will be necessary as a next step towards an implementation of our novel car IT architecture with these principles. Also other classes of sensors like radar or video have to be taken into account.

References

[1]　Robert Bosch GmbH; Adaptive Fahrgeschwindigkeitsregelung ACC; Kraftfahrzeugtechnik, 2002.

[2]　Müller-Rathgeber, et al., "A unified Car-IT Communication-Architecture: Design Guidelines and prototypical Implementation" IEEE Intelligent Vehicles Symposium, 2008.

[3]　Schneider, M. "Automotive Radar – Status and Trends" GeMiC 2005.

[4]　Wenger, J. "Automotive Radar – Status and Perspectives" CSIC Digest 2005.

[5]　Profiling Tool: http://oprofile.sourceforge.net/

Stefan Holzknecht, Erwin Biebl
Technische Universität München
Fachgebiet Höchstfrequenztechnik
Arcisstr. 21
80333 München
Germany
holzknecht@tum.de
biebl@tum.de

Hans-Ulrich Michel
BMW Forschung und Technik GmbH
Hanauer Straße 46
80788 München
Germany
hans.michel@bmw.de

Keywords:　laser, scanner, lidar, fusion, radar, data rate, computing power, network

Lane Departure Warning and Real-time Recognition of Traffic Signs

N. Luth, R. Ach, University of Applied Sciences Amberg-Weiden

Abstract

The paper represents a vision-based lane departure warning system, as well as, a driver assistance system for the automatic recognition of traffic signs. It could inform the driver about actual speed limits or passing restrictions, which could be ignored in difficult traffic situations. The system is implemented partially on a new multi-core processor, developed by the Infineon Technologies AG.

1   Introduction

Camera-based driver assistance systems (CDAS) are one of the emerging markets for automotive electronics. First applications like a "lane departure warning" (LDW), a monitoring of blind spots or an improved vision during nights ("night vision") are already introduced and improve driver's comfort and safety. Pedestrian detection, traffic sign recognition, automatic parking assistance or strategies for collision avoidance will complement this trend with a forward looking camera as one of the key sensor technologies. Today CDAS are offered as options where the price plays an important role for customer acceptance and therefore for achievable market penetration. A LDW-System warns the driver if the vehicle moves out of its lane. The approach is based on finding the edges of the road markings. Based on these measure points a linear lane model is derived over several frames. In the next step we use an extended Kalman-Filter and a clothoid model to follow the road and get an appropriate prediction.

The reuse of a LDW for further applications like traffic sign recognition reduces the overall system costs and increases the attractiveness for customers. An automatic recognition of traffic signs could inform the driver about actual speed limits or passing restrictions. For that purpose an efficient detection of traffic sign candidates (denoted as traffic sign detection − TSD) and a successive reliable classification of these candidates are required.

Colour is often used to realize a fast TSD approach [1], whereas greyscale images are sufficient for classification [1,2]. Unfortunately the colour representation of objects in an image depends on the environment and is not consistent over different light situations and is especially not available at night. Because camera systems for LDW do not require colour and traffic sign recognition should be a functional extension of these systems, this paper presents a TSD approach based on greyscale images.

The proposed TSD is based on a cascade of trained classifiers. This approach was already successfully implemented for traffic signs [2,3]. One drawback of a learning approach is the required generation of a large set of training data. This data set has to cover a sufficient sample of all possible signs including all possible variations (rotation, size, illumination), which must be considered in order to identify a real robust classifier. To avoid time consuming training set generation from real images this paper proposes the use of a virtual database generated by computer graphics. For this purpose a simulation of traffic scenes was developed, which considers the perspective and the projection parameters of a dedicated real camera.

The next section presents the virtual simulation of traffic scenes. Section 3 illustrates two proposed traffic sign detection approaches based on a cascade of classifiers. The classification of the detected traffic-sign candidates is explained in section 4. The LDW-System is presented in section 5. The paper ends with a summary and outlook in section 6.

2   A Virtual Simulation of Traffic Scenes

The developed prototype driver assistance system should fulfil two main constraints: A high rate of detection and recognition of traffic objects using the real-time processing. Beside a powerful computational hardware a robust, efficient and adaptive image processing approach is required. The implementation of robust methods for image processing requires the collection of a wide variety of events and extensive image data material. This data needs to cover thousands of case studies. In this context collecting dynamic outdoor-scenes is a big challenge. Due to the diverse composition of static and moving rigid and non-rigid objects especially in different lighting situations and variable atmospheric and weather conditions. The image acquisition with traditional technical optical systems cannot fulfil these requirements within realistic time and cost parameters. The generation of synthetic pseudo-realistic data leads to a significant remedy of these problems.

The paper presents a traffic sign detection based on a trainable classifier-cascade. It requires a large set of positive and negative training data (several thousand images). Because the final system has to handle different weather and illumination conditions, as well as, combinations of rotation, scale, occluding, pollution etc. the training data has to include all these variables in a sufficient number of samples. Obviously, the generation of this training set from "real" images is time consuming and sometimes it is even impossible to capture all required traffic sign variations.

Therefore, this paper proposes the use of computer-graphics generated training images instead of "real" images. Computer graphics generated images means the generation of artificial images by computer graphics in such a way that the quality of the artificial images is similar to real ones. A very high expenditure of effort must be expexted, so that computer-generated images are indistinguishable from real images. A large number of polygons are necessary to recreate the 3-dimensional image of the natural environment. Many photos and relief textures, as well as, complex lighting calculations are needed to simulate complex surfaces and lighting effects in a realistic way. The smoothing of sharp edges can be achieved with the help of a computerized intensive low-pass filter. Air pollution and other atmospheric effects are simulated using fog, rain and snowfall with the help of particle systems.

A 3D simulation of traffic scenes based on the modelling of all necessary objects, different illuminations and weather conditions allows a parameter-based rendering of the virtual images. The parameter-based rendering process could also include the simulation of the projection behaviour of a real camera.

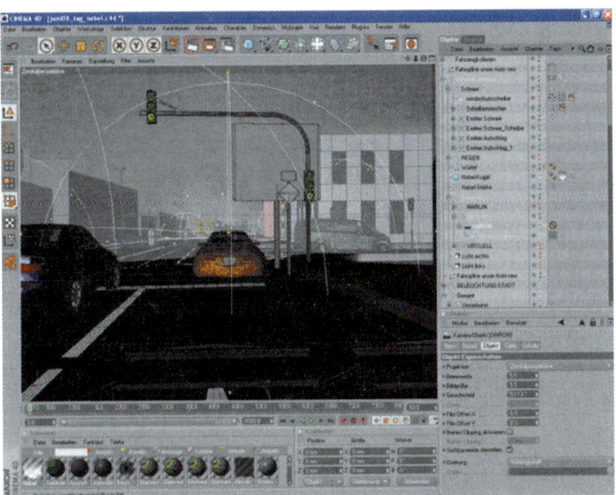

Fig. 1.   Computer-graphics generated traffic scene

To achieve this, a camera model was derived from the evaluation of real camera images. Fig. 2 compares a real camera image and the computer-graphics generated image. Influences such as depth of field blurring, colour adjustment or lense effects are considered.

Fig. 2.　Comparison of real image (left) and a computer-graphics generated image (right)

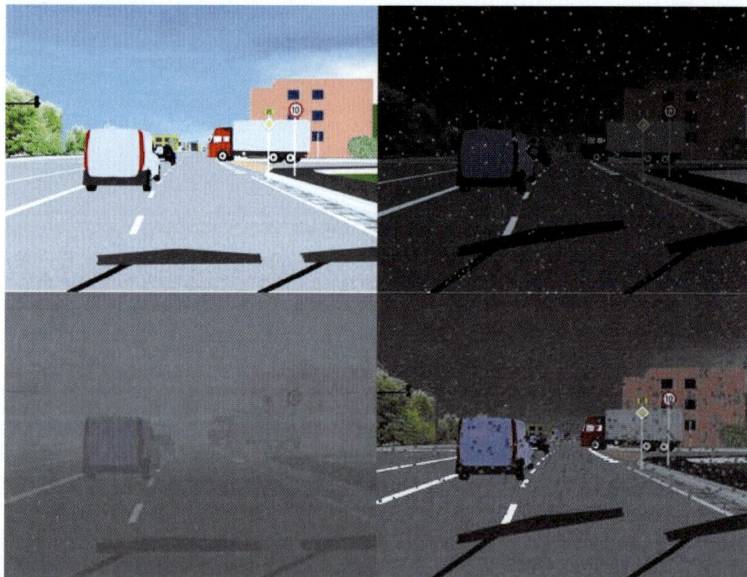

Fig. 3.　Examples of computer-graphics generated traffic scenes under different environmental conditions: day (top left), snow(top right), fog (bottom left), rain (bottom right)

This approach allows the generation of a sufficient set of training data even under extreme weather or illumination conditions (see Fig. 3). The artificial training data is currently used for vehicle detection, for detection and tracking road lanes, as well as, for the classification of traffic signs, as is discussed in this paper. The combination of computer graphics and computer vision enables new and simple ways to verify processing approaches with a wide set of test material or to adjust them based on new requirements (e.g. adjustment to country-specific traffic signs).

The adaptive parameter-based system for 3D simulation was developed. The functionality of this system could be characterized as follow: Preparation of 3D scenes for urban, country and highway traffic situations including different kind of vehicles, buildings, trees, lamp posts, as well as, traffic signs.

- ▶ A parameter-based simulation of artificial and atmospheric illumination
- ▶ A parameter-based simulation of different weather conditions (snow, fog, rain and clear visibility)
- ▶ A parameter-based simulation of traffic flow
- ▶ A parameter-based simulation of real cameras mounted in a car and its movement

The computer-graphics generated traffic scenes were successfully used for the training, as well as, the detection phase using an AdaBoost classifier cascade and for the classifier of traffic signs using neural network. To train the neural network a special image database is used. The database consists of artificial images generated with the help of computer graphics.

This approach allows the generation of thousand of traffic sign patterns with different characteristics and appearances. It is important to note that the generation process, which simulates the projection of a real camera from a 3D world into the 2D image plane, results in very realistic set of training patterns. This is a precondition to achieve a high classification rate for real traffic signs based on training with an artificial training set. Fig. 4 shows the set of the traffic signs used for the current implementation of the traffic sign classifier. Fig. 5 presents some examples of the generated artificial training patterns. Fig. 6 shows some examples of "real" traffic signs for a comparison of the quality of the artificially generated images.

Fig. 4.   The set of traffic signs for speed limits and their abolishment, which are currently recognized by the classifier

Fig. 5.   Examples of the generated training patterns based on the simulated 2D projection from a 3D traffic scenario model. The training set of each sign covers about 2600 images in a resolution range from 16x16 to 45x45.

Fig. 6.   Examples of detected traffic sign candidates

This approach of computer generated images allows a simple and fast adjustment to country-specific traffic signs, to other camera sensors or any camera position. Actually we use traffic-sign types used in Germany but it is possible to extend the training set for the neural network with additional traffic signs based on other specifications.

3   Detection of Traffic Signs based on Haar-Features

The detection of traffic signs in 2D images is based on a cascade of trainable classifiers. To detect signs inside an image a search window is scanned over the image. At every position of the search path, the image is analysed through a cascade of classifiers. If one of the classifiers fails, the analysis process is stopped and the search window is moved to the next position. A positive classification of a traffic sign is only possible, if all cascades return a positive result. This approach is time efficient because negative patterns could be discarded early without the computation of all required features. Fig. 7 illustrates the concept of the cascade classifier.

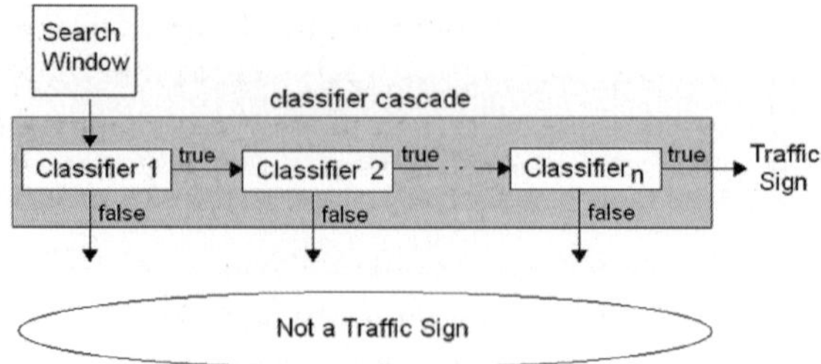

Fig. 7.   Illustration of the concept of a classifier cascade

Each Classifier consists of one ore more features. It is possible to use different types of features, for example, Haar-Features, Histogram-Features, Gabor-Features, Pixel-Features, et cetera.

To detect traffic signs at different scales, the search window including the set of trained features is scaled over a range of defined steps. For the TSD implementation the image resolution is reduced internally from 640x480 to 320x240 pixels (only every second image column is considered). The search area is set to a region of 160x200 pixels related to the 320x240 image size (see Fig. 8).

Fig. 8.   Illustration of the search area definition related to an image size of
320x240.

3.1 Haar-Features

This approach for object detection was introduced by Viola and Jones [4] for the detection of faces and extended by Lienhart [5]. Their method uses simple classifiers, which are based on so-called Haar-like features. To accelerate the computation of the features a so-called integral image is used. The type of Haar-like features, as well as, their size and position related to the search window is determined through a learning process. Every classifier of the cascade uses one or more features.

The classifier-cascade consists of 18 single classifiers and overall about 200 Haar-like features are used. Together this requires 15 kBytes for memory. For every classifier the number of features, the feature parameters and a trained threshold are stored. The feature parameters contain the number of rectangular areas, the rectangle parameters, an indicator for rotated features, a crossover value for further features and a further specific threshold. Finally, the rectangle parameters consist of an offset related to the search window, the size, as well as, the weighting factor for the different scales and pointers to the integral images.

For the verification of the proposed traffic sign detection approach, real video sequences were recorded. The camera was mounted in a passenger car and has a framerate of 25 frames per second. Over 150 different signs were captured.

Due to the camera motion each sign is visible over 5-25 frames. A total of 2000 images with traffic signs were available, with 1400 images containing traffic signs in the resolution range from 16x16 to 35x35 pixels. Detailed results of the detection algorithm can be found in Tab. 1. and Tab. 2. See [6] for details.

Haar-Feature-Cascade	
Correct detect traffic signs	*89,2%*

Tab. 1.   Rate of correctly detected traffic signs

	Average per frame	*Maximum per frame*
False detected traffic sign candidates	*0,027*	*2*

Tab. 2.   Rate of falsely detected traffic sign candidates

3.2 Pixel-Features

Beside Haar-Features simple grey value-differences of two pixels can be used to find the traffic signs. As described in [7] even pedestrians can be detected in an image. This method is very fast and doesn't need an integral image like the Viola-Jones approach. For this reason, the memory consumption is much lower. The Pixel-Feature Cascade is nearly two times faster than the Haar-Cascade (approximately 15 ms per Frame).

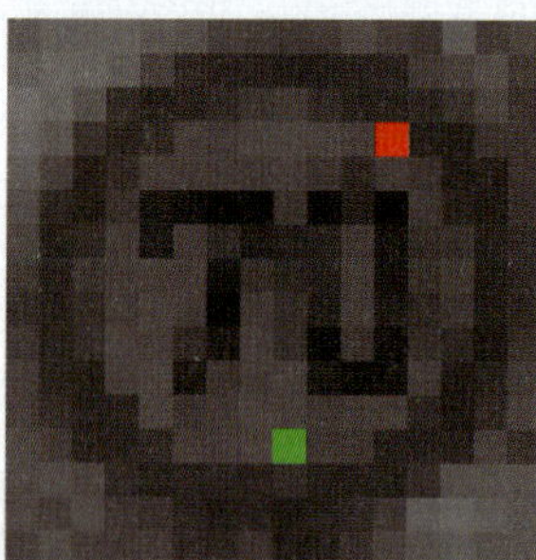

Fig. 9.   The intensity values of two pixels are compared

With respect to several thousand training images the appropriate pixel combinations are learned by AdaBoost [4,5]. Each classifier of the cascade takes one or more Pixel-Features to decide between object and non-object. If the intensity value of the first pixel is bigger than the intensity value of the second pixel one feature returns true otherwise false (if $I(x1,y1) > I(x2,y2)$ return 1 otherwise 0). The actual cascade uses 89 Pixel-Features separated into 6 classifiers. The positive detection results are similar to the results of the Haar-Feature cascade. Unfortunately, the false positives rate is much higher. In addition to the correct detections there are several false findings in every frame. For this reason a support-vector-machine (SVM) is used to verify every object, which is found by the pixel-feature cascade.

Fig. 10. Left: several false positives, Right: verification with the help of a SVM

The binary answers of all 89 Pixel-Features inside the cascade build the feature vector used for the SVM (so each vector is 89-dimensional). The training of the SVM takes place with the help of the free library LibSVM 2.86 [9]. Based on 16000 training-vectors, 894 support vectors are assigned. The SVM reviews every traffic sign candidate which is found by the pixel-feature cascade by computing the following polynomial kernel equation (if the result is > 0 return true otherwise false):

$$\left(\sum_{i=0}^{NoOfSV} coef1[i] * \left(\left(\gamma * \sum_{j=1}^{VecLength} SV[i][j] * FeatureVector[j] \right) + coef0 \right)^{degree} \right) - \rho \tag{1}$$

Detailed results of the detection algorithm can be found in Tab. 3 and Tab. 4

Pixel-Feature-Cascade	
Correct detect traffic signs	*90,4%*

Tab. 3. Rate of correctly detected traffic signs

	Average per frame	*Maximum per frame*
False detected traffic sign candidates	*0,03*	*3*

Tab. 4. Rate of falsely detected traffic sign candidates

4   Classification of Traffic Sign Candidates

The developed traffic sign detection generates one candidate per detected sign, but the quality of the region of interest (ROI), which represents a candidate, differs from sign to sign.

Fig. 11.   Examples of detected traffic sign candidates

This is caused by a changing environment, the condition of the respective sign, as well as, a coarse quantification of the scale space during the detection process. Therefore, normalization concerning contrast, position and size is required. For details see [8]. The final reduced and normalized ROI of every traffic sign candidate is used as the input vector of a neural network classifier.

To assign traffic sign candidates to the correct sign type a multi-layer-perceptron net with a feed forward topology is used. The extracted features are pixel based. Greyscale representations for traffic sign recognition are also used by [1] and [2].

Two separate networks for speed limit signs and their abolishment were trained. According to the still recognizable pattern size of 16x16 pixels, the input vector was defined with the same size. Due to this, the input layer of the neural net contains 256 neurons. The number of output neurons corresponds to the number of signs, which have to be distinguished, plus one additional neuron to indicate the negative class. The output layer consists of 15 neurons for speed limit signs and 16 neurons for the abolishment classes.

To train the neural network a special image database is used. The database consists of artificial images generated with the help of computer graphics. This approach allows the generation of thousands of traffic sign patterns with different characteristics and appearances. Fig. 4 shows the set of the used traffic signs used for the current implementation of the traffic sign classifier. Fig. 5 presents some examples of the generated artificial training patterns. This approach of computer-generated images allows a simple and fast adjustment to country-specific traffic signs, to other camera sensors or any camera position. Actually we used traffic-sign types found in Germany (font Linear-Antiqua, DIN 1451) but it is possible to extend the training set with additional traffic signs based on other specifications.

The average classification rate on 'real' images is about 97.7%. Classification failures are mainly caused by detected candidates having a ROI size smaller than 25x25 pixels. In a final test the positive classification rate of two 2-layer networks on real data was evaluated. One network was trained with artificial data. The other network was trained with real test images. An adequate number of "real" training data was not available for all traffic signs. Therefore, the set of traffic signs was reduced to 5 different types and one additional negative class. Table 5 gives the positive classification rate per traffic sign type. The average positive classification rate is about 97% for the network that was trained with "real" test data. The average classification rate of 73% for the network with artificial trained data results from relatively small, detected traffic sign candidates, as well as, a lower pixel accuracy of the detected ROIs on real data.

Number of positive classifications per sign in %		
speed limit sign	generated signs for training	real signs for training
50 $\frac{km}{h}$	88,0%	98,5%
60 $\frac{km}{h}$	85,5%	96,8%
70 $\frac{km}{h}$	77,7%	98,8%
80 $\frac{km}{h}$	51,3%	99,5%
100 $\frac{km}{h}$	51,1%	94,3%
negative class	84,6%	99,1%

Tab. 5. Comparison neural network trained with computer-generated data and 'real' data

Fig. 12. Detection and Classification results for camera and computer generated images.

5 Lane Departure Warning (LDW)

In order to warn the driver if the vehicle moves out of it road lane a LDW system is introduced. The LDW-System is implemented using a new multi-core processor, developed by Infineon Technologies AG. The approach is based on an edge detector to find the road markings on several measure lines in the image.

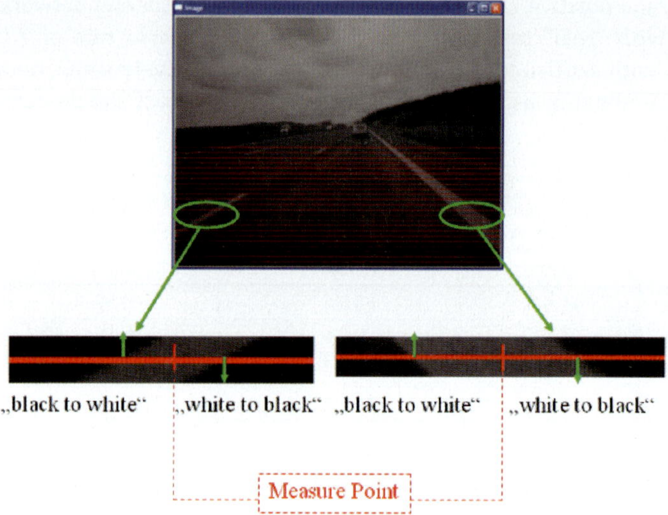

Fig. 13. A lane mark is defined by both a ascending and a descending edge

After some validation frames to confirm the data and gain additional measurements a linear lane model can be derived referencing these lane points.

Fig. 14. Left: validation phase to verify the detected lane measure point; Right: linear lane model

In the next step, an extended Kalman Filter and a clothoid model is used to follow the road and get an appropriate prediction $x(k)$. The predicted state vector considers the distance of the vehicle from the road border (Δx), the clothoid-curvatures of the road (c_0, c_1) as well as the yaw angle (θ) of the vehicle (see Fig. 15 for illustration).

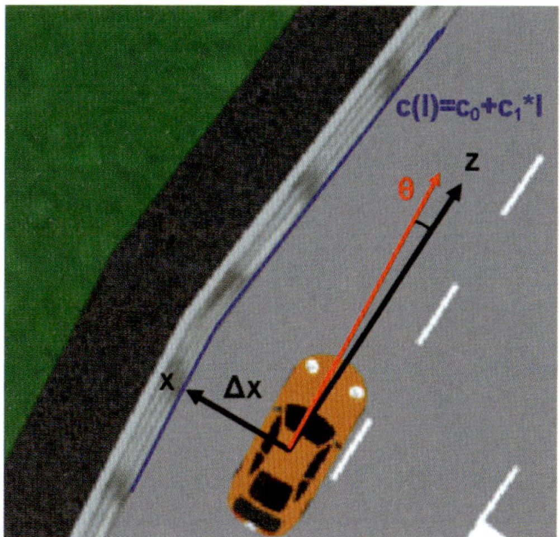

Fig. 15.   Vehicle position related to the road lane (real world x,z-Plane)

By means of a transition matrix A, depending on the velocity, the state vector $x'(k)$ can be estimated. In order to compare the prediction with the actual measurements $y(k)$ this state vector is transferred to $y'(k)$ into the measurement model by a non-linear mapping equation gc (see below). The Prediction is adjusted corresponding to the actual prediction error $e(k)$ and the Kalman-gain-matrix K. The goal is to get a robust prediction after a few seconds where only small adjustments are necessary. In this case the new state vector $x(k)$ relates only on the prediction $x'(k)$. Fig. 16 illustrates this system model.

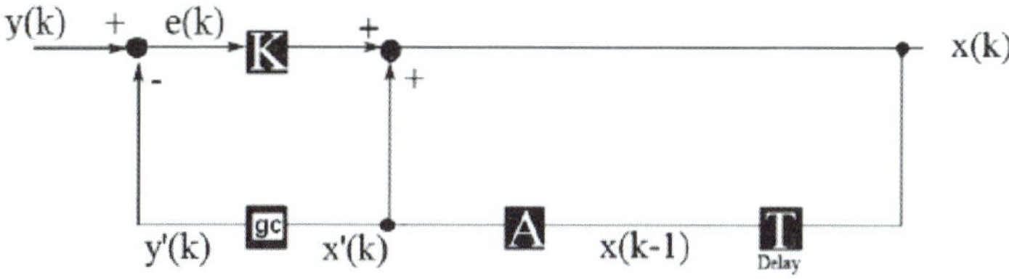

Fig. 16.   Kalman system model

state transition matrix

$$A = \begin{bmatrix} 1 & \frac{1}{2}(v\Delta t)^2 & \frac{1}{6}(v\Delta t)^3 & -v\Delta t \\ 0 & 1 & v\Delta t & 0 \\ 0 & 0 & 1 & 0 \\ 0 & v\Delta t & \frac{1}{2}(v\Delta t)^2 & 1 \end{bmatrix} \qquad (2)$$

v: velocity

state vector:

$$x(k) = \begin{bmatrix} \Delta x \\ C_0 \\ C_1 \\ \Theta \end{bmatrix} \qquad (3)$$

Θ: yaw angle

Δx: distance to road border

C_0, C_1: clothoid curvatures

road lane characteristics:

$$S = \Delta x + \frac{C_0}{2} Z_E^{\,2} + \frac{C_1}{6} Z_E^{\,3} \qquad (4)$$

Z_E: distance in real world [metre]

Transition equation from system-model to measure-model:

$$gc = \cos(\Theta) * S - \sin(\Theta) * Z_E \qquad (5)$$

The measurement $y(k)$ takes place in the two-dimensional image plane (u,v). Provided that the road is flat, it is possible to calculate the distance for each point in the image in real world coordinates, considering the known mounting and intrinsic parameters of the camera (such as the camera height h, the viewing direction, the focal-length f and the sensor dimensions).

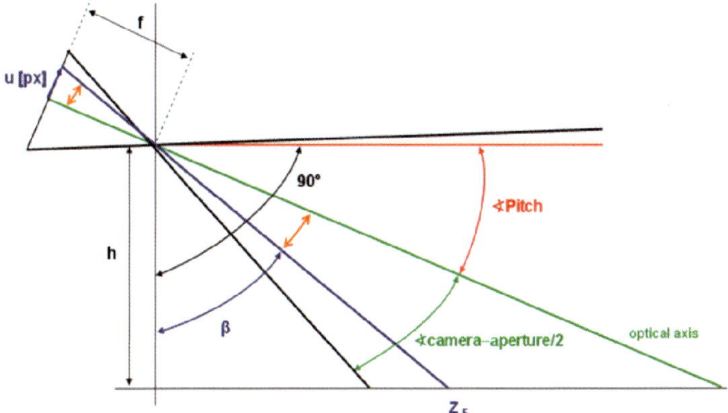

Fig. 17.   Geometric conditions. The pixel (u,v) in the image plane represents a real point in the 3D-World (the distance of this point is named foresight Z_E)

As it is illustrated in Fig. 17 the foresight Z_E [in metres] for a point in the image is computed using the following equations:

$$\beta = 90° - \angle\text{Pitch} - \arctan(\frac{u[px] * \text{Pixelheight}[\mu m]}{f}) \qquad (6)$$

f: focal length

$$Z_E = h * \tan(\beta) \qquad (7)$$

The Kalman-Process for every frame is defined as described below. The state vector $x(k-1)$ and the prediction error covariance matrix $P(k-1)$ need to be initialised with start values.

a.) calculate the real foresight Z_E in metres for every measure point

Fig. 18.

b.) Update the Jacobi-Matrix C (partial derivatives of the non-linear mapping equation gc for every measure-point)

$$C = \begin{bmatrix} \cos(\Theta) & \dfrac{\cos(\Theta)}{2}Z_E{}^2 & \dfrac{\cos(\Theta)}{6}Z_E{}^3 & -\sin(\Theta)*S-\cos(\Theta)*Z_E \\ \cdot & \cdot & \cdot & \cdot \\ \cdot & \cdot & \cdot & \cdot \\ \cdot & \cdot & \cdot & \cdot \end{bmatrix} \tag{8}$$

for every measure point

and calculate new Kalman-Gain Matrix:

$$K(k) = P'(k)*C(k)^T*(C(k)*P'(k)*C(k)^T+R(k))^{-1} \tag{9}$$

R(k): measurement error covariance matrix

c.) for every measure point: compute road lane characteristics S with the new predicted state vector x'(k)

$$S = \Delta x + \frac{C_0}{2}Z_E{}^2 + \frac{C_1}{6}Z_E{}^3 \tag{9}$$

and consider yaw-angle:

$$v_{Act} = \cos(\Theta) * S - \sin(\Theta) * Z_E \tag{10}$$

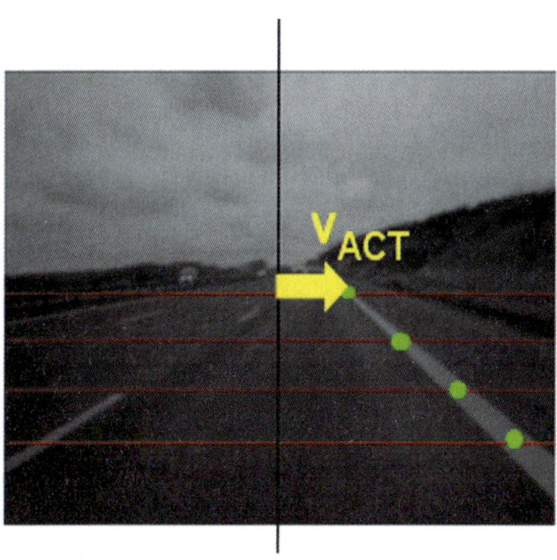

Fig. 19. V_{act}

$$u_{Act} = \sin(\Theta) * S + \cos(\Theta) * Z_E \tag{11}$$

$$v_{Target} = (measurement * PIXELWIDTH[\mu m]) * u_{Act} / f \tag{12}$$

d.) compute the new state vector x (k)

$$v_{Target} = (v * PIXELWIDTH[\mu m]) * u_{Act} / f \tag{13}$$

$$x(k) = x'(k) + K * (\vec{V}_{Target} - \vec{V}_{Act}) \tag{14}$$

e.) Update the prediction error covariance matrix P(k)

$$P(k) = (I - K(k) * C(k) * P'(k) \tag{15}$$

I: identity matrix

6 Summary

The paper presents the implementation of a traffic sign detection system on a multi-core processor. The detection uses a cascade of trained classifiers. This approach is known to be robust and efficient but requires the training of a classifier-cascade with large set of positive and negative training patterns. To overcome the problems of a time-consuming and semi-automatic labelling process to generate training sets from "real" images, a new approach based on computer graphic was developed. This allows a complete automatic process to generate artificial training patterns, which simulate realistic traffic scenarios, as well as, a realistic image capturing process. Therefore, the trained classifier-cascade could be easily adjusted to country-specific traffic signs, to other camera sensors or any camera position. The verification with real test images showed that the developed traffic sign detector is robust and reliable. The detector generates one candidate per sign and only few false regions are generated. Therefore, the proposed approach minimizes the computational effort for the successive classification step. Further improvements could reduce the time-consuming detection step on different scales. A full detection step is only required at lower resolutions. Higher resolutions require only the validation of already detected traffic signs in successive frames. Future work will also cover the combination of the traffic sign recognition with the lane departure warning system. The knowledge of the lane allows a further reduction of the search area.

For the final classification approach a separation of the classifier in two neural nets seems to be reasonable. A training set derived from real images results in an average positive classification rate of 97%. The use of training images generated by computer-graphics results in a positive classification rate of 73,8% for test images generated from real images. The harmonization of the position inaccuracy for computer generated training data and for the detection of traffic sign candidates is currently in process and will improve the classification rate. The implemented extraction of the inner part of a traffic sign candidate is currently not robust enough especially for small sizes. This affects the training data, as well as, the classification results and requires further improvements. In addition the generation of artificial images needs further analysis to model more real world impacts to the image capturing process similar to a rolling shutter or motion blur.

Furthermore a Lane-Departure-Warning System is introduced. The Algorithm uses an edge detector to find the road border. Several measurement points are necessary to build a linear lane model. After that a Kalman-Filter is introduced to predict the road course based on a clothoid model.

References

[1]   A.Broggi, P.Cerri, P.Medici, P.Porta, G. Ghisio, Real Time Road Signs Recognition, IEEE Intelligent Vehicles Symposium, Page(s) 981-986, 2007.

[2]   C. Bahlmann, Y. Zhu, R. Visvanathan, M. Pellkofer, T. Koehler, A system for traffic sign detection, tracking and recognition using color, shape and motion information, IEEE Intelligent Vehicles Symposium, Page(s) 255-260, 2007.

[3]   B. Alefs, G. Eschemann, H. Ramoser, C. Beleznai, Road Sign Detection from Edge Orientation Histograms, IEEE Intelligent Vehicles Symposium, 2007.

[4]   P. Viola, M.Jones, Rapid Object Detection using a Boosted Cascade of Simple Features, IEEE CVPR, Volume I, Page(s) 511-518, 2001.

[5]   R. Lienhart, A. Kuranov, V. Pisarevsky, An Empirical Analysis of Boosting Algorithms for Rapid Objects with an Extended Set of Haar-like Features, Intel Technical Report MRL-TR-July02-01, 2002.

[6]   A. Techmer, N. Luth, R. Ach, Real-Time Detection of Traffic Signs on a Multi-Core Processor, IEEE Intelligent Vehicles Symposium, Page(s) 307-312, 2007.

[7]   L. Leyrit, T. Chateau, C. Tournayre, J. Lapreste, Association of AdaBoost and Kernel Based Machine Learning Methods for Visual Pedestrian Recognition, IEEE Intelligent Vehicles Symposium, Page(s) 67-72, 2007.

[8]   A. Techmer, N. Luth, R. Ach, T. Schinner, S. Walther, Classification of Traffic Signs in Real-Time on a Multi-Core Processor, IEEE Intelligent Vehicles Symposium, Page(s) 313-318, 2007.

[9]   C. Chang, C. J. Lin, A Library for Support Vector Machines, http://www.csie.ntu.edu.tw/~cjlin/libsvm, 2001.

Nailja Luth, Roland Ach
University of Applied Sciences Amberg-Weiden
Faculty of Electrical Engineering and Information Technology
Kaiser-Wilhelm-Ring 23
92224 Amberg
Germany
n.luth@haw-aw.de
r.ach@haw-aw.de

Keywords: driver assistance, traffic sign detection, Haar-feature, lane-departure-warning system, Kalman-filter

Test-bed for Unified Perception & Decision Architecture

L. Bombini, St. Cattani, P. Cerri, R. I. Fedriga, M. Felisa, P. P. Porta
Università degli Studi di Parma

Abstract

This paper presents the test-bed that will be developed for a Unified Perception & Decision Architecture (UPDA). Due to the increasing demand of ADAS systems to be mounted on cars, it is more and more important to develop a unified architecture that can communicate and share information between these systems. This is the aim of an ERC-founded project and to develop and test such architecture a car has been set up with many different sensors.

1   Introduction

VisLab is undertaking highly innovative research within its ERC-founded European project, whose topic is the development of an open standard for the perception and decision subsystems of intelligent vehicles. The subject, gathering the attention of car industries all over the world, was first outlined in 2006 by the German Research Foundation under the name of "Cognitive Automobiles" [1]. Currently, many commercial vehicles include sophisticated control devices like ABS, ESP, and others. These control equipments have been independently developed by car manufacturers and suppliers. Generally, they also act independently, and are singularly tuned. Nevertheless, new methods to improve overall performance are currently under development, exploiting communication and cooperation of these devices: the recently introduced Unified Chassis Control (UCC) is an example. The deployment of the UCC in the mass market requires to adapt and rethink all control subsystems to provide communication, data fusion, and an overall tuning: namely to integrate all of them together. From the car manufacturers′ and suppliers′ point of view, the introduction of the UCC requires the redesign of each single block (ABS, ESP, ...) meaning an additional financial effort, besides the obvious delay in reaching the market. Had a complete UCC architecture been defined well in advance with respect to the development of each single block, its implementation would have been straightforward, less costly, and would have reached the market earlier. Perception and decision modules are in an earlier development stage than control ones: the advanced driver assistance systems that are currently available on the market are only basic ones (Lane Departure Warning, Blind Spot Monitoring, Automatic Cruise Control, Collision Mitigation), independently developed by different car manufacturers and suppliers. The state of the art of advanced driver assistance systems, in fact, has not yet defined a complete architecture that would allow fusion of all these basic blocks and benefit from their integration. The availability of such architecture would allow to define a standard module interface so that the following research efforts could be more focused in providing modular systems, already engineered to fit into this architecture.

In order to develop these concepts, VisLab is setting up BRAiVE, (Fig. 1) a prototype vehicle (three more are expected for 2009) integrating various sensing technologies. A new architecture (UPDA, Unified Perception & Decision Architecture) is meant to take as input all the data coming from the different perception modules, fuse them, and build a more sophisticated world model to be used in the following decision level. In doing this, a standard interface will be defined for each module, allowing different providers to integrate their own perception system into this standard architecture, which will also boost competition.

Fig. 1.   BRAiVE, VisLab's test-bed vehicle for the UPDA architecture

Providing a UPDA involves the understanding of what has to be perceived and classified by the vehicle. We make a list of the main items in the following:
- *pedestrians*: still or moving and, if moving, how fast;
- *vehicles*: and their speed;
- *other obstacles*: and their speed;
- *lanes*: any sort of roadways demarcation line –temporary or permanent, continuous or dashed;
- *stop lines*;
- *junctions*: crossings layout;
- *parking lots*: available parking areas delimited by horizontal demarca-

tion lines;

▶ *tunnels and bridges*;

▶ *road signs*: danger, priority, prohibitory, mandatory and indication signs;

▶ *traffic lights*.

▶ *free space*: surrounding environment areas available for the vehicle to move and steer;

▶ *blind spots*: areas of the road that cannot be seen while looking forward or through either the rear-view or side mirrors;

▶ *light and visibility conditions*: daylight, nighttime, dazzle, rain, fog, snow, etc.;

▶ *road slope*: road plane inclination;

▶ *environment*: can be urban or extra-urban; if the latter, no lanes will be available for the vehicle to find its way.

The advanced driver assistance systems currently available on the market can be described in terms of the items just listed: Lane Departure Warning involves the detection of *lanes* and *free space*; Blind Spot Monitoring the detection of *vehicles*; the ACC functionality is allowed by detecting the preceding *vehicle* and Collision Mitigation involves the detection of *vehicles, other objects, tunnels and bridges, free space.*

BRAiVE will be provided with driver assistance systems similar to those already available on the market but with improved performances, plus other major capabilities, listed in the following:

▶ **Crossings Assistance**: lateral perception is important when facing an intersection, and as junctions may have very different layouts, the lateral system must be able to perceive traffic coming from different angular directions. During a traffic merging manoeuvre, the vehicle is allowed to pull into traffic only when oncoming traffic leaves a gap of a sufficient time interval (usually at least 10 seconds). Hence the vehicle, regardless of the junction layout, needs to perceive the oncoming traffic as well as estimate cars speed from long distance.

▶ **Overtaking Assistance**: when driving on a road with multiple lanes, the lane change manoeuvre is a common task in order to overtake slower traffic. Two rear cameras installed acquire images of the road behind and on the vehicle side. The cameras are installed so that they can frame the area close to the vehicle and extend their vision over the horizon. This system can overcome some LIDAR limitations, like its inability to provide useful results when moving along dusty roads, where the clouds produced by the vehicle itself negatively affect the laser beams. This is a problem mainly of the rear LIDAR since dust clouds are produced by the vehicle in motion; sometimes it affects also the front ones especially during sudden stops when dust clouds overtake the car. Despite this problem, LIDAR data have been used to refine the detected obstacles position measurement, thus reducing the errors introduced by vehicle pitching, since distance estimation is performed by a monocular system.

▶ **Lane Keeping**: navigation in a urban environment has to be very precise. The vehicle must be able to detect obstacles and lane markings with high confidence on the detection probability and position accuracy; moreover the system must detect stop lines, to allow a precise positioning of the vehicle at intersections, and lane markings in sharp curves. The perception system must therefore cover a distance range useful for detections at driving at medium-to-high speeds. The vehicle includes two stereo systems (one on the front and one at the back) which provide precise sensing in its close proximity.

▶ **Parking Assistance**: the system aims at supporting the driver during parking manoeuvres.

▶ **ACC Stop-and-Go**: this functionality is to be used for speed adjustment in respect of the preceding vehicle. Stop-and-Go is for queue-driving, when the vehicle speed is lower than 15 km/h in a urban-like environment (whereas ACC operates at higher speeds, like on highways). Once engaged, gas and brake control are handled by the system until some driver's action on the pedals disables the automatic speed control of the car.

To meet these requirements, a sensors belt has been set up around the vehicle. The choice to primarily use passive sensors is dictated by the will of producing a car suitable for mass production: active ones are likely to interfere with each other, thus potentially degrading the performance of multiple autonomous vehicles operating in close proximity. The vehicle is equipped with a variety of devices, including sensors for world perception, navigation and control systems, localizations devices, computers, displays, batteries, wiring. A lot of effort must be spent to obtain a high integration level, making the perception devices barely visible from outside the vehicle and the controls and displays integrated into the dash board, headrest and armrest. Fig. 2 shows the overall equipment schema.

BRAiVE will be used for testing both UPDA and ADAS applications.

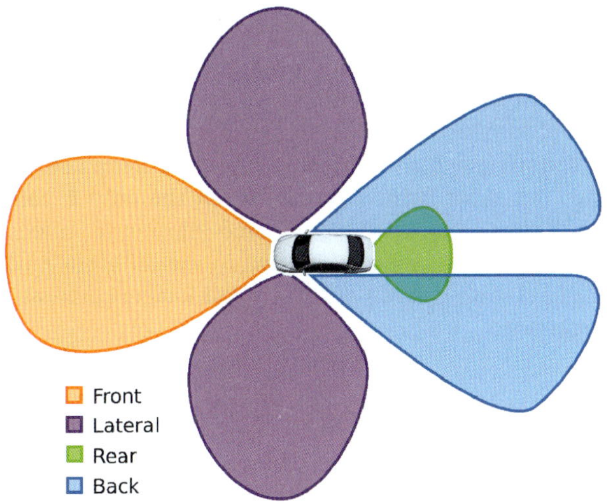

Fig. 2.   BRAiVE's perception modules, sensing its surrounding areas

2   Hardware Architecture

As introduced in the previous section, the aim of the architecture proposed is to give BRAiVE an all-round view in order to let it perceive the environment and its evolution before taking any decision about its own movement. The vehicle must be able to sense its surroundings, understand situations and react quickly; a service-based system architecture coupled with efficient path planning, trajectory algorithms and behaviour modes ensures that sensor data are processed in a timely manner, while robust low level control and X-by-Wire components enable the vehicle to manage different situations.

The complete perception apparatus integrates 10 cameras, 3 laserscanners, a laser, a GPS unit connected to an Inertial Measurement Unit, and 4 PCs; the whole car has been set up for drive by wire, and can be controlled via CAN messages. The UDPA architecture to be implemented on BRAiVE receives, as input, information coming form laser, laserscanners, cameras, GPS antenna, inertial navigation system and vehicle state. All these information are gathered to create, and continuously update, a perception map of the environment where BRAiVE is called to move, on the basis of which the implemented driver assistance actions are evaluated and taken.

To get into the architecture details, in the following the description will be split in 4 subsections, each referring to a monitored area of Fig. 2 and describing a perception module: front, lateral, rear, and back. The items listed in section 1 will be associated with each monitored area and, to give an idea about what each module's output will look like, similar systems previously developed by VisLab and their respective outputs will be cited. Those systems will be the prototype vehicle TerraMax T2, which participated to the 2007 Urban Challenge [2], a market available Hyundai Grandeur, demonstrated at IV 2008 conference [3]., TerraMax, which was between the only 5 vehicles to have completed the 2005 Grand Challenge course [4], and a Volvo truck. In the following sections' images, devices involved in the different modules perception are highlighted with colored ovals: red to indicate cameras and green to indicate laserscanners.

2.1 Front

The vehicle frontal perception must cover the orange area in Fig. 2. Specifically, referring to the items listed in section 1, the head-on system must detect *pedestrians, vehicles, other obstacles, lanes, stop lines, parking lots, tunnels and bridges, road signs, traffic lights, free space, the road slope* together with the kind of *environment* around and its *light and visibility conditions*.

For this purpose, BRAiVE head-on system includes 2 stereo pairs –of which one with 2 color cameras and the other with 2 b/w cameras– placed behind the windscreen, 2 single layer laserscanners with a 270 degrees scanning range and a sensing range from 0.1 to approximately 60 meters, another laserscanner with 4 parallel and simultaneous scanning layers, a scanning range of 100 degrees and a sensing range from 0.3 to 200 meters, and finally a laser with 16 beams, a 16 degrees scanning range and a sensing range up to 200 meters.

Some of VisLab's previously developed systems for frontal area monitoring are shown in Fig. 3, 4, 5.

Fig. 3.   TerraMax T2 head-on area was equipped with a trinocular color cameras system and two laserscanners for obstacles and lane detection [2]: system output is shown in the right image

Fig. 4.   The Hyundai Grandeur had a single b/w camera installed behind the windscreen and a laserscanner with a 90 degrees scanning range and an extended sensing range up to 80 meters to detect road signs [5] and pedestrians [3]; images on the right give some examples of its outputs: road signs detection and pedestrian detection during nighttime visibility conditions

Fig. 5.   On TerraMax was installed another trinocular color cameras system aided by a single laserscanner with four scanning layers, a scanning range of approximately 150 degrees and a sensing range of approximately 80 meters [4]; right images show some examples of its output: obstacle detection (bridge pillars and small poles) in extra-urban environments

2.2   Lateral

Perception on the vehicle's lateral sides must be able to detect *pedestrians, vehicles, other obstacles, junctions* and *free space*. BRAiVE lateral system, monitoring the violet areas in Fig. 2, includes two color cameras installed near the car's hood, above the tires and at the bumper's both ends.

Some of VisLab's previously developed systems for lateral areas monitoring are shown in Fig. 6, 7.

Fig. 6.   TerraMax T2 was equipped with a lateral system made of two color cameras placed on a bar in front of the ventilation grid with the purpose of detecting cars approaching at traffic junctions [2]; right image shows an output example of T2 lateral system: cars driving towards the junction are framed by a red bounding box

Fig. 7.   The Hyundai Grandeur had a single b/w camera installed near the wing mirrors for parking assistance purposes; both images on the right show examples of free space detections: green areas (red in the bird's eye vieew underneath) are suitable for parking the car

2.3 Rear

BRAiVE rear perception system, monitoring the green area in Fig. 2, is committed to the detection of *vehicles* and *other* obstacles, *parking lots, free space* and *blind spots*. For these purposes, two color cameras are placed in both wing mirrors lodgings besides the respective mirrors. Some of VisLab's previously developed systems for rear areas monitoring in Fig. 8, 9.

Fig. 8.   TerraMax T2 rear view system consisted of two color cameras placed on top of the truck cabin, pointing backwards [2].; right

images show some approaching cars detected by red U-shaped lines

Fig. 9. The Hyundai Grandeur rear view system was made of two b/w cameras installed in the wing mirrors lodgings; right image shows an example of the system's output: lanes with a blue stripe and an overcoming vehicle with a red bounding box, are correctly detected; the traffic lights image indicates if overtaking is permitted or not

2.4 Back

To monitor areas at the back, blue colored in Fig. 2, BRAiVE is equipped with a stereo pair of two color cameras and a laserscanner between rear bumper and number plate. Aim of the system is the detection of *pedestrians, vehicles, other objects, parking lots* and *free space*.

Some of VisLab's previously developed systems for frontal area monitoring, but convertible for back areas, are shown in Fig. 10, 11.

Fig. 10. TerraMax T2 frontal system for near obstacles detection consisted of a color cameras stereo pair and two laserscanners (described in Fig. 2.1 caption) [2]; the right images roadway lanes and a preceding car being correctly detected

Fig. 11.   Another color cameras stereo system installed by VisLab on the front of a Volvo truck cabin, for near obstacles detection [6]; examples oft he system's output are shown in the right image

3   Final Considerations

The project is aimed at the development of autonomous vehicles and supervised driving systems with the ultimate goal of defining a common open architecture which will be proposed as a standard to the automotive sector. Besides providing clear advantages on safety for road users, the availability of an open architecture will encourage and make possible the sharing of knowledge between public and private research communities (academic and automotive industry) and thus speed up the design of a standard platform for future vehicles. Further research steps will be eased -and therefore made more effective- thanks to the common and open architectural layer proposed. The project is divided into the following two main milestones:

1. the development of fully autonomous vehicles and,
2. extension towards driving assistance systems, namely systems able to supervise a driver and to intervene when necessary.

Fig. 12 shows the evolution of driving assistance systems leading to autonomous driving. The first four steps have a human being as vehicle main leader: starting from a set of independent warning systems (step 2) like lane departure warning, to independent active systems (step 3) like adaptive cruise control, collision avoidance, and finally a unified perception and decision architecture (step 4) devoted to perform an active cooperative driving. From step 5 onward the vehicle leader changes from the human being to the electronic pilot increasing the sensing capabilities. First with autonomous driving and then (step 6) with supervised driving, in which the human instructs and directs the manoeuvre while the main control is owned by the electronic pilot. Human contribution, in this case, is treated like one of the other sensing devices, and thus overridable. This project will reach step 3 and step 4, using the new concept architecture to make all the systems cooperate together.

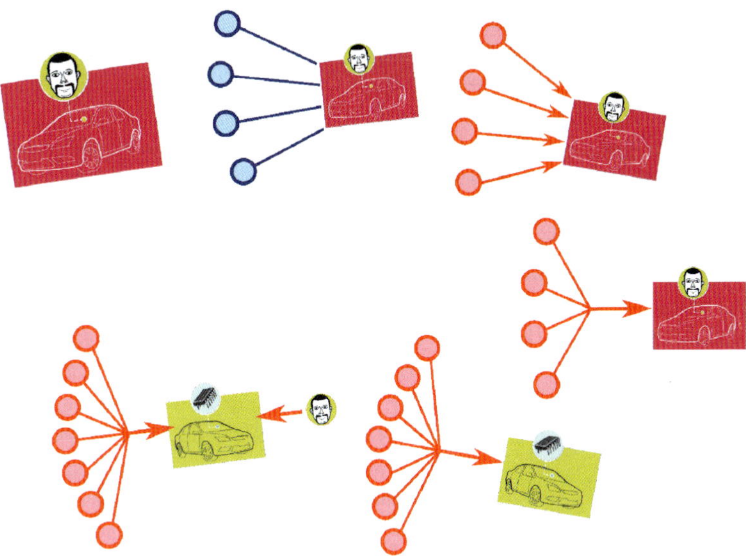

Fig. 12.   Intelligent vehicles evolution

The first milestone will be a demonstration of fully autonomous vehicles able to cope with real scenarios and not only with controlled environments. This requires to develop both an extended perception system able to build an accurate world model and a sophisticated decision system. This part will be based on the work already developed by VisLab for other projects (like ARGO [2] and the DARPA Challenges [7]) and its experience as a primary player in this field.

In the second stage, leading to the second milestone, a perception module able to analyze the driver's intentions as well as a Human Machine Interface will be added in order to enable driving assistance features. As mentioned, this requires to extend both the perception and decision functionalities in order to integrate new inputs.

During both stages, the logical architecture of the vehicle (i) autonomous system and (ii) supervisory system will be designed, tested, and validated thanks to intermediate tests on its completeness, feasibility, and scalability by means of the test bed described.

References

[1]   Details at http://www.kognimobil.org

[2]   Yi-Liang Chen, et al., TerraMax: Team Oshkosh Urban Robot, Journal of Field Robotics, 25(10):841-860, October 2008.

[3]   A. Broggi, et al., Localization and Analysis of Critical Areas in Urban Scenarios, In Procs. IEEE Intelligent Vehicles Symposium 2008, pages 1074-1079, Eindhoven, Netherlands, June 2008.

[4]   D. Braid, et al., The TerraMax Autonomous Vehicle, Journal of Field Robotics, 23(9):693-708, Sept 2006.

[5]   A. Broggi, et al., Real Time Road Signs Recognition, In Procs. IEEE Intelligent Vehicles Symposium 2007, pages 981-986, Istanbul, Turkey, June 2007.

[6]   A. Broggi, et al., StereoBox: a Robust and Efficient Solution for Automotive Short Range Obstacle Detection, EURASIP Journal on Embedded Systems - Special Issue on Embedded Systems for Intelligent Vehicles, June 2007, ISSN 1687-3955 2007.

[7]   A. Broggi, et al., Automatic Vehicle Guidance: the Experience of the ARGO Vehicle. World Scientific, Singapore, April 1999.

Luca Bombini, Stefano Cattani, Pietro Cerri
Rean Isabella Fedriga, Mirko Felisa, Pier Paolo Porta
Università degli Studi di Parma
VisLab – Dipartimento di Ingegneria dell'Informazione
Via G.P. Usberti 181
Parma
Italy
bombini@vislab.it
cattani@vislab.it
cerri@vislab.it
fedriga@vislab.it
felisa@vislab.it
porta@vislab.it

Keywords: ADAS, autonomous vehicle, unified perception and decision architecture

Real-Time Camera Link for Driver Assistance Applications

M. Römer, Inova Semiconductors GmbH
T. Heimann, Helion GmbH

Abstract

With the increasing demand on driver information and multimedia content, displays and video signalling are getting more and more attention in the automotive industry. The requirement of transmitting video signals includes applications like infotainment displays, dashboard and head-up displays, but also driver assistance systems requiring real time video streams. The Automotive Pixel Link (APIX) Technology has been specifically designed to address the different requirements for video transmission in automotive applications. Optimized for low EMI, the APIX technology offers the ability to combine high speed real-time video data, full-duplex sideband channels and the power supply over a single cable. Especially on multi-camera applications, the video information needs to be provided with minimum delay, with special requirements on synchronization and image processing. This article provides an introduction on the APIX technology, followed by the discussion of various requirements of implementing camera based driver assistance systems in the automotive environment.

1   Introduction

The term "driver assistant systems" describes different applications, offering the driver functionality, higher comfort as well as they provide higher security for critical situations. Applications like automatic parking systems or adaptive cruise control provide higher comfort, while systems like blind spot detection, night vision or topview systems provide higher safety, helping the driver to react in critical situations. Each of these systems requires one or multiple sensors or cameras, of which the signals need to be evaluated and prepared for the user interface or to even send the messages to the main control unit for immediate action.

Especially looking at multi-camera based driver assistance systems, the video signals need to be provided with high reliability and at lowest latency, in order to be able to combine all images with minimized processing effort. In addition, the car environment requires lowest electromagnetic interference and to reduce cabling requirements to a minimum.

Based on the example of a multi-camera based topview system, this article shall show the challenges and requirements for the implementation of a multi-camera system and how the designer may implement the camera link including control and power transmission using the APIX technology.

2 APIX Overview

2.1 APIX Architecture

The Automotive Pixel Link (APIX) architecture is designed to act as a single interface to a display or to a remote digital camera solution. In order to support high transmission speeds at distances of up to +15 m (1 GBit/s mode) and up to +40 m (500 MBit/s mode), the data are serialized and transmitted via Current Mode Logic technology.

The APIX link provides three independent channels for data transfer, which allow full duplex transmission in both directions.
- ▶ a high speed downstream pixel channel with up to 1 GBit/s
- ▶ a downstream sideband channel with up to 26 Mbit/s
- ▶ an upstream sideband channel with up to 20 Mbit/s

The pixel channel and the downstream sideband channel are multiplexed and commonly transmitted over the downstream link. The upstream sideband channel can either be established over the same pair of cable as the downstream link or over a separate pair of cables (see Fig. 1).

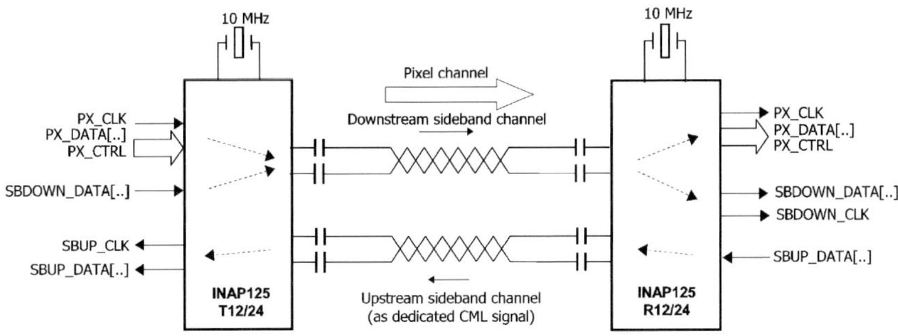

Fig. 1.   APIX Architecture

The high speed downstream channel acts as transparent gateway for the parallel video interface. Video data are transmitted real-time without compression, minimizing the latency for time critical applications. The video interface supports any parallel data format like RGB or YUV, with four different bit widths of 10, 12, 18 or 24 bit. In addition to the pixel data interfaces, three pixel control signals are implemented on the video interface.

In parallel to the video interface, the APIX architecture offers the so called sideband data channels, providing full-duplex downstream and upstream capabilities. The side-band data are sampled at dedicated pins on either transmitter or receiver and are transparently provided at the respective pin on the remote side. Since the APIX link does not expect any special formatting, this transparent sampling allows flexible 'extension' of interfaces like I²C, UART or SPI over a long distance.

Due to the fact that transmitter and receiver are driven by local reference crystals, of which the high speed differential link is derived, the differential link is independent of the video pixel clock and therefore allows the transmission of the sideband data even without pixel clock available. The ability of carrying full-duplex information without pixel interface offers a significant benefit for point-to-point applications. Since the sideband data channel is available as soon as the link is powered up, it can be used as the main configuration interface of the camera or display implementation. Even with no video data available, a camera can still be configured and synchronized. With using the APIX as a single system interface even powering the remote device (see Section 2.3), the system cost of the implementation can be reduced significantly.

2.2 Features

The APIX link has been designed to be able to support different application scenarios for cameras or displays in the automotive environment. These requirements include the need of low electromagnetic interference (EMI), as well as the support of cables in various lengths and types.

In order to be able to adapt the line driver to different cables and lengths, the nominal output swing of the differential signal, as well as optional pre-emphasis is configurable. The longer the cable, the higher the impact of reflections and the cable capacitance. Adjusting the nominal output swing allows longer cable length or reducing the power consumption, while pre-emphasis reduces the drive level to optimize the switching characteristics.

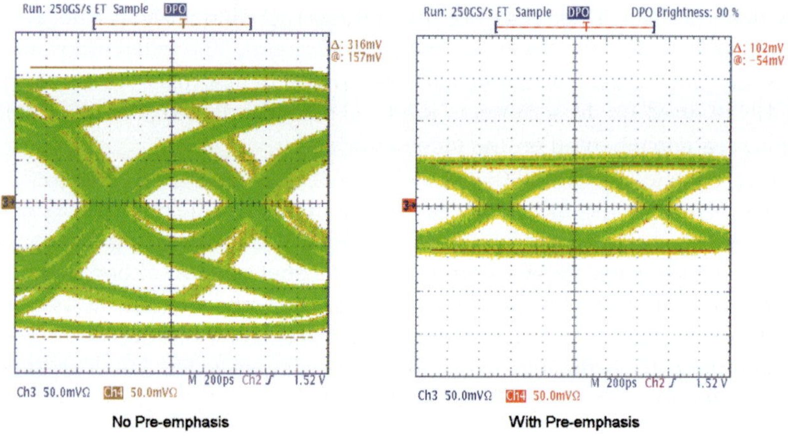

Fig. 2.   APIX CML eye pattern after 20 m of cable without and with pre-emphasis enabled

A major issue in the automotive area is electromagnetic interference (EMI), which can have severe impact on the overall system. EMI includes immunity as well as emissions, which both need to be optimized for acceptable performance.

The APIX current mode logic signalling together with the optimized line coding offers a very stable transmission, providing high immunity against external influences. However, since EMI is evaluated at system level, the designer needs to take care of several aspects for the design, including the power supply design and the differential signal design.

A critical source of EMI is the wide pixel interface, toggling at speeds up to 62 MHz. A typical approach to reduce the emissions of this interface is to allow a certain jitter on the reference clock spreading the overall emission spectrum. In order to support this method, the INAP125T12/24 transmitter accepts a wide range of jitter on the pixel clock, with no impact on the serial transmission. In addition, the INAP125R12/24 receivers use staggered outputs, to reduce the overall switching noise.

2.3   Power over APIX

Applications like cameras or rear-seat displays typically are just connected to one main unit, requiring a video interface as well as a configuration interface. In addition, the devices need to be supplied with power, which typically requires additional cabling and may also induce ground offset issues.

By using the APIX link, the interface to these 'extension' devices can be implemented by just two pairs of wire, transporting the video data and configuration data but also providing the power supply over the same cable. In addition, this supply eliminates the problem of ground offsets.

The power supply is realized by using either one or both existing wire pairs, using inductors, acting as low pass filters to the signal lines. Since high DC transients may couple into the chip or may appear as common mode noise, the design needs to be carefully designed to avoid high emissions. Fig. 3 illustrates several examples of transmitting the power supply over the data lines. Version a) would be the solution with minimum effort, but might cause common mode to differential mode conversion at high dynamic currents. Version b) requires 2 pairs of wires, but eliminates the conversion effects.

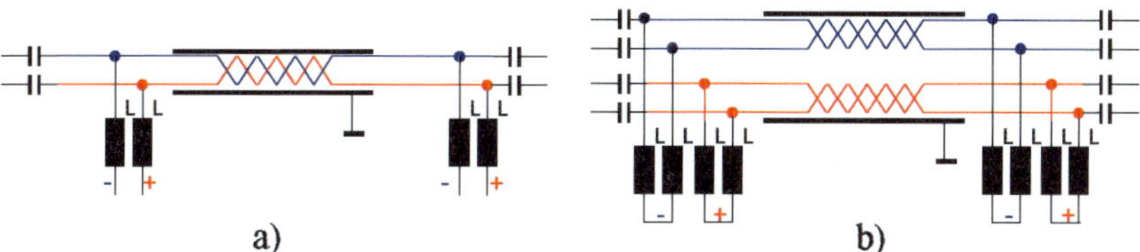

Fig. 3.   Power over APIX configuration examples

2.4 APIX IP Licensing

The APIX technology is offered today by Inova Semiconductors as discrete transmitter and receiver devices INAP125T12/24 and INAP125R12/24, as well as through licensing partners Fujitsu, Toshiba and Xilinx. Integrated in graphics controllers or as hardware block in a FPGA, the APIX technology allows further reduction of system costs by eliminating the need of external serializer or deserializer devices.

Fig. 4 shows a possible implementation of a display using the MB88F332 graphics controller from Fujitsu microelectronic. The graphics controller incorporates the APIX receiver IP, acting as single interface to the complete display implementation. Video data as well as configuration data and the power supply are transmitted over the four wire cable, carrying the APIX signal.

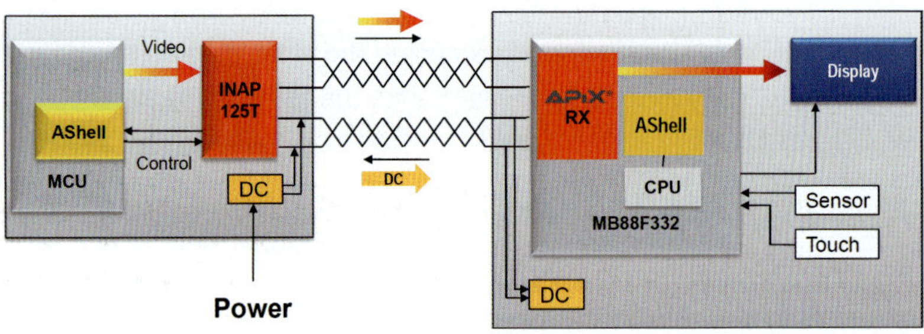

Fig. 4.   Application example with Fujitsu Microcontroller MB88F332, incorporating the APIX IP

With the possibility of implementing the technology into graphics controllers, sensors or displays, APIX has the potential to become a standard interface in automotive digital camera and display connections.

3   Camera Application

3.1 Digital Camera Systems and Camera Requirements

Today's vehicles, especially trucks, do not have a clear view to the surrounding area. There are fields of view that are not visible using conventional mirrors. E. g. while maneuvering (backing into a parking space), the surrounding area (directly behind) the vehicle (e. g. truck or SUV) is difficult to observe. Until today there are a lot of mirrors and displays to observe while moving a vehicle. Often the driver has information overflow and looks at the wrong display or mirror.

This issue can be solved with a camera-display system that presents the whole vehicle and the surrounding area in topview on one display to the driver (see figure 5). Additional information could be displayed on the screen within the image like the estimation of driveway etc.

Fig. 5.   Topview example, truck maneuvers on a parking space (simulated image)

The camera system itself has to fulfill some special requirements. The objective of the system is to observe the whole surrounding area of the vehicle especially in bright sunlight with some shadow areas. Since this environment reflects a huge dynamic range, the camera system needs to provide an even higher dynamic range. One approach to address this is the use of the so-called HDR-technology.

Fig. 6 to 8 show the difference between a standard CMOS camera with over- and underexposed areas (Fig. 6 and 7) and the solution with a combination of HDR-capture and the BLENDFEST algorithm from Helion (Fig. 8). It starts with capturing the HDR image row data. The captured data width is bigger than 8 bit per pixel per color so the data has to be compressed in order to be viewed on a display, which has in the best case 8 bits per pixel per color. Generally the algorithm to compress the data down to 8 bit is called "tone mapping". There are different possible ways to do that. A very convenient way has been developed by Helion and is called BLENDFEST. The results are visible in Fig. 8, too.

Fig. 6.   Image with overexposed areas

Fig. 7.   Image with underexposed areas

Fig. 8.   BLENDFEST HDR-image

To get a topview of the surrounding area of the vehicle a single camera is not sufficient. Four cameras with wide angle lenses would be a good decision to use. By mounting them on the edges of the vehicle all surrounding areas can be observed even on a semi-trailer truck while driving through a curve. To get a topview picture out of four single cameras it is necessary to stitch the single images together.

Before stitching the images they have to be distorted and the lens shading has to be corrected. An additional requirement for a good stitched image is to have the same exposure time for all four cameras. Therefore it is necessary to control the exposure times from the central unit.

3.2 Camera Implementations using APIX

As explained in the section before, it is strongly recommended to control the configuration of all cameras in a central unit. Only the central unit has the data of all cameras and can decide which configuration would be the best to use for all image sensors, to get the best looking overall image. To control the image sensors the communication between the central unit and the cameras outside is necessary. Also the power supply for the cameras is needed.

The issue of accessing and communicating between the four camera modules and the central unit can be solved in different ways. The conventional way is to have a pair of wire for each function:
  ▶ Power supply (two wires)
  ▶ Image data transfer (two wires)
  ▶ Control data transfer (two wires)

Each function requires additional devices for transmitting and receiving of the data on the camera as well as on the central unit. Also the EMC-safe design of supplying the cameras with power is a great issue in automotive.

Using the APIX link, one four-wire cable is sufficient to connect the camera device with the central unit including all required functions. Since there is no need for a separate power supply connector, the camera device belongs to the central unit and the whole system (central unit and four cameras) is one electronic control unit (ECU). Even for the system designer, it is easier and more reliable to use only one APIX link for all functions.

Combining all these functions, APIX provides a real-time, loss-free digital video and communication link, which is required to create a high quality topview image. Analog cameras or autonomous cameras cannot be used to fulfill this requirement.

3.3 Application Example

Fig. 9 shows the complete topview system example described in the sections before.

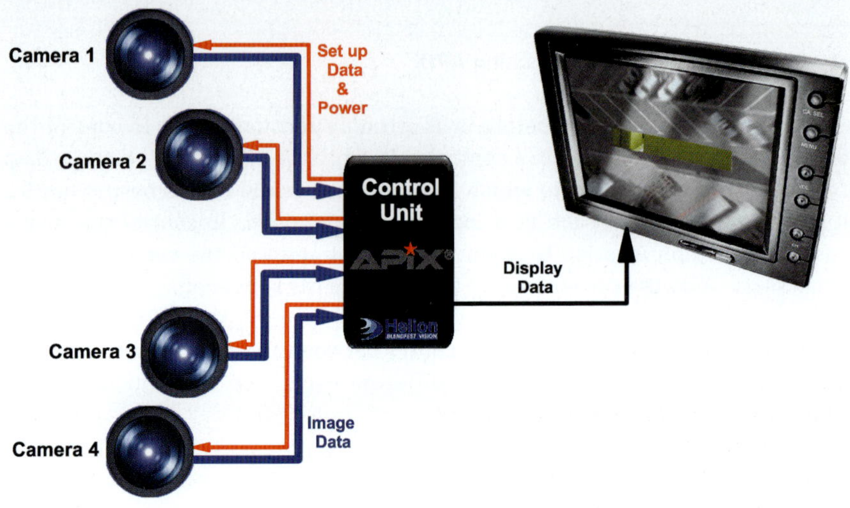

Fig. 9.   Overview over the topview application

With the actual available technology the camera head is based on a three chip solution. These three chips are the image sensor, a really tiny FPGA and of course, the APIX transmitter device. In the future, especially with the licensing model of the APIX technology, it is possible to develop a special device, which integrates all three functions (imager, FPGA, APIX) in one package. The next step is to build one tiny device with a little FPGA and APIX interface inside.

On the other hand the central unit has to control and set up the cameras, receive the image data and to stitch the different images to one resulting image. Of course this unit also has the job to process the resulting image to show the driver a high quality topview image on the display with possibly additional information.

Fig. 10 shows the data flow for one camera. The other three cameras work the same way.

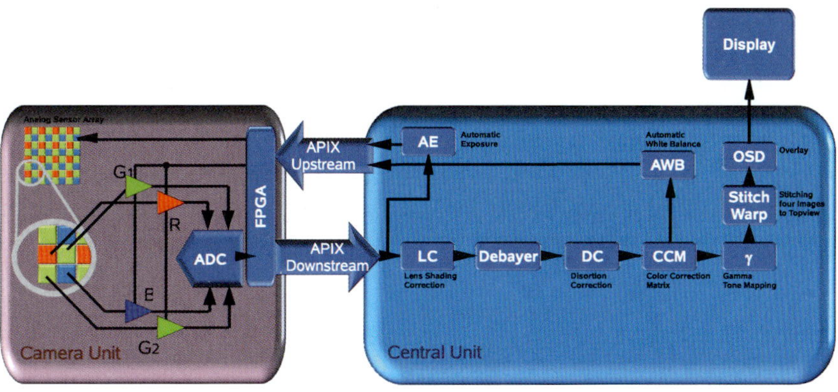

Fig. 10.   Data flow of the topview system using APIX

In detail the set up data flow from central unit to camera consists of the following steps:

- ▶ Create set up data for the cameras, e. g. calculate the exposure time that fits best for the actual scene
- ▶ Transmit the set up data via APIX upstream channel to the camera
- ▶ The FPGA in the camera configures the image sensor and captures an image

The image data flow from the camera back through the central unit to the display consists of these functions:

- ▶ The FPGA in the camera receives the raw image data from the sensor
- ▶ The FPGA prepares the raw data for transmitting and sends them via APIX downstream channel
- ▶ The central unit receives the raw data and starts to process them
- ▶ Analysing the image data regarding the exposure, calculation of the exposure time for the next image to capture
- ▶ Lens shading correction
- ▶ Debayering (only in color mode)
- ▶ Distorting correction
- ▶ Color processing (only in color mode, color correction, auto white balancing (equal white balance for all four images)
- ▶ Tone mapping, gamma correction to reduce the data to 8 bit per pixel
- ▶ Stitching and warping the images to get the resulting topview image
- ▶ Overlay driveway and additional information (optional)
- ▶ Displaying the resulting image in topview

Most of the work needs to be performed by the central unit, so the cameras can be very small and tiny. Besides the optimized processing and design, lower camera cost will also reduce the cost of repair to the owner of the car in case of damage due to an accident, since they typically need to be mounted within bumpers or mirrors.

Also a very important requirement is the latency between capturing and displaying the image, which needs to be kept at a minimum. Since only FPGA's or optimized hardware implementations are able to meet this requirement, the solution above has been described using FPGA's for the time critical path.

4   Summary

Automotive camera based driver assistance and safety applications require advanced video technology. These systems are able to provide higher comfort and safety to the driver. The topview system is the ideal application to assist the driver while maneuvering the vehicle and also reduces the risk of damages or even accidents.

The APIX transmission technology enables the developer to create cost effective and powerful solutions. APIX licensing offers the possibility to create multi-interface ASICs or even to design a one-chip-camera with integrated APIX interface. However, even with a dedicated APIX transmitter device the system costs are reduced by using only one cable for video data, control data and power supply.

Markus Römer
Inova Semiconductor GmbH
Grafinger Str. 26
81671 München
Germany
mroemer@inova-semiconductors.de

Thorsten Heimann
Helion GmbH
Bismarckstr. 142
47057 Duisburg
Germany
thorsten.heimann@helionvision.com

Keywords: automotive pixel link, APIX, camera, driver assistance, real-time video interface, topview, image stitching, HDR

Components and Generic Sensor Technologies

Miniaturised Sensor Node for Tire Pressure Monitoring (e-CUBES)

K. Schjølberg-Henriksen, M. M. V. Taklo, N. Lietaer, SINTEF
J. Prainsack, M. Dielacher, Infineon Technologies
M. Klein, M. J. Wolf, J. Weber, P. Ramm, Fraunhofer IZM
T. Seppänen, Infineon Technologies SensoNor

Abstract

Tire pressure monitoring systems (TPMS) are beneficial for the environment and road and passenger safety. Miniaturizing the TPMS allows sensing of additional parameters. This paper presents a miniaturized TPMS with a volume less than 1 cm^3, realised by 3D stacking and through-silicon via (TSV) technology. Suitable technologies with low electrical resistance and high bond strengths were evaluated for stacking the microcontroller, transceiver, pressure sensor and bulk acoustic resonator (BAR) in the TPMS. 60 µm deep W-filled TSVs with resistance 0.45 Ω and SnAg micro bumps with a bond strength of 53 MPa were used for stacking the transceiver to the microcontroller. TSVs through the whole wafer thickness with resistance 6 Ω were used for the pressure sensor. Au stud bumps were used for stacking the pressure sensor and BAR devices. The final TPMS stack was packaged in a moulded interconnect device (MID) package.

1   Introduction

Tire pressure monitoring systems (TPMS) are beneficial for minimised fuel consumption and optimised road and passenger safety. Driving with correctly inflated tires can reduce the fuel consumption by 2-6% [1], which again will reduce greenhouse gas emissions substantially. The reduced emissions are very favourable for the environment. In addition, incorrectly inflated or damaged tires can result in a variety of safety risks. Under-inflated tires are more prone to stress damage, have less lateral traction, a shorter tread life, and are more vulnerable to flat tires and blow outs. Furthermore, under inflated tires can increase the distance required for a vehicle to stop, especially if the surface is wet. With a TPMS, the driver is notified in case of erroneous tire pressure, and the dangerous situations above can be avoided.

The design of TPMS modules is very challenging. As the battery can not be changed once the system is installed in the tire, TPMS sensor nodes must have the lowest power consumption possible. New developments try to make use of energy scavenging systems, which gain energy from vibrations. Another challenge lies in the environmental conditions like a very high temperature range (– 40. . .125 ^{o}C) and mechanical stress, which can reach values up to 1000 g (tire dimension: 235 = 45R17, speed: 250 km/h) only considering static acceleration being induced by the circular movement [2].

Today's TPMS have a volume above 20 cm^3 and must be mounted either strapped on or attached to a valve of the tire. Attaching the node on the inner liner of a tire instead will allow sensing of important additional parameters such as road condition, tire wear out, temperature, tire friction, side slip, wheel speed, and vehicle load. The information may be used for improved tracking and engine control. However, to mount the TPMS on the inner liner, a substantial miniaturisation down to a volume less than 1 cm^3 is desired. Fig. 1 shows a possible placement of a miniaturised TPMS on the inner liner of the tire, and the communication of measurement data to the vehicle's electronic control unit.

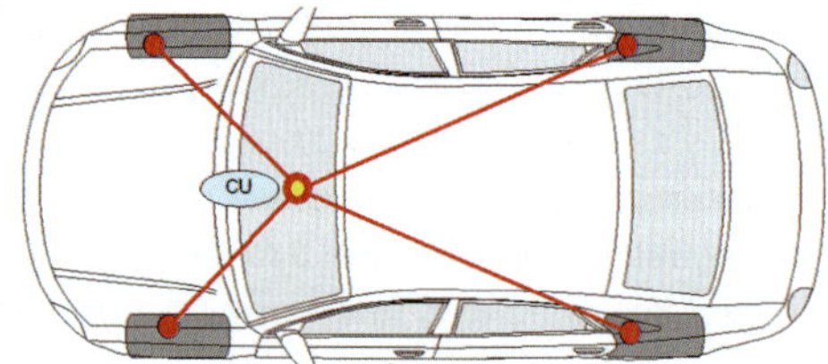

Fig. 1.   Left: possible placement of miniaturised TPMS on the inner liner of the tire. Right: Measurement data from all tires of the vehicle will be communicated to the electronic control unit.

Miniaturisation of sensor nodes for wireless sensor networks is the aim of the EU-funded e-CUBES project [3]. A TPMS is one of the demonstrators in the project. Fig. 2 shows the proposed, miniaturised TPMS consisting of a micro-controller, a transceiver, a bulk acoustic resonator (BAR) and a pressure sensor. To realise the TPMS in Fig. 2, the different chips of the TPMS must be stacked on top of each other as bare dice. Interconnections between the dice are realised by a combination of various through-silicon-via (TSV) and interconnection technologies. The stacking is referred to as 3D integration.

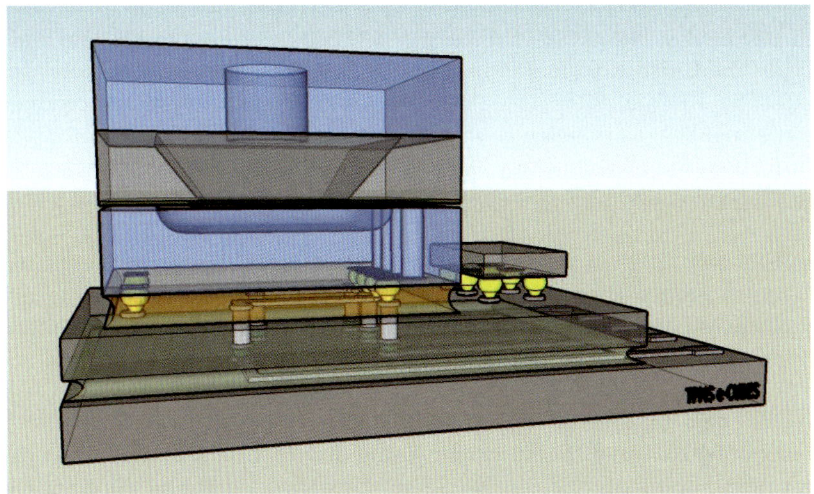

Fig. 2.   The TPMS has a microcontroller at the bottom, and a transceiver
in the middle layer. On top of the transceiver, the BAR is placed to
the left and the pressure sensor to the right in the sketch.

TSVs and interconnection technologies are the two main requirements for 3D
integration. Within the CMOS community and its sub suppliers, substantial
research has been carried out with the objective of realizing 3D integrated
chips. Several suppliers provide TSV and interconnection technology services,
and a variety of materials is available. Copper is one of the most popular materials. Tezzaron's SuperVias and FaStack [4], Rensselaer Polytechnic Institute
(RPI) [5], Fraunhofer IZM [6], Infineon [7], IBM [8], and IMEC [9] all use Cu as
conductive material in the TSVs. Tezzaron [4], IMEC [9], and MIT [10] also use
Cu-Cu bonding to realise the conductive interconnections, whereas RPI uses
Cu interconnections in combination with partially cured BCB [4]. Infineon and
Fraunhofer IZM have investigated Cu-Sn-Cu bonding called the SLID technology [11]. Micro bumps may also be used to realise interconnections, as pre-
sented by IBM [8], Fraunhofer IZM [12] and IMEC [9].

W is another interesting material for conductive material in the TSVs, and has
been used by Fraunhofer IZM [6] and ZyCube [13, 14]. Cu-Sn-Cu SLID technology may also be combined with W-TSVs [5]. ZyCube applies In-Au micro bumps
for realising the interconnections between the TSVs [14].

The TSVs in reference [4-14] all require thinning of the silicon wafer. Thinning
may not be a viable option for devices with mechanical structures, like pressure sensors and resonators. A few via technologies suitable for wafer thick-
nesses of 200-300 µm are presented in the literature. PlanOptik [15] produces
glass wafers with conductive silicon vias. Silex [16] manufactures silicon wafers

with silicon vias, isolated by a dielectric. SINTEF has presented hollow vias in 300 μm thick silicon wafers where the conductive layer is doped polysilicon isolated by thermal SiO_2 [17]. Filled vias with doped polysilicon conductor and thermal SiO_2 as isolation have been presented by groups at Stanford [18], Georgia Institute of Technology [19], and VTT [20]. TSVs through thick wafers with Cu conductors is also possible, and have been published by DIMES [21].

The TSVs through thick wafers can be combined with the interconnection technologies presented above: Cu-Cu bonding with or without BCB, SLID technology, or micro bumps. Other alternatives are the Ni-based conductors as offered by Ziptronix' DBI process [22], and conventional Au stud bump bonding.

The four chips included in the 3D integrated stack of the TPMS shown in Fig. 2 are manufactured on different wafers, using different technologies. The differences in requirements and considerations to be taken called for different technologies for TSVs and interconnections for each chip. For each chip, various technologies were evaluated before a final choice was made for each layer. This paper describes the evaluated alternatives and the final technology choices. Structures for process control monitoring (PCM) of each stacking procedure were designed so that the performance of each procedure could be tested sequentially. The assembly and packaging of the full TPMS is also presented.

2   TPMS Building Blocks

In the TPMS shown in Fig. 2, the transceiver, pressure sensor, and BAR wafers are diced and stacked on top of the microcontroller. The microcontroller was kept as a complete wafer level throughout the 3D-stacking process, serving as a handle wafer. The assembly of the TPMS was performed using chip-to-wafer bonding technologies.

2.1   Microcontroller Chip

The microcontroller chips were delivered as fully processed and functional tested wafers with diameter 200 mm and thickness 0.7 mm. After analysing the topography at critical regions like the scribe line and coil regions by atomic force microscopy (AFM), the wafers were planarized by deposition of sub-atmospheric chemical vapour deposition (SACVD) oxide and etch-back. The pads of the microcontroller chips were opened, and an AlSiCu re-distribution layer metallization was deposited by a physical vapour deposition (PVD) process. The structured re-distribution layer was passivated by deposition of an oxide/nitride layer stack. The chip itself plus area needed for stack assembly and dicing resulted in a footprint of 4.0×3.6 mm^2.

SnAg micro bumps were formed on wafer scale by electroplating. The bumps were plated on Cu under-bump metallization (UBM) pads with diameter of 26 µm. Barrier and seed layers of 200 nm TiW and 300 nm Cu were sputtered on the whole wafer surface. The bump areas were defined in photo resist and 6 µm thick Cu and eutectic SnAg solder was electroplated. Finally, the photo resist and seed layers were removed.

A total number of 621 bumps were deposited on each microcontroller chip. Only 57 of these bumps were needed as electrical interconnects. The remaining 564 bumps were deposited to ensure mechanical stability of the TPMS stack (see Fig. 2). Bumps had to be placed on the microcontroller in the areas where the pressure sensor and BAR devices were connected to the transceiver chip, and in the areas where the transceiver chip was to be wire bonded. This way, there would be metal between the microcontroller and transceiver back side underneath all the transceiver top side pads that would experience mechanical pressure.

AuSn could have been an alternative solder material for the electroplated bumps. However, SnAg has lower melting point, more ductile behaviour, and lower yield stress than AuSn [23]. Therefore, SnAg solder bumps were preferred.

2.2 Transceiver Chip

The transceiver ASICs had a footprint of 3.8×3.3 mm^2 and were fabricated on 200 mm diameter wafers. A "Post Backend-of-Line (BEOL) Via First" approach was applied for the TSVs, meaning that the TSVs were fabricated in predefined areas on the circuit after the standard manufacturing of the ASIC.

TSVs areas of 3×10 µm^2 were defined with accuracy better than 300 nm using an i-line stepper. The TSV structures were etched through the intermetal dielectric layers, consisting of several isolation layers with different layer composition and an overall thickness of about 8 µm, using a special developed etch sequence on different equipments. Subsequently, 60 µm deep reactive ion etching (DRIE) was performed by an STS Pegasus tool, realising TSVs with an aspect ratio 20:1. The TSVs were isolated by depositing a SACVD oxide. Conductive interconnects were formed applying a W chemical vapour deposition (CVD) process including a thin TiN seed layer. A cross-section of the TSVs is shown in Fig. 3. For this application, the system requirements and overall reliability aspects for the MEMS integration recommended the use of W metallization instead of Cu, which is also commonly used [24].

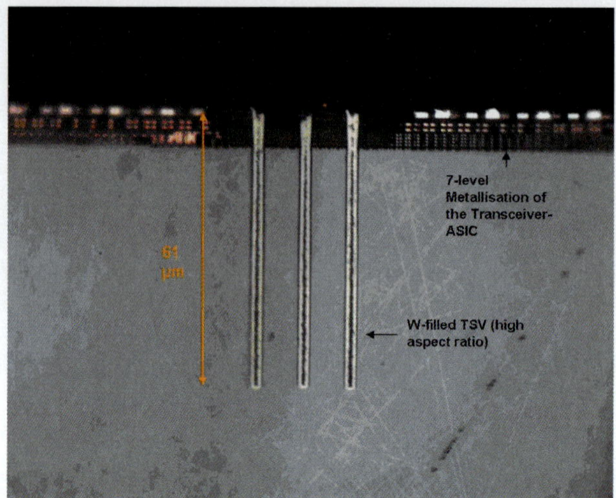

Fig. 3.   Cross section of W-filled TSVs of the transceiver ASIC.

After manufacturing the TSVs, a metal re-distribution layer was made on the front side of the wafer. Selected original pads of the transceiver ASIC were opened, and an AlSiCu-metallization was deposited by PVD and structured. The re-distribution layer hence defined the areas for placing the pressure sensor and BAR devices. This is illustrated in Fig. 4. The front side re-distribution layer was protected by an oxide/nitride layer stack and the bond pads were opened by dry etching.

The transceiver wafers were glued by polymer to handle substrates and ground down to a thickness of about 100 µm. After grinding, they were spin-etched down to 60 µm thickness, and a short CMP step for additional thinning close to the W plugs was applied. A dry etching step using the STS Pegasus finally uncovered the metallization of the TSVs on the back side of the wafers. Then a re-distribution metal layer was patterned, sandwiched between isolating oxide layers which were planarized by CMP. Vias for accessing the signal pads were etched in the upper oxide. Finally, 5 µm Cu bumps were electroplated as UBM for the SnAg flip chip bonding process of the transceiver chip to the microcontroller.

A PCM for the TSVs was integrated into the transceiver ASIC. The PCM was constructed as a daisy chain of eight TSVs, which were connected alternately by the front and back side re-distribution metal layers. This PCM was measured both before and after TPMS stack assembly. A second simple PCM structure for the transceiver-microcontroller stacking was integrated as a single daisy chain link using an array of three parallel connected TSVs and one micro bump for the signal path from the front side of the transceiver ASIC to the redistribution layer of the microcontroller chip.

The transceiver chip contained the bond pads for all required wire bonds. Wires needed to the microcontroller had been rerouted to the transceiver. In order to keep the footprint as small as possible, the bond pads were arranged so that all wire bonds could be made on the same side of the TPMS stack, and not on all four sides.

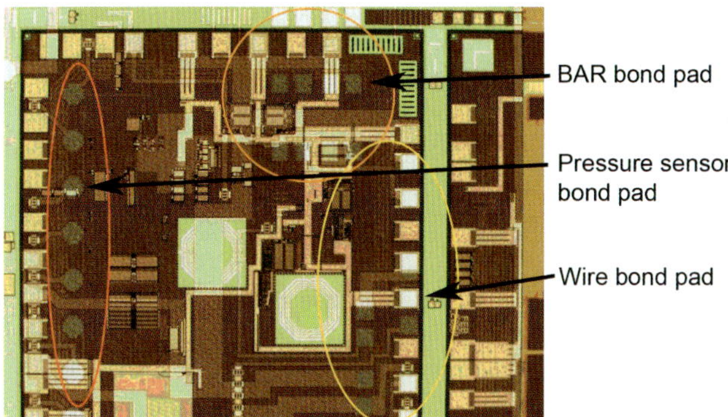

Fig. 4.　Top view of section of transceiver ASIC after processing of TSVs and patterning of redistribution layer. Examples of bond pads for the sensor (red circle), BAR (orange circle), and wire bonds (yellow circle) are indicated.

2.3　Pressure Sensor Chip

The pressure sensor consisted of 3 pcs. 150 mm diameter wafers which were bonded together in a glass-silicon-glass triple-stack. The thickness of the total stack was about 1 mm. The chip had a footprint of 1.8×2.1 mm^2. The cap wafer was a silicon-glass compound wafer from Planoptik [15]. The electrical signal from the sensor was routed through silicon pins insulated by the glass of the cap wafer. Further, the signal was routed through aluminium lines to contact pads patterned on the top surface of the triple stack. A pressure inlet was present on the backside of the triple stack, thus the sensor chips were flipped in the stacking procedure to be able to interact with the environment. The inlet could not see any wet processing, which limited the possibility of post processing of the sensor stack. Special precautions had to be made when the aluminium lines and contact pads were patterned on the top surface of the stack.

An alternative to the chosen TSV solution with a silicon-glass compound wafer could have been tungsten-glass compound wafers as commercially available from Schott [25]. Such wafers were not available early enough for this demonstrator and silicon pins were anyway suitable due to only moderate conductivity demands. If a silicon cap wafer could be acceptable, another alternative could be through silicon vias (TSVs) insulated by dielectrics as offered by Silex [26]. A hermetic seal was required for the reference cavity in the cap wafer of the pressure sensor, and for this application a glass wafer was considered as easier to hermetically bond to the sensor wafer.

Suitable technologies for connecting the pressure sensor to the TPMS were evaluated [27, 28]. Three possible solutions were identified: plated micro bumps, SLID interconnects using Sn and Cu, and Au stud bump bonding. Since plating of the wafers would normally require wet processing, incompatible with the open pressure inlets, Au stud bump bonding was selected. The chosen technology was ideal for the pressure sensor with its low I/O count. Gold stud bumps named "accubumps" [29] with an average diameter of 52 μm were placed on the aluminium pads on the top surface of the triple stack in order to prepare the chips for stud bump bonding (SBB). A total of 32 bumps were distributed along the perimeter of each chip. Only 10 bumps were required for electrical connection, but dummy bumps were added for mechanical stability during flip chip bonding and for process monitoring purposes. A daisy chain of 16 stud bumps was prepared as a PCM structure on every chip by interconnecting every second bump with aluminium routing. The PCM was intended for probing after flip chip bonding. The strength of the SBB was assured by dispensing the adhesive 353ND (Epo-Tek) underneath each pressure sensor chip.

Fig. 5.   The PCM structure on the sensors consisting of 16 bumps in a daisy chain, as indicated by the white box and arrow. The 10 bumps on the other long end of the sensor were used for electrical interconnections to the sensor (Photo: SINTEF).

2.4 Bulk Acoustic Resonator Chip

The BAR was fabricated on 150 mm diameter wafers of thickness 200 µm. The chip had a footprint of 0.8×1.3 mm^2. The material on the I/O pads of the BAR was Au. A total of 16 accubumps with an average diameter of 47 µm were placed on each BAR chip. Two of the bumps were needed for electrical connection whereas the remaining bumps assured a certain mechanical strength of the SBB. There were no designated PCM structures included for the BAR. No underfiller was permitted under the BAR as it would deteriorate its functionality. Even without underfiller, the strength of the SBB compared to the form factor of the chip was sufficient for all subsequent handling, including wafer dicing.

3 Assembly and Packaging

3.1 Building Block Assembly

The transceiver chips, still glued to the handle wafer, were diced according to the overall stacking scheme and assembled to the microcontroller wafer using a flip chip bonder. Soldering was carried out in an oven in activated atmosphere without using liquid flux applying a profile which was especially developed for SnAg solder. The handle chips were removed by a wet chemistry process and underfiller was applied. The underfiller serves two purposes. It is necessary to prevent any residues between the microcontroller and transceiver due to the final sawing of the microcontroller wafer. The underfiller also decreases the stress of the interconnections during the following assembly steps of the module, i.e. bonding of the pressure sensor and BAR devices as well as wire bonding.

An appropriate underfill material was selected for the about 25 µm gap between microcontroller and transceiver chips. After underfilling the pads were cleaned and conditioned by a dry water strip process to remove any residues from the gluing of the handling chip. The mechanical strength of the SnAg bumps was tested by shear testing. A picture of the transceiver ASIC stacked on top of the microcontroller after applying the underfill is seen in Fig. 6.

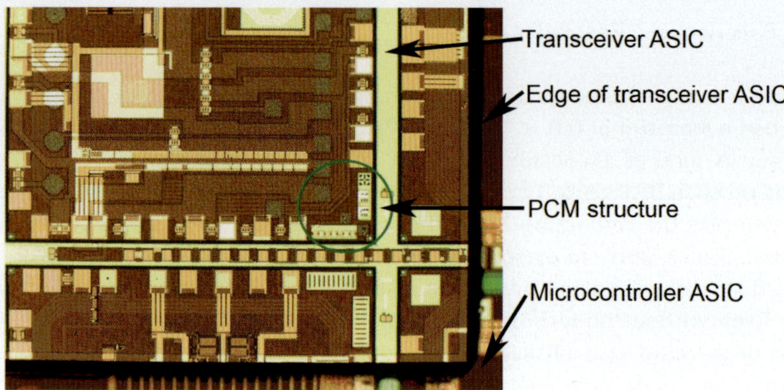

Fig. 6.   Top view of section of the transceiver ASIC stacked on top of the microcontroller wafer. The integrated PCM structure is seen inside the green circle.

The bumped sensor and BAR devices were chip-to-wafer bonded onto the transceiver chips already mounted on the microcontroller wafer. The bonding was performed by Datacon [30]. The bonding parameters had been adjusted in advance on a large number of dummy devices and specially designed test substrates. An example of a test wafer substrate with bonded pressure sensors and BAR devices is seen in Fig. 7.

Thermal cycling, cross sectioning and shear tests had been performed to optimize the bonding procedure [28, 31, 32]. Both thermo compression and thermo sonic bonding was utilized. The chuck temperature was always set to 120-180°C. For thermo compression bonding the tool was heated to 200°C, a tool force of 20-30 N was applied and the bonding lasted for 10 seconds. For thermo sonic bonding the tool was kept at room temperature, the tool force was reduced to 12-20 N and the time was reduced to 2 seconds. The reduced bonding parameters were compensated by the added ultrasonic energy. For both bonding methods, a droplet of 353ND (Epo-Tek) adhesive was dispensed on top of the transceiver chips before bonding the sensors.

Some sensor and BAR devices bonded without adhesive were sheared off during set-up to inspect chip alignment and the fracture surfaces. The resistance of the electrical interconnects were measured on test substrates and TPMS stacks. Pads for manual probing were accessible on the surface of the transceivers.

Final dicing was performed by Infineon Technology SensoNor. Special precaution had to be made as the total height of the stack was several millimetres.

Additionally, the pressure inlets had to be protected to avoid slurry entering the device.

Fig. 7. Pressure sensors and BAR devices bonded onto a dummy transceiver wafer for process optimization of the SBB process (Photo:
SINTEF).

3.2 Moulded Interconnect Device (MID) packaging

The TPMS stacks were mounted onto a four layer printed circuit board (PCB) consisting of Rogers RO4450B material which was chosen because of its suitability for RF applications. The size of the PCB is about 11 mm x 10 mm, and its top-level metallization is shown in Fig. 8. The large area in the upper right corner is the footprint of the TPMS stack. Besides the TPMS stack, the PCB only carried a few capacitors required for voltage regulators and for tuning of the antenna matching network.

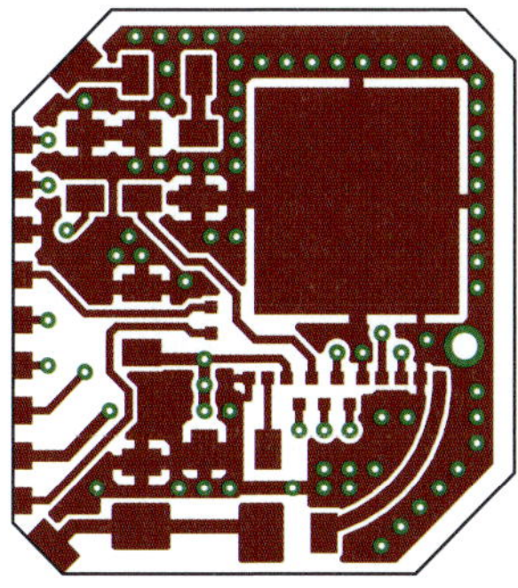

Fig. 8. Top-level metallization of the PCB for the TPMS.

The 3D integrated TPMS stack was assembled together with additionally required parts of a complete TPMS in a three dimensional substrate called a moulded interconnect device (MID). An MID is a plastic part which undergoes a selective surface metallization process after moulding or milling in order to create conductor lines as well as solderable and/or bondable pad surfaces. The MID for the presented demonstrator took advantage of a laser direct structuring method. The surface of the plastic part was selectively activated by a laser beam making it susceptible for subsequent electroless CuNiAu-plating [33]. The 2.4 GHz antenna [34, 35] was also fabricated by this process, placing it directly inside the MID.

The PCB was placed in the MID and electrically connected to the pads using conductive adhesive. Two buffer capacitors to provide the required peak current for the transceiver were mounted onto the PCB, and the MID lid held two small coin cell batteries. By connecting the batteries with a stamped metal spring, unavoidable dimensional height tolerances were accounted for. The lid did not need to undergo a complicated metallization process but could be manufactured as a simple plastic part. The overall concept is illustrated in Fig. 9.

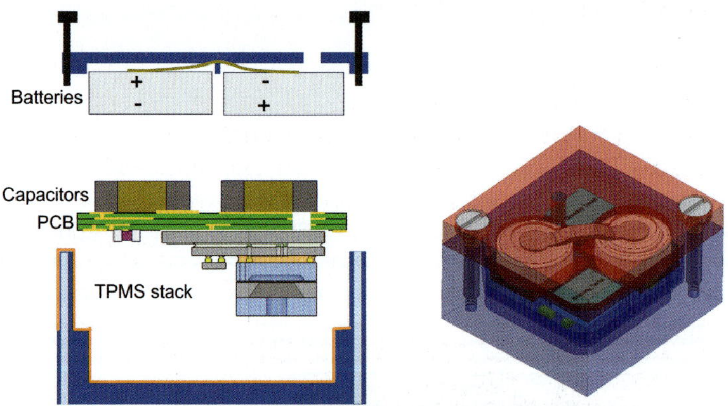

Fig. 9.   Left: Basic concept for the packaging of the 3D integrated stack inside an MID. Right: A 3D model of the package. Screws enable changing of batteries in the test phase. Dimensions of the cube (in cm^3) are 1.2 x 1.3 x 0.64 = 0.998.

4   Results and Discussion

4.1   Microcontroller-Transceiver Stacking

The TSVs were tested after post-processing of the transceiver wafers. Daisy chains of 6 W-filled TSVs were measured from the back side of the wafer. Typical DC electrical resistance for these daisy chains was 2.7 Ω, suggesting 0.45 Ω per TSV. These values were obtained on more than 220 devices from three different wafers, and show that the yield and process repeatability are both highly satisfactory.

After stacking the transceiver on top of the microcontroller, the daisy chain consisting of 8 W-filled TSVs could be measured. Typical DC electrical resistance for these daisy chains was 3.6 Ω, again indicating 0.45 Ω per TSV. The PCM for testing the connection between the microcontroller and the transceiver was also measured. 77 devices were measured, and the mean value for the DC resistance between the front side of the transceiver and the redistribution layer on the microcontroller was 0.47 Ω.

Fig. 10 shows the cross section of a reflow soldered transceiver chip with the glued-on handling chip bonded to the microcontroller wafer. The mechanical strength of the SnAg bumps was tested by shear testing. Seven transceiver chips were shear tested. The average force was 1742 $\pm$ 92 cN, corresponding to a bond strength of 53 MPa. The bumps had fractured within the solder, indicating good wetting of the SnAg solder to the Cu UBM. Fig. 11 shows the top view of an underfilled transceiver chip assembled on the microcontroller wafer.

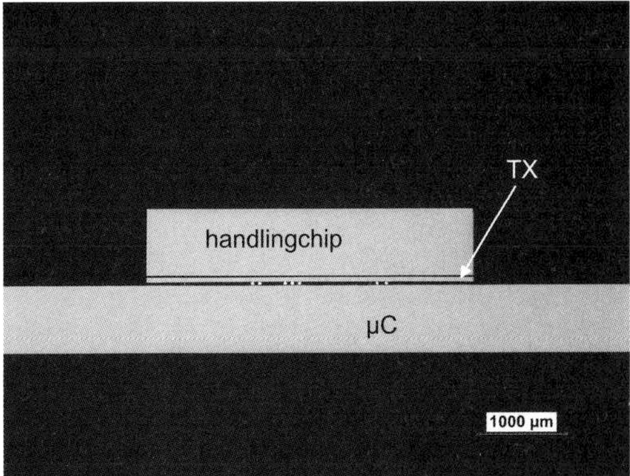

Fig. 10.   Cross section of a reflow soldered transceiver chip with handling chip on the microcontroller wafer.

Fig. 11. Light microscopy image of an underfilled transceiver chip. The underfiller is seen as the thin dark line around the transceiver chip.

4.2 Pressure Sensor and BAR stacking, and Final Dicing

The Au SBB process was approved when the sheared, inspected bumps showed fracture within the bumps rather than within the chips or at the bump/chip interface. The stand-off height between the chip and substrate after bonding was inspected by cross sectioning comparable dummy devices having seen the same bonding parameters. The stand-off height was between 10 and 15 µm.

The resistance between the pressure sensor and transceiver chip was measured on 25 sensors bonded to a dummy transceiver wafer as shown in Fig. 7. The average contact resistance was 6 Ω. When probing the PCM structures for the sensors on the TPMS stacks, a contact resistance in the range of a 6 Ω or less was therefore considered as an indication of a well aligned and reliable bonding.

A total of 58 TPMS stacks with pressure sensor and BAR devices were successfully mounted, distributed on two different microcontroller wafers. Half of the stacks had pressure sensors and BARs bonded with thermo compression, and the rest had thermo sonically bonded pressure sensors and BARs.

Both microcontroller wafers with completely mounted TPMS stacks were successfully diced. When dicing the first wafer, the standard setting for the water jet proved to be too high for the relatively weakly bonded BAR devices. This was probably because the BAR did not have any adhesive that improved the mechanical bond strength. The water force was reduced for the following wafers, letting the BARs remain in place during dicing as shown in Fig. 12.

Fig. 12. The 3D integrated stack for the automotive demonstrator before adjusting the dicing parameters (left, BAR was removed; Photo: SINTEF) and after process optimization (right, BAR in place; Photo: SensoNor).

4.3 MID packaging and full TPMS

Fig. 13 shows an empty MID and its lid with complete metallization. A step designed to carry the PCB can be seen around the inside-walls of the MID. On one side of this step, there are metal contacts from which circuit tracks lead to the outside. On the PCB (shown in Fig. 8) one can also see a row of metal contacts which exactly matches the contacts inside the MID. The 2.4 GHz antenna is seen as a larger metallized loop along the bottom of the MID in Fig. 13. Using the laser direct structuring method proved to be a very cost-efficient method for manufacturing the antenna. Fig. 14 shows an assembled TPMS sensor node with the PCB mounted into the MID.

Fig. 13. Empty MID package and its lid.

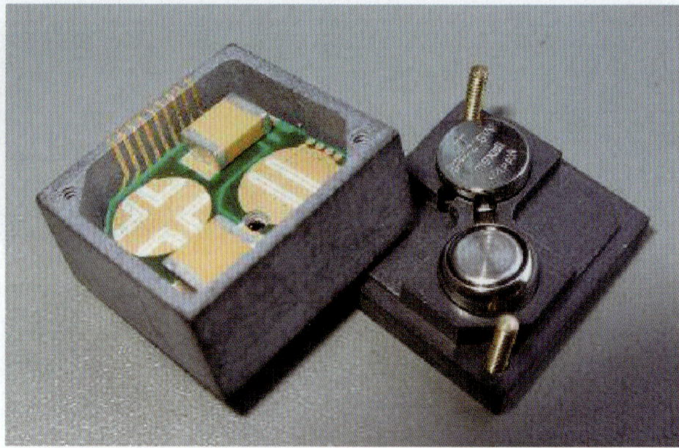

Fig. 14. PCB mounted into MID. On the backside of the PCB one can see the buffer capacitors as well as the contact areas for the batteries. The hole which goes through the lid of the MID and through the PCB is the pressure inlet for the sensor.

5   Conclusions

Through-silicon via (TSV) and bonding technologies for miniaturizing a tire pressure monitoring system (TPMS) have been presented. With these technologies, a TPMS volume less than 1cm^3 was obtained. The applied TSV technologies showed sufficiently low resistance. The selected bonding technologies had both high bond strengths and low electrical resistance. The final TPMS stack was packaged in a moulded interconnect device (MID) package. Testing of the full packaged TPMS is underway. So far, the mechanical and electrical results are promising and show that miniaturized TPMSs realised by 3D integration are feasible.

References

[1]   Pearce, J.M., Hanlon, J.T., "Energy conservation from systematic tire pressure regulation", Energy Policy, Vol. 35, pp. 2673-2677, 2007.

[2]   Fischer, M., "Tire Pressure Monitoring", Verlag Moderne Industrie, 2003.

[3]   www.ecubes.org

[4]   Gupta, S., Hilbert, et al. ""Techniques for producing 3D ICs with high-density interconnect", Proceedings of the 21st International VLSI Multilevel Interconnection Conference, Waikoloa Beach, HI, USA, September 2004.

[5] Gutmann, R.J., et al. "Damascene-patterned metal-adhesive (Cu-BCB) redistribution layers", MRS Symposium Processings, Vol 970, pp. 205-214, 2007.

[6] Ramm, P., Klumpp, et al. "3D system integration technologies", MRS Symposium Processings, Vol 766, pp. 3-14, 2003.

[7] Benkart, P. et al. "3D chip stacking technology using trough-chip interconnects", IEEE Design and Test of Computers, Vol. 22, pp. 512–518, 2005.

[8] Knickerbocker, et al. "Development of next-generation system-on-package (SOP) technology based on silicon carriers with fine-pitch chip interconnection", IBM Journal of Research and Development, Vol. 49, pp. 725-753, 2005.

[9] deMoor, P., "3D integration technologies for imaging applications", Nuclear Instruments and Methods in Physics Research A, Vol.591, pp. 224-228, 2008.

[10] Tan, C.S., et al. "Silicon layer stacking enabled by wafer bonding", MRS Symposium Processings, Vol 970, pp. 193-204, 2007.

[11] Klumpp, A., Merkel, R., Ramm, P., Weber, J., Wieland, R., "Vertical system integration by using inter-chip vias and solid-liquid interdiffusion bonding", Japanese Journal of Applied Physics Part 2, Vol. 43, Issue 7a, pp. L829-L830, 2004.

[12] Wolf, M.J., et al. "3D Process Integration – Requirements and Challenges", To be published in MRS Symposium Proceedings, Vol 1112.

[13] Bonkohara, M., et al. "Current and future three-dimensional LSI integration technology by chip-on-chip, chip-on-wafer and wafer-on-wafer", MRS Symposium Processings, Vol 970, pp. 35-45, 2007.

[14] Park, K.T., et al. "A 3-dimensional wafer-level stacking technology with precise vertical interconnections to MEMS applications", Proceedings of Transducers'01/ Eurosensors XV, Digest of technical papers, pp. 1590-1593, Munich, Germany, June 2001.

[15] http://www.planoptik.com/produkte/produkte_en.html

[16] http:/www.silex.com.au

[17] Lietaer, N., Storås, P., Breivik, L., Moe, S., "Development of cost-effective high-density through-wafer interconnects for 3D microsystems", Journal of Micromechanics and Microengineering, Vol. 16, pp.S29-S34, 2006.

[18] Chow, E.W., Chandrasekaran, V., Partridge, A., Nishida, T., Sheplak, M., Quate, C. F., Kenny, T.W., "Process compatible polysilicon-based electrical through-wafer interconnects in silicon substrates", Journal of Microelectromechanical systems, Vol. 11, pp. 631- 640, 2006.

[19] Ok, S. J., Kim, C., Baldwin, D.F., "High density, high aspect ratio through-wafer electrical interconnect vias for MEMS packaging", IEEE Transactions on advanced packaging, Vol. 26, pp. 302 - 309, 2003.

[20] Ji, F., Leppävuori, et al. "Fabrication of silicon based through-wafer interconnects for advanced chip scale packaging", Proceedings of Eurosensors XX, Contribution M3C-P1, Gothenburg, Sweden, 2006.

[21] Tian, J., et al. "RF-MEMS wafer-level packaging using through-wafer interconnect", Sensors and Actuators A, Vol. 142, pp. 442-451, 2008.

[22] Enquist, P., "High density bond interconnect (DBI) technology for three dimensional integrated circuit applications", MRS Symposium Processings, Vol 970, pp. 19-24, 2007.

[23] Klein, M., et al. "Development and Evaluation of Lead Free Reflow Soldering Techniques for the Flip Chip Bonding of Large GaAs Pixel Detectors on Si Readout Chip", Proceedings of the 58th ECTC, pp. 1893-1899, Orlando, Florida, 27-30 May 2008.

[24] Ramm, P., et al. "Through Silicon Via Technology –Processes and Reliability for Wafer-Level 3D System Integration", Proceedings of the 58th ECTC, pp. 841-846, Orlando, Florida, 27-30 May 2008.

[25] http://www.schott.com/epackaging/english/auto/others/hermes.html

[26] http://www.silexmicrosystems.com/pages/

[27] Taklo, M.M.V., et al. "Technologies enabling 3D stacking of MEMS", Proceedings of the IEEE workshop on 3D System Integration, München, 1-2 October 2007.

[28] Lietaer, N., Taklo, M.M.V., Klumpp, A., Ramm, P., "3D Integration Technologies For Miniaturized Tire Pressure Monitor System (TPMS)", To be presented at IMAPS 5th International Conference and Exhibition on Device packaging, Scottsdale, Arizona, 10-12 March 2009.

[29] www.kns.com

[30] Datacon semiconductor, Innstrasse 16, A-6240 Radfeld, Austria

[31] Ramm P., et al. "3D Integration Technologies for Wireless Sensor Systems (e-CUBES)", IMAPS Proceedings of 4th International Conference and Exhibition on Device packaging, Scottsdale, Arizona, 17-20 March 2008.

[32] Taklo, M.M.V., et al. "MEMS Sensor/IC Integration for Miniaturized TPMS (e-CUBES)", Proceedings of SEMATECH workshop: Manufacturing and reliability challenges for 3D ICs using TSVs, San Diego, California, 25 - 26 September 2008.

[33] Wissbrock, H., "Laser direct structuring of plastics - A new addition to MID technologies", Kunststoffe-Plast Europe, Vol. 92, p. 101, 2002.

[34] Cheng, S. et al. "Gain and Efficiency Enhanced Flip-Up Antennas for 3D Integrated Wireless Sensor Applications", 2nd European Conference on Antennas and Propagation (EuCAP 2007), Edinburgh, UK, 11-16 Nov 2007, pp. 1-5.

[35] Cheng, S., et al. "Inverted-F Antenna for 3D Integrated Wireless Sensor Applications" 3rd IEEE International Workshop on Antenna Technology, Cambridge, England, 21-23 March 2007, pp. 447-450.

Kari Schjølberg-Henriksen, Maaike Margrete Visser Taklo, Nicolas Lietaer
SINTEF
Department of Microsystems and Nanotechnology
P.O. Box 124, Blindern
0314 Oslo
Norway
Kari.Schjolberg-Henriksen@sintef.no
MaaikeMargrete.VisserTaklo@sintef.no
Nicolas.Lietaer@sintef.no

Josef Prainsack, Markus Dielacher
Infineon Technologies Austria AG
Babenberger Straße 10
8020 Graz
Austria
Josef.Prainsack@infineon.com
Markus.Dielacher@infineon.com

Matthias Klein, Jürgen Wolf
Fraunhofer Institut Zuverlässigkeit und Mikrointegration
Gustav-Meyer-Allee 25
13355 Berlin
Germany
Matthias.Klein@izm.fraunhofer.de
Juergen.Wolf@izm.fraunhofer.de

Josef Weber, Peter Ramm
Fraunhofer Institut Zuverlässigkeit und Mikrointegration
Hansastraße 27d
80686 München
Germany
Josef.Weber@izm-m.fraunhofer.de
Peter.Ramm@izm-m.fraunhofer.de

Timo Seppänen
Infineon Technologies SensoNor AS
P.O. Box 196
3192 Horten
Norway
Timo.Seppanen@infineon.com

Keywords: MEMS, tire pressure monitoring system, 3D integration, through-silicon via

Development of a Single-Mass Five-Axis MEMS Motion Sensor

R. M. Boysel, L. J. Ross, Virtus Advanced Sensors Inc.

Abstract

MEMS (Micro-Electro-Mechanical Systems) are being included in an increasing number of applications in automotive systems, particularly because of their small size and cost. However, most automotive applications require more information than can be provided by a single MEMS motion sensor. We report on the development of a MEMS motion sensor that can detect acceleration along three axes and angular velocity around two or three axes using a single proof mass, enabling the integration of multiple applications into a single device. The sensor consists of four support springs patterned into the device layer of an SOI (Silicon on Insulator) wafer and a single proof mass formed by etching through the handle layer using DRIE (Deep Reactive Ion Etching). We present the fabrication and packaging process and modeling results for the sensor.

1   Introduction

Much has been written about the growth of MEMS (Micro-Electro-Mechanical Systems) sensors in automotive safety and performance systems.[1,2,3] MEMS sensors have been adopted not only because of their low cost and small size, but also for their performance and reliability. The automotive industry has been fertile territory for MEMS motion sensors with higher functionality (i.e. degrees of freedom) and sensitivity in particular, starting with crash sensor accelerometers for air-bag deployment and expanding into multi-axis accelerometers and gyroscopes for automatic braking systems (ABS) and electronic stability control (ESC). In fact ESC is being mandated for all US vehicles by 2012, and is being adopted even more quickly in Europe. There is additionally a growing trend to integrate MEMS motion sensors into GPS-based navigational systems and Inertial Measurement Units. It has been estimated that by 2012 the automotive market for MEMS inertial sensors (accelerometers and gyroscopes) and pressure sensors will grow to $2B.[2] There are a number of well-established automotive MEMS sensor manufacturers including Bosch, VTI, Denso, ADI, and ST Micro. Currently automotive systems employ around 30

MEMS sensors of which about half are inertial sensors. These sensors consist of one-, two-, and three-axis accelerometers and single-axis gyros assembled into around half a dozen sensor clusters. Some of the systems and the required inertial sensors are listed in Table 1. [1, 2]

Given that there are only six independent degrees of freedom, there is obviously some redundancy among these MEMS inertial sensors. The cost and size savings achieved by using MEMS sensors is somewhat reduced by the need to integrate multiple sensors into sensor clusters, all of which need to be aligned to the vehicle axes. The cost of the sensor cluster can be two to three times the cost of the individual MEMS sensors. As an example, the accelerometer and yaw sensor in an ESC cluster cost around $15, but the cluster costs $30-$40. [2] Thus a cost savings can be achieved in two ways. First the integration of multiple degrees of freedom into a single chip could reduce chip count. But additionally, the replacement of multiple sensor clusters by a single inertial cluster or single sensor could provide an even larger savings.

An additional advantage of a single MEMS inertial sensor is its size. The best place in the vehicle to position the sensor is near the center console inside the vehicle. However space is at a premium inside the passenger compartment, so having fewer and smaller sensors is desirable. A single small MEMS IMU would be ideal in that it could provide the six degrees of freedom functionality in a small volume. Obviously it will not be possible to replace all the MEMS sensors with a single sensor. Multiple crash detectors will probably still be required. The single sensor might not have a range which matches all the sensor requirements. However, many of the sensor ranges are redundant, in the 1-3 g range. Finally, and more importantly, safety systems will require some redundancy as well as self-test.

To address this opportunity Virtus is developing a 5- and 6- axis MEMS motion sensor that is roughly the same size as or smaller than current MEMS accelerometers and gyros. Although the MEMS structure for the two sensors is the same, Virtus is leading with the 5-axis sensor because the drive and sense electronics are simpler. The sensor is designed to provide inertial measurements in the +/- 3 g and +/- 300 deg/sec range important for navigation and automotive safety and performance applications. The MEMS transducer is 4 mm square and 1.2 mm thick and can be bonded directly to a custom ASIC leading to a very small 5 or 6 axis unit. This paper describes the architecture, fabrication process, and progress on fabrication and modelling.

Function	Accelerometer	Gyroscope
Airbag deployment	*X,Y*	*Yaw*
Automatic Braking System	*X*	
Electronic Stability Control	*X,Y*	*Yaw*
Roll Stability	*Y,Z*	*Yaw, Roll*
Active Suspension	*Z*	
Navigation		*Yaw*
Electronic Parking Brake	*X*	*Pitch*
Cruise Control		*Yaw*
Anti-theft	*X,Y*	

Tab. 1. Automotive Safety and Performance Systems Using MEMS Inertial Sensors

2 Device Description

2.1 Bulk Micromachined SOI (Silicon on Insulator) Architecture

MEMS inertial sensors have been attractive for all these applications because of their small size, and because they provide relatively high performance at low cost. However, the addition of more degrees of freedom has come with size and cost penalties that arise from microfabrication methods used to make the chips. The dominant approach to fabricating MEMS inertial sensors has been surface micromachining, an additive process in which layers of thin films such as silicon dioxide and polysilicon are deposited and patterned. Finally some of the layers are selectively removed (typically the oxide) leaving a released mechanical structure. This approach has been widely adopted because the MEMS can be monolithically integrated with CMOS detection circuitry. However, some compromises must be made to do this integration.

First a noise penalty is paid. Because accelerometers are so small, the proof mass is subject to Brownian motion. The Brownian Force, expressed in units of N/√Hz, is given by:

$$F_{noise} = \sqrt{4k_B T_K b} \tag{1}$$

where k_B is Boltzmann's constant, T_K is the Kelvin temperature, and b is the damping coefficient. [4] This produces a random motion of the proof mass equivalent to that due to an acceleration given by:

$$a_{noise} = \frac{\sqrt{4k_B T_K b}}{m} \tag{2}$$

where m is the mass of the proof mass. Thus this acceleration noise density is minimized by low damping and high mass. Because surface micromachined sensors use thin films, typically polysilicon films a few microns thick, the mechanical noise density is higher than that achievable with bulk-silicon sensors with proof masses typically the thickness of the wafer, 500 – 675 microns. The first MEMS accelerometers, fabricated in the late 1970's, actually were bulk micromachined Si devices. These devices used the entire thickness of the wafer for the proof mass but used anisotropic wet etching to form the mass and supports. Thus the devices were sensitive because of the large mass, but were relatively simple because of the resolution of the wet etches. Consequently, surface micromachined sensors using comb drives and sense capacitors came to dominate the industry.

Second, a compromise in chip area is made when the MEMS transducer is integrated with the CMOS circuitry. Typically the MEMS surface micromachining processes, though compatible with the CMOS processes, are not identical to them. For example the low stress polysilicon used for MEMS may not be the same as gate polysilicon. The high MEMS polysilicon deposition temperatures generally require a specialized CMOS process to accommodate the additional thermal budget, so standard CMOS foundry processes cannot be utilized. The CMOS needs to be protected from the aggressive MEMS release chemistries, typically HF acid or vapor. This typically means that the CMOS is fabricated first in the periphery of the chip and the MEMS sensor is fabricated afterward in a space reserved for it at the center of the chip. Thus silicon area dedicated to CMOS cannot be used for MEMS and vice versa. This in the end is a real opportunity cost. While the silicon wafer is going through the CMOS process, the MEMS area, which can be 20-40% of the die area, is not being processed, although the cost of processing is equivalent to processing an entire wafer of CMOS. The CMOS process cost is, in effect, 20-40% higher.

Finally, because surface micromachining is a planar or two-dimensional process, in general as sensor degrees of freedom are increased, additional proof masses must be designed in. So although 2D accelerometers have been fabricated with a single proof mass and x/y detection, some 3D accelerometers can require two masses (although some do not). Adding gyroscopes requires one proof mass (and sometimes more) per degree of freedom. These extra masses take up additional space on the silicon chip that cannot be used for CMOS detection circuitry.

In the last few years, however, there has been a reduction in the cost of, and an increase in the availability of, both SOI (Silicon-On-Insulator) wafers and Deep Reactive Ion Etching (DRIE) for controlled, high rate bulk etching of silicon wafers. This has already enabled Virtus to commercialize an inexpensive bulk-micromachined SOI 3-axis accelerometer (VAS340A) fabricated by Asia Pacific Microelectronics. The VAS340 is a 3-axis piezoresistive accelerometer that is fabricated in SOI and bonded directly over the ASIC (Fig. 1), producing a packaged device 4 mm x 4 mm x 1.4 mm (Fig. 2). This SOI architecture and stacked chip approach is being applied to the development of the 5- and 6- axis motion sensor.

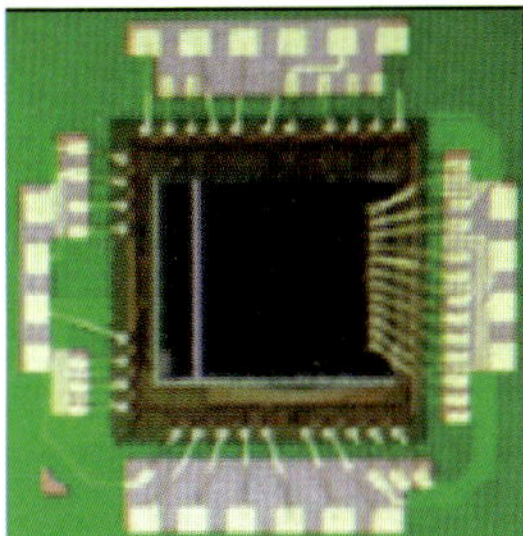

Fig. 1.   VAS340A 3-axis accelerometer stacked bonded over ASIC chip

Fig. 2.   Packaged VAS340A 3-axis accelerometer.

2. 2 Design and Fabrication

An exploded sketch of the MEMS layers of the motion sensor is shown in Fig. 3. The device is fabricated in an SOI wafer, in this case with a 10 µm thick single crystal silicon (SCS) device layer, a 1 µm thick buried oxide (BOx), and a 380 µm thick handle layer. The springs are positioned in the SCS layer along the device diagonals, and the four-lobed mass is etched into the thick handle layer using DRIE. The primary mechanical properties of the motion sensor, the spring constant and the mass can thus be controlled separately. This permits straightforward tuning of device parameters such as sensitivity, range, reso-nant frequency, etc., by independently changing mass and spring thickness and lithographic dimensions. The geometries of the springs and mass are not challenging. The baseline split (there are several design splits on the wafer) has springs 600 microns long and 600 microns wide. The etch gaps are wide ($\sim$100 microns) to avoid RIE etch lag while patterning the large openings under the springs

SOI is an attractive material because the springs can be fabricated in single crystal silicon. The electrical and mechanical properties of single crystal are well known and can be modeled. Furthermore, SCS is less prone to the process variations that can stress and deform thin films. The proof mass, on the other hand, can take up the entire thickness of the wafer. These wafers are 380 microns thick to minimize the DRIE etch time. Even so, a typical proof mass is about 1-3 milligrams, some three orders of magnitude greater than typical surface micromachined proof masses.

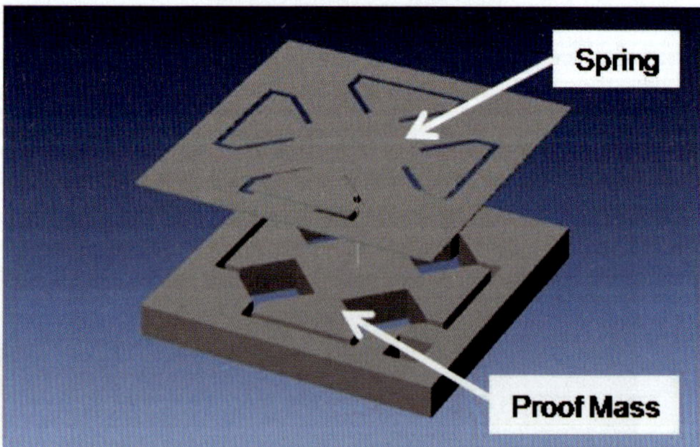

Fig. 3.   Exploded view of the 5-DOF MEMS sensor.

A cross-sectional process engineering sketch (not to scale) of the full device is shown in Fig. 4, showing the MEMS wafer bonded to the cap wafers which contain the capacitor gap and electrodes, which are fabricated separately. Our MEMS foundry partner is Silex Microsystems AB, Järfälla, Sweden. Processing proceeds roughly as follows. Contacts are etched between the SCS layer and handle to electrically connect them and are filled with polysilicon. The springs are next patterned in the SCS. Meanwhile, the caps are processed in parallel with the MEMS. The capacitor gaps (~4 μm deep) are etched into one side of the cap wafer. The sense and drive electrodes are next patterned in the cap wafers using a Silex proprietary trench DRIE etch and refill process. The MEMS wafer is fusion bonded to the top cap wafer and the mass is patterned in the handle (backside) using DRIE. Finally the MEMS wafer and the bottom cap wafer are fusion bonded, hermetically sealing the MEMS sensor under vacuum. The wafer stacks are then metalized and diced.

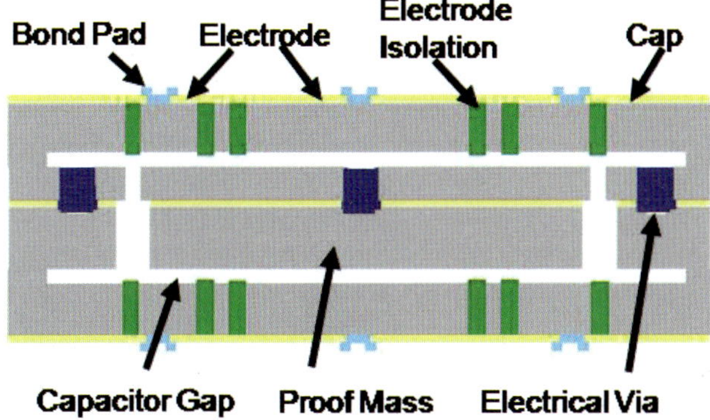

Fig. 4.   Process engineering cross-section of sensor.

2. 3 Principle of Operation

The principle of operation of the device, shown in Fig. 5 with a top view of the actual layout for comparison, has been described in an earlier paper. [5] However, a brief description is useful. When the sensor body accelerates in the -z direction, the proof mass translates a distance Δz relative to the frame in the +z direction given by Newton's second law:

$$F = ma = k\Delta z \tag{3}$$

where m is the mass of the proof mass and k is the restoring spring constant. The sensor response has been modeled analytically using well-known formulas for mechanical structures by treating pairs of springs as a beam of length 2L, moment of inertia I, and Young's modulus E, rigidly supported at each end with a concentrated force, ma, at the center.[6] The deflection is given by:

$$\Delta z = \frac{maL^3}{192EI} \tag{4}$$

and the resulting capacitance between the mass and the sense electrode with area A is:

$$C = \frac{\varepsilon_0 A}{(d \pm \Delta z)} \tag{5}$$

where d is the rest gap of 4 µm. The capacitance of C_{1top} increases, while the capacitance of $C_{1bottom}$ decreases. To first order in Δz,

$$C_{1top} - C_{1bottom} = \frac{2C_0 \Delta z}{z_0} \tag{6}$$

where C_0 and z_0 are the rest capacitance and rest gap (~1.2 pF and 4 µm, respectively).

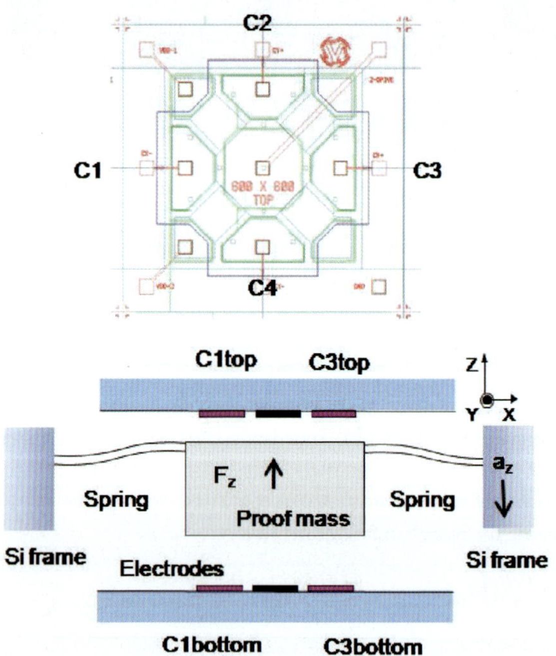

Fig. 5. Principle for sensing z acceleration. (a) Top view of the sensor showing the four tope sense electrodes. (b) Side view of the sensor showing the z deflection of the proof mass.

The method for detecting acceleration in the x or y directions is shown in Fig. 6. When the motion sensor body accelerates in the x direction, the center of mass (COM) of the proof mass of thickness T moves in the $-x$ direction a distance Δx. This deflection produces a torsional moment:

$$M = ma\frac{T}{2}2 \tag{7}$$

on the springs and the mass tilts. The deflection Δx can be related to the tilt angle which can in turn be determined by measuring $Cx1$-$Cx3$. This mode of operation was modeled analytically by decomposing the moment along the springs which are at $45°$ to the principle axes. These components in turn produce a bending moment along the spring and a torsional moment along the perpendicular spring. The net resultant moment produces a tilt angle:

$$\theta_x = \frac{\alpha}{1+\alpha}\frac{m\frac{T}{2}L}{8EI}a_x \tag{8}$$

where α is the ratio of bending to torsion constants for the springs:

$$\alpha = \frac{4EI}{KG} \tag{9}$$

and E is Young's modulus for silicon, I is the spring moment of inertia, K is the spring torsional compliance, and G is the torsional modulus.

The capacitance can then be calculated by integrating over the tilted gap between the trapezoidal electrode and the mass, a straightforward geometrical integration in principle.

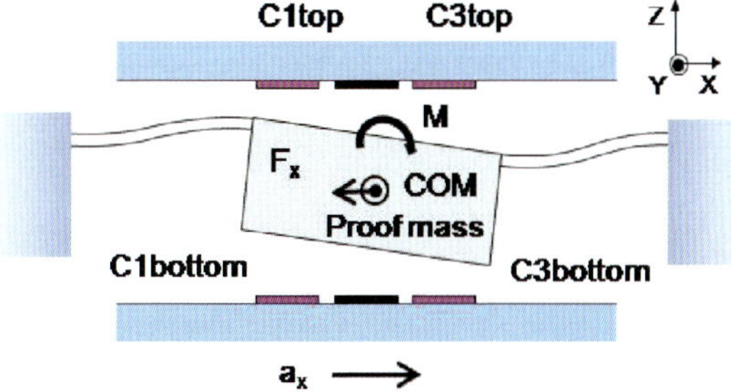

Fig. 6.   Sensing x acceleration.

Finally, the angular velocity measurement (here Ω_y) is shown in Fig. 7. Angular velocity is measured using the Coriolis force, F_{Coriolis}, a fictional force perceived at right angles to the velocity v and angular velocity Ω, and given by their cross product:

$$\vec{F}_{Coriolis} = 2m\vec{v} \times \vec{\Omega} \tag{10}$$

The mass is driven into resonance by driving the z electrodes at the resonant frequency with an amplitude of around 2 µm, or half the rest gap. The Coriolis force (acceleration) is measured in the same way as the x or y acceleration, but in quadrature with the resonant drive, that is, in phase with the resonant velocity. The maximum deflection due to angular velocity occurs at the maximum resonant velocity, which occurs at the zero-point of the motion of the center of mass. This is modeled in the same manner as the x or y acceleration, with the replacement of the specific force $F_x = ma_x$ by the Coriolis force.

$$F_{Coriolis\,x} = -2mv_z \times \Omega_y \tag{11}$$

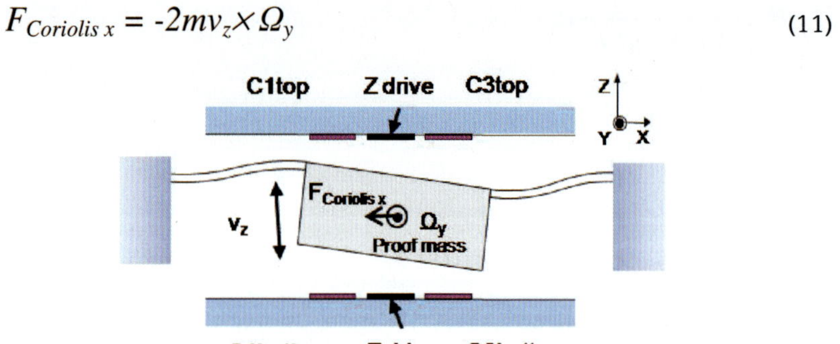

Fig. 7.   Sensing y angular velocity.

In this way acceleration along x, y, and z and angular velocity around the x and y axes can be measured with a single proof mass. By adding the electronic capability to drive the x electrodes resonantly and measure motion in the y direction we should be able measure angular velocity around the z axis as well without changing the MEMS design. Because of the additional electronics complexity we are reserving this addition of the sixth degree of freedom for a follow-on development phase, although that will be part of the characterization of the MEMS prototypes currently in fabrication.

This architecture enables in a projected package size of approximately 6 mm x 6 mm x 2 mm a 5 or 6 axis Inertial Measurement Unit (IMU) with the functionality of much larger sensor clusters that require the assembly of multiple accelerometers and gyros with their accompanying additional assembly costs.

Fig. 8 is a plot for many of the inertial sensors currently on the market of the packaged sensor size (in cm^3) against the sensor degrees of freedom, a measure of functionality. Gyros (blue diamonds) are typically single axis and large. Accelerometers (green triangles) range from one to three degrees of freedom, and have become quite small, largely through the pressure of consumer electronics.6-axis IMUs (magenta circles), although high-functionality, are still quite large because of the fixturing required for the multiple integrated motion sensors. The Virtus motion sensors (red squares), because of the single proof mass design, maintain the size benefits of single accelerometers or gyros, but have the functionality of the larger IMUs. Thus they should find wide applicability in space-sensitive automotive (and other) applications as described in the Introduction.

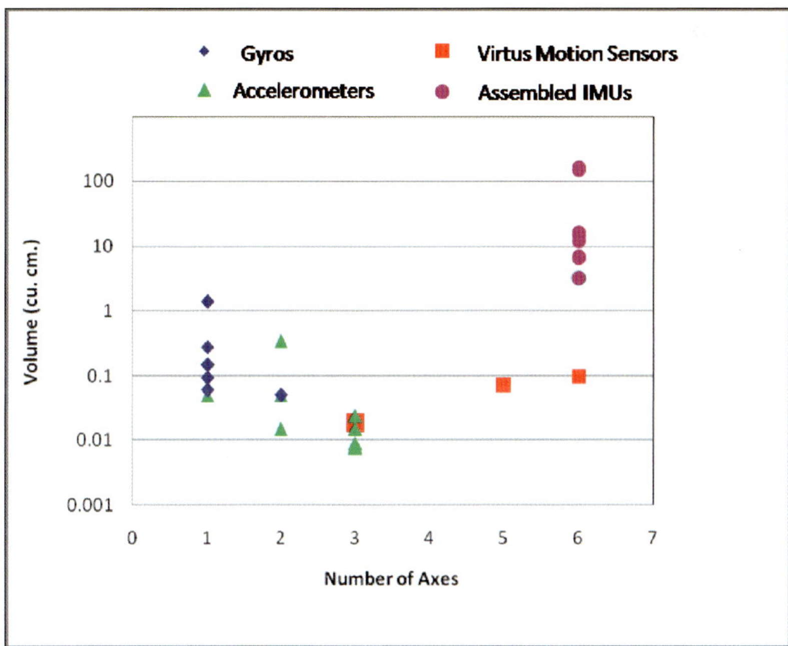

Fig. 8.   Plot of sensor volume vs. number of degrees of freedom for several commercial sensors.

3 Device Modeling

3.1 Linearity and Sensitivity

In order to understand the sensitivity and linearity of the sensor we started by modeling the change in capacitance in each of the electrodes due to an acceleration constrained to be only along that axis. As mentioned in section 2.3, an acceleration in the x direction causes a tilt θ_x of the mass given by:

$$\theta_x = \frac{\alpha}{1+\alpha} \frac{m\frac{T}{2}L}{8EI} a_x \tag{8}$$

Fig. 9 shows the geometry used for calculating the capacitance. For acceleration in the x (or y) direction, the

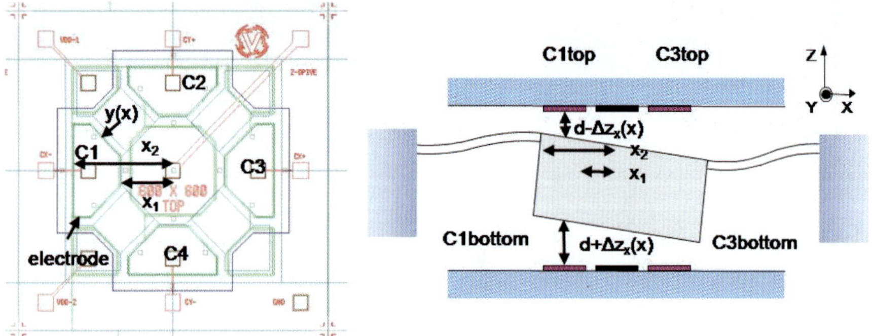

Fig. 9.   Geometry used for calculating response of capacitance to acceleration.

capacitance between the cap electrode and the tilted mass lobe can be calculated from:

$$C_{1top} = \frac{2\varepsilon_0}{d} \int_{x_1}^{x_2} \frac{y(x)}{1-\dfrac{\Delta z}{d}}\, dx \tag{12}$$

where x_1 and x_2 denote distance from center of the leading and following edges of the electrode and $y(x)$ describes the shape of the electrode, in this case including a tapered and rectangular section. For the bottom electrode the sign in the denominator becomes a +. Δz_x is the tilted gap width, the maximum of which is given by:

$$\Delta z_{max} = x_2 tan\theta_y \approx x_2\theta_y \qquad (13)$$

which is proportional to a_x. The range of acceleration over which the capacitance is linear can be determined by expanding the denominator in terms of $\Delta z_x/d$:

$$C_{1top} \approx \frac{2\varepsilon_0}{d} \int_{x_1}^{x_2} y(x)[1+\frac{\Delta z}{d}+(\frac{\Delta z}{d})^2 + \cdots]dx \qquad (14)$$

So if the dependence of capacitance on acceleration is to be linear to some criterion, usually around 1% of full scale, then we need $\Delta z_{xmax}/d$ to be <0.1. Thus from Eq. 7 and Eq. 12 the maximum linear x (or y) acceleration is:

$$a_{x\,max} = 0.1d\,\frac{1+\alpha}{\alpha}\frac{16EI}{TLx_2} \qquad (15)$$

A similar expression for the maximum linear z acceleration can be obtained in the same way:

$$a_{z\,max} = 0.1d\,\frac{48EI}{ML^3} \qquad (16)$$

The capacitances for each electrode can be calculated over these ranges using Equation 11. The zero state capacitance can be removed by measuring the capacitance differentially (Equations 16 and 17 below). The C_{1top}-$C_{1bottom}$ and C_{1top}-$C_{1bottom}$ terms remove the dependence on C_0 , the rest capacitance. C_{1top}-C_{3top} and $C_{3bottom}$-$C_{1bottom}$ remove any offset due to z acceleration (refer back to Fig 6).

$$C_x = C_{1top} - C_{1bottom} - (C_{3top} - C_{3bottom}) \qquad (17)$$

$$C_z = 4(C_{1top} - C_{1bottom}) \qquad (18)$$

Tab. 2 shows the spring dimensions, mass, maximum linear acceleration (acceleration range), and sensitivity for each of the design splits on the wafer calculated numerically with Equation 1 using MathCad. The baseline split (A) targets +/- 3g and +/- 300 deg/sec. The other splits include spring dimension splits for stiffer and more compliant springs and a sensor with reduced area. These results indicate that the sensor can be operated open loop over a large linear range of acceleration and angular velocity. Virtus' first devices will not use feedback, although there may be other issues such as bandwidth control and active damping that may eventually make closed loop operation attractive.

Split	*A* *Baseline*	*B* *Stiff*	*C* *Compliant*	*D* *¼ Area*
Spring length (um)	600	325	905	255
Spring width (um)	600	509	175	185
Proof mass (mg)	2.589	1.528	3.304	0.902
Linear (1% FS) x(y) acceleration range (g)	10.17	34.73	1.65	39.18
Linear (1% FS) z acceleration (g)	26.28	237.7	1.75	303
Cross-coupling (1% FS) y acceleration (g)	3.8	13.1	14.8	0.62
X,Y Sensitivity (fF/g)	33.39	4.82	322.6	3.085
Z Sensitivity (fF/g)	84.81	10.04	917.6	8.331
Resonant frequency (Hz)	2858	8591	737	9700
Ω Sensitivity (aF/deg/sec)	4.239	1.841	10.58	1.331

Tab. 2.   Predicted Sensor Response

3.2 Cross-axis Coupling

Because of the 3-D nature of the proof mass a potential concern often raised is cross-axis coupling. Once again we turn to Equation 12 for capacitance as a function of tilt angle, but include all three axes:

$$C_{1top} = \frac{2\varepsilon_0}{d} \int_{x_2-y(x)}^{x_1} \int^{y(x)} \frac{1}{1-(\frac{\Delta z_x}{d} + \frac{\Delta z_y}{d} + \frac{\Delta z_z}{d})} \, dxdy \tag{19}$$

And once again we expand the denominator and look at the terms beyond those linear in $\Delta z_x/d$.

$$C_{1top} \approx \frac{2\varepsilon_0}{d} \int_{x_1}^{x_2} \int_{-y(x)}^{y(x)} [1 + \frac{\Delta z_x}{d} + (\frac{\Delta z_y}{d} + \frac{\Delta z_z}{d}) + (\frac{\Delta z_x}{d} + \frac{\Delta z_y}{d} + \frac{\Delta z_z}{d})^2 + \cdots] dxdy \qquad (20)$$

The $\Delta z_y/d$ term integrates to zero in the y integral, since the y deflection is antisymmetric in y ($\Delta z_y/d = y\,\tan\theta_y$). The $\Delta z_z/d$ term remains through the integration, but it will be removed with the differential measurement (Eq. 17). So the main contributions to cross-coupling are the quadratic terms. The $(\Delta z_x/d)^2$ term is not part of cross-coupling and the $(\Delta z_z/d)^2$ term will also be removed with the differential measurement since it is the same for all sets of electrodes. So we are left with seven cross terms of order $(\Delta z_y/d)^2$. This sets a slightly more rigorous limit on the accelerometer range. If we want to keep the cross-coupling contribution to 1% or less, we would need to set a_{max} to $\sqrt{7} = 2.65$ times the value estimated in the linearity calculation. For design split A, the x acceleration range becomes 3.8 g. So we can expect the cross-coupling to be less than 1% over the +/- 3g range of baseline design.

3.3 Dynamic Response

Since the device is driven in resonance for the angular velocity measurements it is necessary to estimate the damping and noise response. These were modeled using well-known expressions for mechanical and electrical noise. [4, 7] The results of this section are summarized in Table 3 at the end of the section. The squeeze-film damping coefficient b is given by:

$$b = \frac{96\mu A_t^2}{\pi^4 d^3} \qquad (21)$$

where A_t is the total proof mass area, d is the gap width, and μ is the pressure-dependent squeeze film viscosity. μ is given by:

$$\mu = \frac{\mu_{atm}}{1 + 9.638 K_n^{1.159}} \qquad (22)$$

where μ_{atm} is the viscosity of air at atmospheric pressure and K_n is the Knudsen number, the ratio of the pressure-dependent mean free path to the gap. The quality factor at resonance, Q, can then be calculated from the damping coefficient using

$$Q = m\omega/b \qquad (23)$$

The damping coefficient and quality factor of the baseline design, A, are plotted as a function of the log of the ambient pressure in Fig. 10. At atmospheric pressure the mass is heavily damped with $Q<1$. However, the motion sensor process flow includes the fusion bonding of the MEMS and cap wafers, which occurs under vacuum, at a few milliTorr. At these pressures, $Q>100$, so reasonable resonant drive should be achievable. In fact by perforating the mass to permit air flow, an additional 2-3 orders of magnitude increase in Q is possible, with some sacrifice in mass and electrode area. We have included a perforated design split, but do not expect that this will be required in the final device.

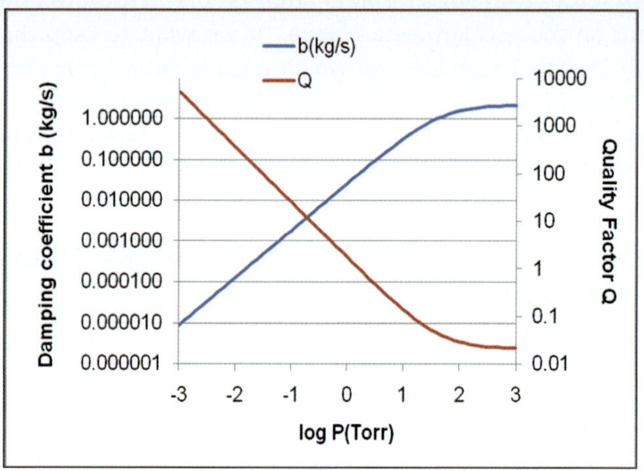

Fig. 10. Pressure dependence of the damping coefficient and quality factor for the baseline MEMS sensor design.

The maximum resonant drive voltage (defined by the pull-in voltage at collapse, which occurs at about d/1.73) can be estimated by:

$$V_r = \sqrt{\frac{8}{27}\frac{Kd^2}{C_d}\frac{2}{Q}} \tag{24}$$

where K is the effective spring constant and C_d is the drive electrode capacitance.[8] A plot of the resonant drive voltage as a function of pressure is shown in Fig. 11. We estimate that the required 2 μm drive amplitude should be achievable with about 3 volts at 10 mTorr.

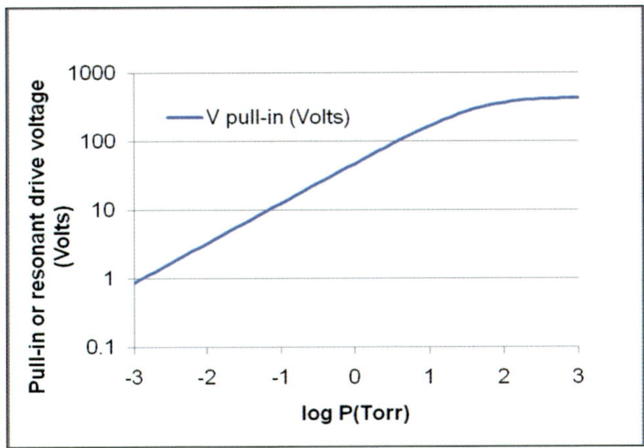

Fig. 11.   Resonant drive voltage as a function of pressure for the baseline MEMS design.

As described in Section 1, the mechanical noise density due to Brownian motion can be represented as a noise equivalent acceleration:

$$a_{noise} = \frac{\sqrt{4k_B T_K b}}{m} \tag{2}$$

where k_B and T_K are Boltzmann's constant and the Kelvin temperature respectively. Again a dependence upon pressure arises through the damping coefficient as seen in Fig. 12. At the expected package pressure of around 10 mTorr, this noise effective acceleration is less than 0.1 μg/rt-Hz. Since the damping coefficient depends on the square of the area, which cancels the area dependence of the mass, the noise depends only on the thickness of the mass.

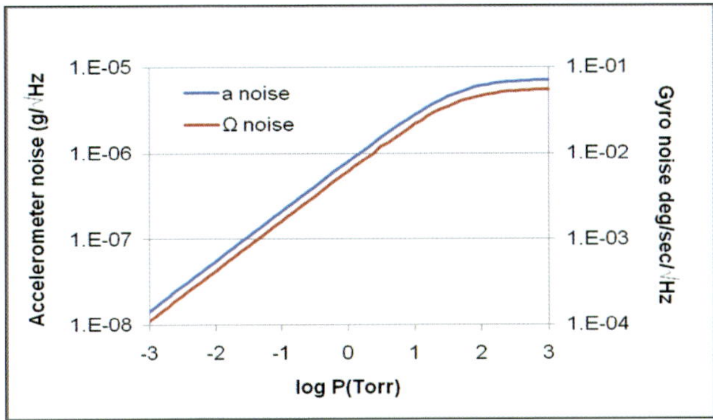

Fig. 12.   Accelerometer noise density due to mechanical or Brownian noise.

A similar calculation of the noise-effective angular velocity which replaces m with 2mv gives a noise effective angular velocity at 10 mTorr of 0.5 m-deg/sec/rt-Hz. This corresponds to an angle random walk (ARW) of 0.03 deg/rt-hr. In practice this extremely low noise will be dominated by IC noise, although one could envision developing a custom low-noise IC for high resolution applications. Nonetheless the low mechanical noise demonstrates a potential advantage of the 3D bulk Si approach, the large proof mass, on the order of 2 milligrams. This is some 3 orders of magnitude more massive than the surface micromachined proof-masses which are only a few microns thick and on the order of a microgram. The bulk proof mass has more inertia against the collisions from ambient gas molecules that cause the mechanical noise.

Split	*A* *Baseline*	*B* *Stiff*	*C* *Compliant*	*D* *¼ Area*
Damping coefficient (g/sec)	0.128	0.044	0.208	0.015
Quality factor	364	1858	73.7	3554
Accelerometer noise density ($\mu g/\sqrt{Hz}$)	0.0546	0.0546	0.0546	0.0546
Velocity Random Walk ($\mu m/sec/\sqrt{hr}$)	32.13	32.13	32.13	32.13
Gyroscope noise density (mdeg/sec/$\sqrt{Hz}$)	0.427	0.142	1.656	0.126
Angle Random Walk (milli-deg/sec/$\sqrt{hr}$)	25.65	8.527	99.38	7.552
Resonant Drive Voltage (Volts)	3.264	3.708	4.054	3.924

Tab. 3.   Predicted sensor dynamic properties at 10 mTorr for MEMS design splits

4   Status

As of this writing in January 2009 the 5-degree-of-freedom MEMS sensor is nearing completion at Silex. The process described in Section 2.2 has been completed through final wafer bonding and cap thinning. Fig. 13 is an SEM (Scanning Electron Micrograph) of the MEMS sensor from the backside or mass side just prior to final bonding with the bottom cap wafer. Visible are the 380 micron thick four-lobed proof mass, the 100 micron-wide trenches, and, at the bottom (actually topside) of the mass, the 600 micron by 600 micron spring of the baseline design. As one can see from the SEM scale, the entire mass is about 2.5 mm across.

Fig. 14 is an SEM of the inside (sensor side) of the top cap prior to bonding to the MEMS wafer. The gray area is the gap area which is recessed 4 microns. The corner polygons are the electrical feedthroughs to the MEMS sensor. The "D"-shaped structures are the four top capacitor electrodes. The central octagon is the drive electrode for the x and y angular velocity measurements. All are fabricated in heavily doped single crystal silicon and each is surrounded by an insulator-filled trench 20 microns wide.

Fig. 15 is an acoustic microscope scan of the bonded and thinned wafers from (a) the top and (b) the bottom. The dark area is the bonded area and a solid color is an indication of a robust bond. The corner feedthroughs are dark indicating that there is good contact to the MEMS. The absence of dark spots in the sensor area is an indication that there has not been any collapse or stiction during the release and bonding processes, although this remains to be confirmed by probing. The pitch of the sensors is 4 mm, which will be the final die size. The final contact and metallization steps are underway and completion of the devices is expected in early February.

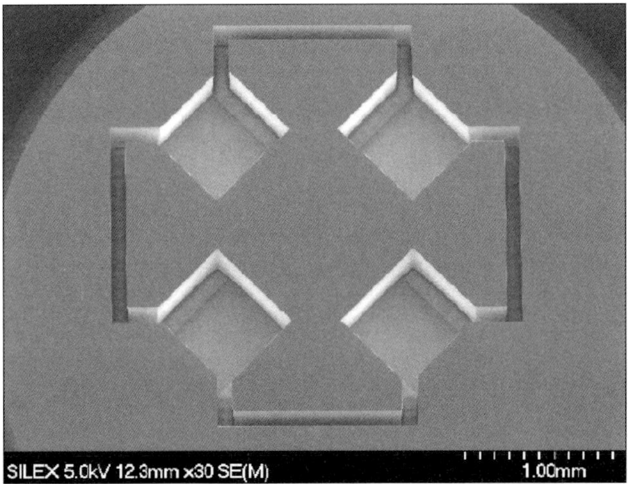

Fig. 13.   SEM photo of the baseline MEMS design from the bottom side.

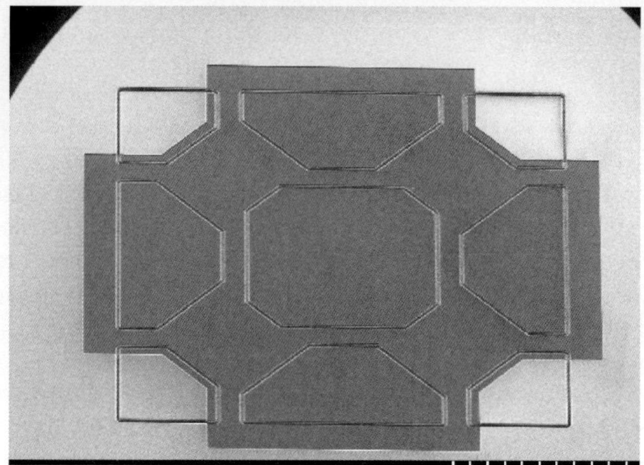

Fig. 14.   SEM photo of the baseline top cap design from the sensor side.

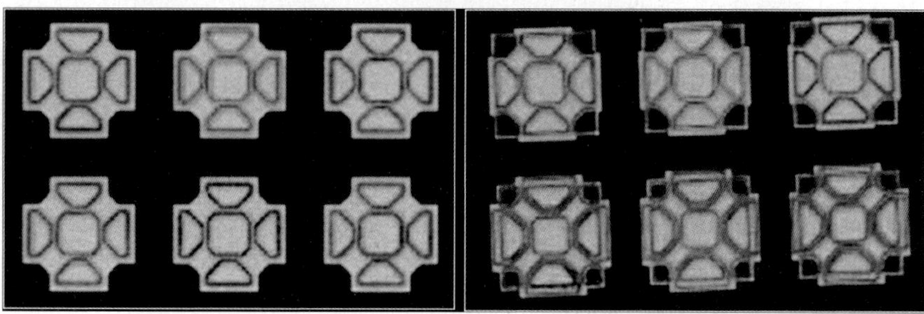

Fig. 15.   Acoustic microscope scans of (a) the top and (b) bottom of the
bonded cap and MEMS wafers.

5   Conclusion

Virtus is developing a MEMS motion sensor that can measure 3 axes of acceleration and 2 axes of angular velocity in a device 4 mm x 4 mm x 1.2 mm. The sensor is preliminarily specified to have a ranges of +/- 3g and +/- 300 deg/sec. Modeling indicates that the baseline design will be linear within this range to less than 1% of full scale, and cross-axis coupling should be less than 1%. Sensitivity is modeled to be 33 fF/g in x and y and 85 fF/g in z with an angular velocity sensitivity of 4.2 aF/degree/sec. With a target of a package pressure of 10 mTorr, we expect accelerometer noise on the order of 0.05 $\mu g/\sqrt{Hz}$, an ARW of 25 milli-degree/sec/$\sqrt{hr}$, and a resonant drive voltage of around 3 Volts. Fabrication will be complete in early February, 2009.

References

[1]   Weinberg, H., "MEMS Sensors Are Driving the Automotive Industry", Sensors, online article, http://www.sensorsmag.com/sensors/article/articleDetail. jsp?id=361648, Feb. 1, 2002.

[2]   Dixon, R., Bouchaud, J.,"MEMS in automotive: How regulatory issues will reshape the market", Advanced Automotive Electronics 2008, online article, http://www. aae-show.com/Presentations/dixon_slideswithnotes.pdf. Oxfordshire, UK, Sept. 23, 2008.

[3]   Layton, M., "Inertial Sensors Enable Automotive Safety", EC Magazine, online article, http://www.ecnmag.com/Inertial-Sensors-Enable-Automotive-Safety.aspx, Oct. 01, 2007.

[4]   Yazdi, N., et al, "Micromachined Inertial Sensors", Proceedings of the IEEE, v. 86, no. 8, pp 1640-1659, August, 1998.

[5]   Watanabe, Y., et al, "SOI Micromachined 5-Axis Motion Sensor Using Resonant Electrostatic Drive and Non-resonant Capacitive Detection Mode", Sensors and Actuators A, v. 130/131, pp. 116-123 (1995).

[6]   Young, W., Roark's Formulas for Stress and Strain, McGraw-Hill, 6th Edition (1989).

[7]   Berny, A., "Substrate Effects in Squeeze-Film Damping of Lateral Parallel-Plate Sensing MEMS Structures", online article http://robotics.eecs.berkeley.edu/~pister/245/project/Berny.pdf.

[8]   Seeger, J., Boser, B., "Parallel-Plate Driven Oscillations and Resonant Pull-In, Proceedings of the Solid State Sensor, Actuator and Microsystems Workshop, pp. 313-316, Hilton Head Island, South Carolina, June 2-6, 2002.

R.M. Boysel, L.J. Ross
Virtus Advanced Sensors, Inc.
707 Grant St., Suite 3200
Pittsburgh, PA
USA
boysel@virtusensors.com
ross@virtusensors.com

Keywords: MEMS, motion sensor, inertial measurement unit, accelerometer, gyroscope

MEMS Sensors for non-Safety Automotive Applications

M. Ferraresi, S. Pozzi, ST Microelectronics

Abstract

With the launch of the AIS326DQ, its first AECQ100-qualified 3-axis accelerometer, STMicroelectronics is addressing non-safety applications like vehicle alarms, tracking and monitoring, black-box systems, and navigation assistance. Government regulations, insurance requirements, and a general increasing user demand for car security are the main driving factors of this high-CAGR market.

Fully leveraging the economies of scale of a proven 200 mm wafer technology (STMicroelectronics was the first company in the world to start MEMS volume production on 8″ wafers), advanced features such as user selectable full scale ranges, 12-bit usable resolution, together with the mounting freedom provided by the 3-axis sensing, make AIS326DQ the recommended choice when customers look for the best trade-off between performances and price.

STMicroelectronics will continually be enriching its offer by complementing the portfolio with more innovative solutions: the new generation of accelerometers is reported to show a state-of-the-art thermal stability below 0.2 mg/°C, while, after a successful launch in the consumer segment, ST gyroscopes will find their way into automotive applications.

1   Introduction

In the automotive market, MEMS (Micro Electrical Mechanical Systems) first found their entry in the 90's as pressure and flow sensors (MAP, BAP) and accelerometers for airbag deployment.

In the following years, the penetration of MEMS sensors in powertrain, chassis, and body applications grew at an amazing pace. It is estimated that today the TAM in units is close to 600 million, with an average of 10 per vehicle, reaching 30 in the case of high-end models. [1]

A stable growth, partially offset by the deep crisis hitting the car industry and the expected sensor fusion, is basically guaranteed in the future by safety mandates that will force car companies to implement features like stability control or tire pressure monitoring in all their fleet.

Safety, however, is not the only area in the car where MEMS can find a place. There is another emerging and rapidly growing field that is shaped by non-safety and security applications, such as anti-theft systems, black boxes and event recording, car alarms, embedded navigation system with dead reckoning capabilities, vehicle tracking, road tolling, and others.

This area positions itself somewhere between the consumer and industrial worlds and the automotive world, taking features from both of them. Consumer-oriented semiconductor manufacturers might find it difficult to comply with the stringent requirements imposed by the automotive environment, while purely automotive-oriented ones might not have the dynamics and efficiency required by such applications: what is needed is a balanced blend of features, as shown in Fig. 1:

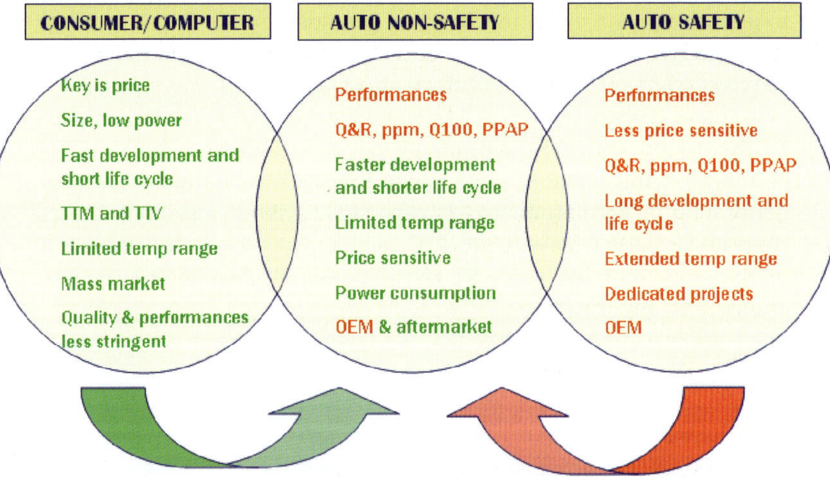

Fig. 1.   Overlapping of Automotive and non-Automotive worlds

STMicroelectronics, leveraging a consolidated and stable MEMS manufacturing machine (more than 200 million sensors shipped in 2008 to computer, industrial, and consumer applications) and as a long-term player in automotive, nominates itself to fill this gap and quickly become a recognized supplier in the segment.

Government mandates will also boost sensor demands for non-safety applications.

For instance, according to Resolution 245 issued on the 27th July 2007 by the Brazilian government, all new vehicles (both 2- and 4-wheelers) leaving the factory, produced in Brazil or imported, starting 24 months from the date of publication of the resolution, may only be sold if they are equipped with an anti-theft and electronic tracking device. The equipment features are the following:

▶ Immobilization
▶ Calculation and storage of geographic positioning
▶ Integrated communication module to send information to a monitoring services center and to receive commands from it

After the resolution's launch, due to the turmoil in the industry, (the 100% applicability) by August 2009 was changed to a more gradual introduction rate. All vehicles currently do require at least one accelerometer. [2]

A second example is the "Pay-as-you-drive" project (pay for the use, not for the ownership), approved by the House of Representatives in July 2008, that will be deployed in the Netherlands starting 2011. Each vehicle must be equipped with a suitable GPS system that will feature, among other things, an accelerometer and gyroscope, both of which must feature high precision, quality, and compliance with the Q100 [6] standards.

We foresee that other countries will follow with similar mandates, both in the field of security and of road tolling.

2   Current Products Line-Up

On October 1st 2008, STMicroelectronics announced the AIS326DQ, its first 3-axis accelerometer qualified for automotive applications.

The AIS326DQ is a three axis capacitive accelerometer that includes a sensing element and an IC interface able to process the information acquired from the sensor and to provide the measured acceleration signals to the external world through digital serial interface. The sensing element is manufactured using a dedicated process, THELMA [3], developed by ST. The IC interface, on the other hand, is manufactured using a CMOS 0.25 µm process with a high level of integration to design a dedicated circuit factory trimmed to better match the sensing element characteristics. The whole system is stacked and assembled in plastic Quad Flat Package No-lead surface mount (QFPN) using a low-stress molding resin with silicone-based die attach material, and it is specified over a temperature range extending from -40 °C to +105 °C, eligible for non-safety automotive applications.

Here is a brief excerpt of the main features:
- ▶ 3.3 V supply
- ▶ User selectable full scale (± 2 g / ± 6 g)
- ▶ SPI interface
- ▶ 12 bit accuracy
- ▶ Programmable interrupt

The sensor structure is made of two separate regions: the first one to sense in-plane accelerations (xy), the second one to detect accelerations orthogonal to the package (z):

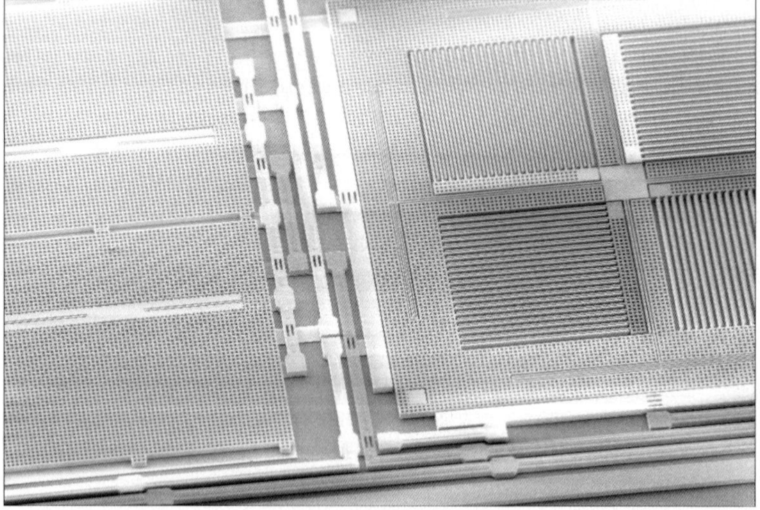

Fig. 2.　The sensing masses of AIS326DQ (z structure on the left, xy on the right)

To be compatible with the traditional packaging techniques, a cap is commonly placed on top of the sensing element to avoid blocking the moving parts during the molding phase of the plastic encapsulation and to prevent further particle contamination:

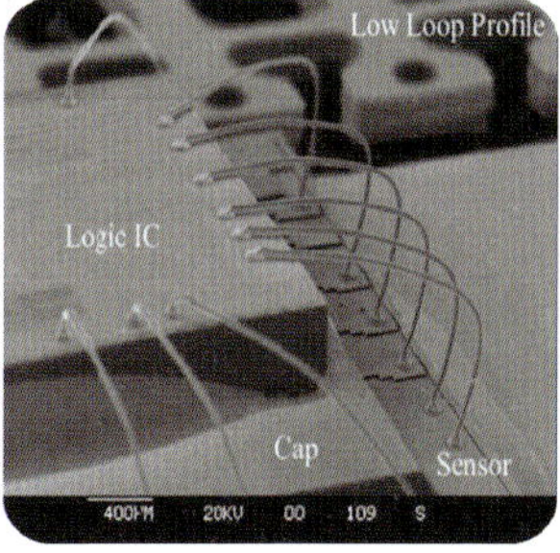

Fig. 3.   MEMS and ASIC stacking in AIS326DQ, before and after molding

The quality process in STMicroelectronics puts a great emphasis on the hermeticity of the cap sealing. This issue is really important for molded packages since the resin injection applies a significant mechanical stress, which in turn may induce delamination of the interconnection between cap-die and active MEMS, with dramatic consequences on device reliability. In Fig. 4 is shown a section of an accelerometer in stacked configuration, with the areas more subjected to stresses clearly visible:

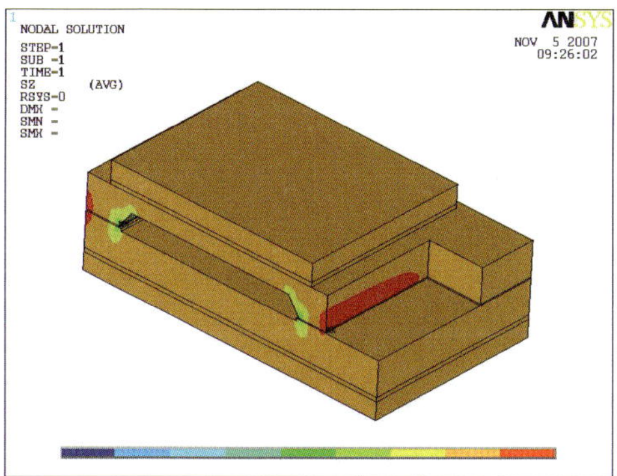

Fig. 4.   MEMS die Section Stress Simulation

Specific process inspections are implemented on 100% of the wafers to prevent the possibility that a potential weakness of the glass-frit junction could degenerate during the assembly steps. The Scan Acoustic Microscope (SAM), for instance, is used for the detection of early delamination or bonding defects.

If we look at the device block diagram, after a stage needed to amplify the small signal coming from the sensor (capacitance variations in the range of a few fF/g), the core of the reading chain is built on a triple high-precision $\Sigma\Delta$ converter and a low-pass digital filter (Fig. 5):

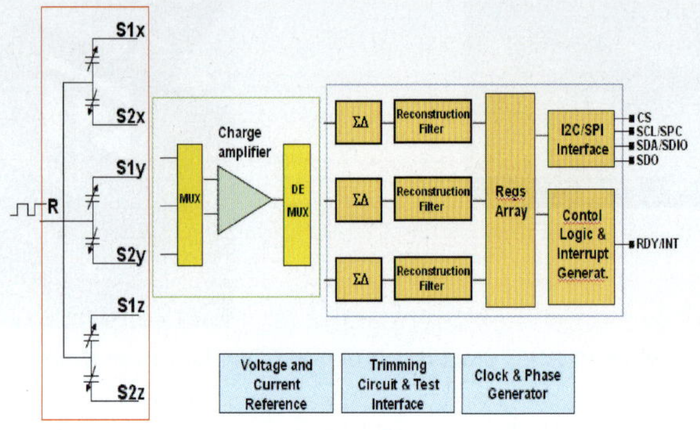

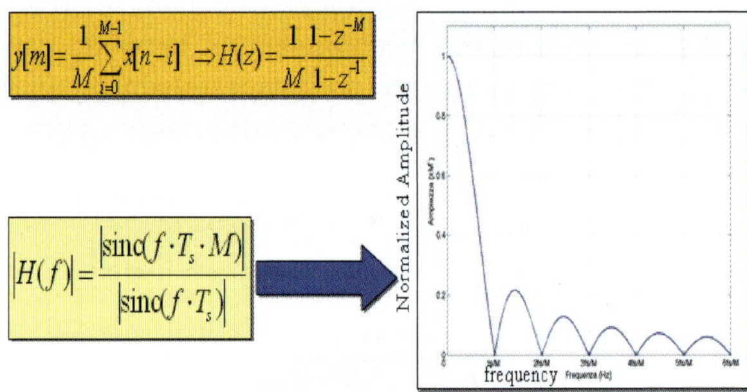

Fig. 5.   AIS326DQ: block diagram and transfer function of the embedded digital filter. Sinc decimation filter is realized by calculating the mean value of the latest M samples

After only three months from its official launch, the AIS326DQ had already recorded a few design-ins, and it is currently in pre-production in at least three major first tier customers.

While a 3-axis accelerometer gives the customer complete mounting freedom, a 2-axis one is less flexible in terms of mounting, but introduces a sensitive advantage in terms of cost. The AIS226DS, which is now in sampling stage, is the response of STMicroelectronics whenever a trade-off flexibility-price is required.

The ASIC of the AIS226DS is the same of the 3-axis version, while the sensor is different; it is a dual axis implementation, also designed for harsh environments, with excellent hermeticity, mechanical robustness, and more resistant to delamination. The side-by.side assembly in a standard SOIC16 package wide-body with reverse frame forming also increases its EMI robustness.

In Fig. 6 is shown the sensor layout, package, along with the main features at 3.3V typical conditions:

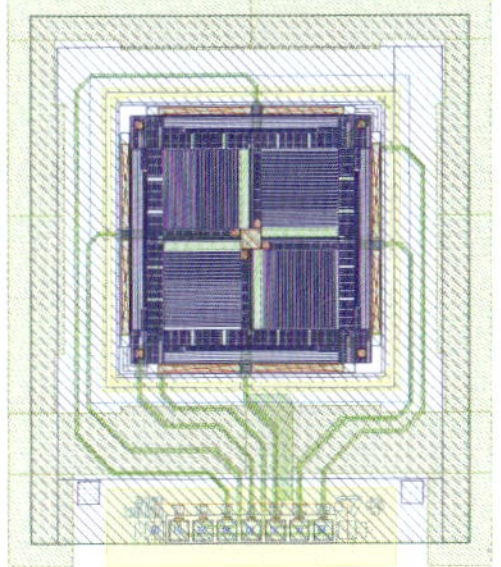

Parameter	Typ Value
C_0 *[pF]*	*0.60*
Sensitivity [fF/g]	*30*
$F_{res\ xy}$ *[kHz]*	*1.6*
K [N/m]	*1.5E-8*
g range [g]	*±2, ±6*
Cross-axis [% FS]	*2.5*
Non linearity @2 g [%FS]	*0.5*
0 g offset drift [mg/°C]	*<0.8*

Fig. 6.   The AIS226DS: sensor layout, package, and main mechanical and electrical characteristics

A modern navigation system integrates the information coming from GPS-based and inertial navigation systems (INS). By properly combining the information from an INS and the GPS system (GPS/INS), the errors in position and velocity are stable. Furthermore, INS can be used as a short-term fallback if GPS signals are unavailable, for example when a vehicle passes through a tunnel.

An INS module is based on accelerometers and gyroscopes: the former measure the linear acceleration of the system in the inertial reference frame, but in directions that can only be measured relative to the moving system (since the accelerometers are fixed to the system and rotate with the system, but are not aware of their own orientation); the latter measures the angular velocity of the system in the inertial reference frame.

By tracking both the current angular velocity of the system and the current linear acceleration of the system measured relative to the moving system, it is possible to determine the linear acceleration of the system in the inertial reference frame.

The AY515DT, a yaw-axis gyroscope with 140°/s of full scale, is the ideal companion of the accelerometers described above for effectively implementing a dead-reckoning function in INS, tracking a car in anti-theft devices, or a road toll payment in pay-per-usage applications.

The AY515DT is a low-power single-axis yaw rate sensor, with both analog and digital output. It includes a sensing element and an IC interface able to provide the measured angular rate to the external world through an analog output voltage and SPI/I2C digital interfaces. The sensing element, capable of detecting the yaw rate, is manufactured using the same technology as linear accelerometers (THELMA), while the IC interface is manufactured using a CMOS process that provides the high level of integration needed to design a dedicated circuit that is trimmed to better match the sensing element characteristics.

The analog output of AY515DT is capable of measuring rates with a -3 dB bandwidth up to 88 Hz, while the digital output can reach 1 kHz of Output Data Rate.

The Coriolis effect describes the deflection of a mass which is moving - in a rotating system – along the radius. The deflection motion is a tangential motion and perpendicular to the motion along the radius.

In the MEMS implementation, a tiny mass experiences a linear backwards/forwards motion (driving mass, see Fig. 7). If this oscillating mass is exposed to an external rotation, a force that is perpendicular to the oscillating direction is measurable (Coriolis Acceleration) with an on-chip, high resolution accelerometer (sensing mass, see Fig. 7).

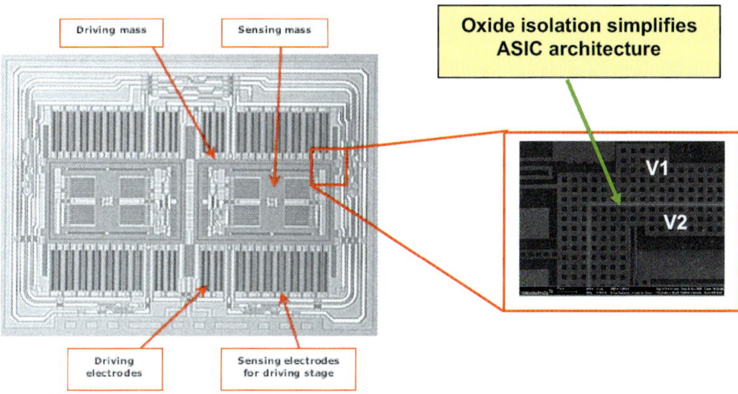

Fig. 7.   AY515DT: mechanical sensor with driving and sensing mass, with zoom on the oxide isolation

The AY515DT has already been tested in the field with promising results. In Fig. 8 is presented the setup of a real navigation test performed in a high-density urban area in cooperation with a third party. The test aimed at comparing the output from STM gyro with a similar device of the competition:

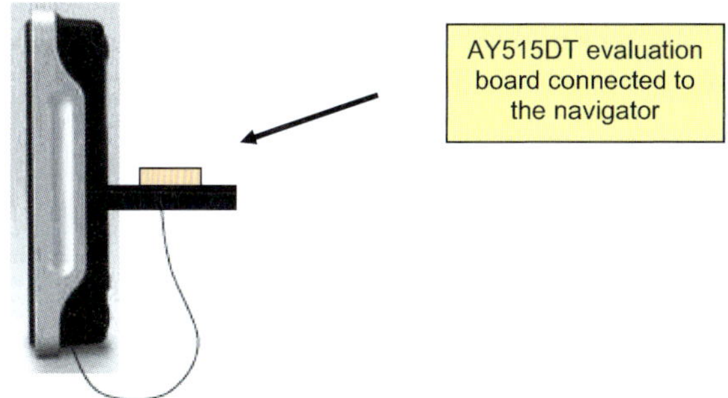

Fig. 8.   The AY515DT: test setup in the field

Two tests were performed in sequence:

▶ 1-minute sensitivity calibration (indoor), where the navigation module was rotated of 90 degrees and then rotated back, to verify the output:

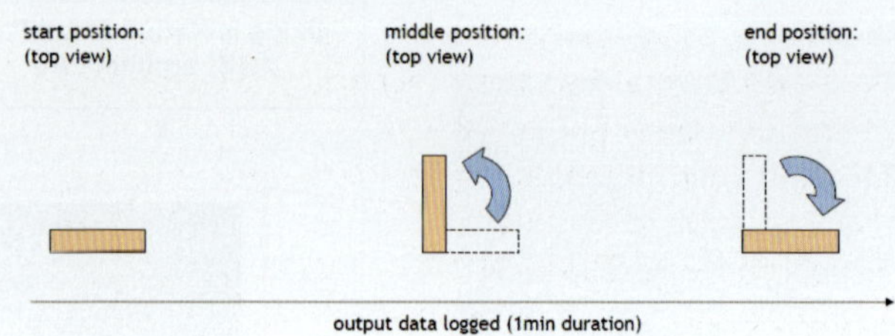

Fig. 9.   Test procedure

▶ 10 minutes real drive outdoor, where data coming from the gyro embedded in the navigator were compared with those ones coming from AY515DT, mounted externally.

By plotting the angle measured by the gyroscope vs. time and, using basic trigonometric algebra, by converting the angle into a 2D view (x-y plot), the driven trajectory can be calculated.

The results (not reported here) show a close match between the actual driven route and the output data from STM gyro. Route reconstruction based on the AY515DT data is more accurate, indicating better time stability and offset drift performance, both crucial for a reliable inertial navigation.

3   New Products in the Pipeline

The market trend, which is clearly visible from both first-tier system suppliers and aftermarket manufacturers, is to better address their designs with specific care on power consumption and low-drift in temperature, without compromising the accuracy needed by the application and by keeping, at the same time, an attractive cost.

Other than the 2-axis analog accelerometer for automotive applications that will be launched this year to complement the product portfolio, efforts are now addressed towards the definition of the next generation of low-power/low-drift 3-axis digital accelerometers for automotive and of multi-axis combo devices, integrating acceleration, rate, and magnetic sensors

The interface IC of the new accelerometer is able to read signals from a 3-axis mechanical element with a typical current consumption of 10 μA @ 10 Hz ODR (Output Data Rate). This ASIC communicates with the environment via SPI or

I2C and is able to provide signals for motion detection and sleep/wake-up functions. Follows is an excerpt of the main electrical characteristics:

▶ 1.8 V to 3.6 V single supply operation
▶ ±2 g / ±4 g / ±8 g dynamically selectable full-scale, with 2 independently programmable interrupts
▶ Power consumption of 10 μA @ 10 Hz ODR
▶ Offset drift in temperature below 0.2 mg/°C
▶ 12-bit guaranteed accuracy
▶ Sleep to wake-up function to optimize power consumption

The main parameters of the mechanical element are shown in the following table:

	X	*Y*	*Z*
C_0 *[fF]*	*400*	*400*	*400*
Non linearity [%]	*1*	*1*	*1*
Sensitivity [fF/g]	*7.5*	*7.5*	*8*
F_{res} *[kHz]*	*2*	*2*	*1.7*

Tab. 1.   Sensor element parameters

The sensor element is coupled with an ASIC developed in HCMOS BiCMOS technology available in ST. The system obtained has shown excellent performance both in terms of power consumption and stability over temperature.

Fig. 10.   Digital output ASIC for three-axis ultra low power accelerometer

Measures on the bench confirm the simulation results. Fig. 11 shows the 0 *g* level drift measured on 22 samples exposed to the following temperature cycle: [+25°, -40°, -20°, 0°, +25°, +55°, +85°, +25°]. The DUTs were soldered on boards and always powered at 2.5 V; acquisition was done after a temperature stabilization of 2 minutes:

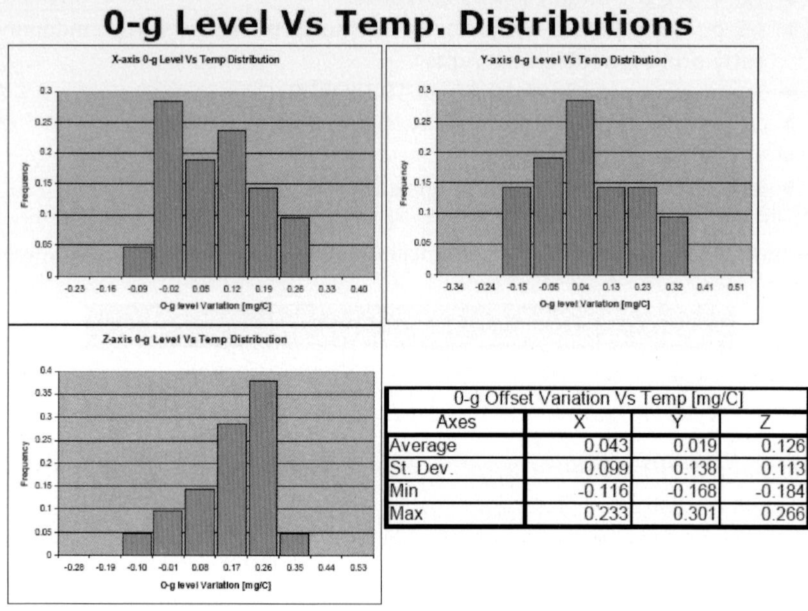

0-g Offset Variation Vs Temp [mg/C]			
Axes	X	Y	Z
Average	0.043	0.019	0.126
St. Dev.	0.099	0.138	0.113
Min	-0.116	-0.168	-0.184
Max	0.233	0.301	0.266

Fig. 11.   Offset drift in temperature of the new ST accelerometer

The new high-performance digital accelerometer is therefore matching the industry requirements to have high-accuracy, low power consumption, low temperature drift, and attractive cost.

4   Conclusions

After becoming an undisputed leader in MEMS for consumer electronics, mobile, and computer, STMicroelectronics is strongly investing to quickly gain acceptance and market share in the more challenging automotive environment.

MEMS growth in automotive is guaranteed by two main driving forces:
- ▶ an increasing number of applications that require sensors
- ▶ the proliferation of government mandates that compel the implementation of certain functions in the car. These mandates are both in the field of safety and non-safety, and STMicroelectronics has made the decision to address both areas.

In this paper, it is explained how the company, by leveraging the economies of scale of a proven 200 mm wafer technology, intends to become a recognized MEMS player in new emerging automotive applications, by fulfilling the technical and economical needs of more and more demanding customers.

References

[1]   R.Dixon, J.Bouchaud, "Driver Safety: Key to Robust Automotive MEMS Growth", ISupply Market Tracker, July 2008.

[2]   Conselho Nacional de Transito – CONTRAN, "Resolução Nº. 245, Brazil, July 2007.

[3]   Vigna B., "More than Moore: micro-machined products enable new applications and open newmarkets", Invited Talk, Proc. International Electron Devices Meeting, Washington, DC, USA 2005.

[4]   www.st.com/mems

[5]   Oboe R, Antonello R, Lasalandra E, et al. Control of a z-axis MEMS vibrational gyroscope. IEEE/ASME Transactions on Mechatronics 10:364-370, 2005.

[6]   (http://www.verkeerenwaterstaat.nl/english/topics/mobility_and_accessibility/roadpricing/index.aspx)

Marco Ferraresi, Stefano Pozzi
Via Tolomeo, 1
20010 Cornaredo (MI)
Italy
marco.ferraresi@st.com
stefano.pozzi@st.com

Keywords: MEMS, non-safety, accelerometer, gyroscope, low power, low drift

Fault-tolerant ASIC Design for High System Dependability

G. Schoof, M. Methfessel, R. Kraemer
Innovations for High Performance Microelectronics

Abstract

Fault-tolerant devices are becoming more and more important in safety-critical applications. In addition, because of further decreased geometries, integrated circuits are becoming more susceptible to induced interference. This paper presents new methods and design concepts to make application specific integrated circuit (ASIC) devices fault-tolerant to effects generated in the harsh automotive environment, especially to single event effects (SEEs). We describe how to mitigate single event effects which can immediately affect the function of electronic components. ASICs provided with this technique will increase the reliability and dependability while simultaneously maintaining the full real-time behaviour of the system.

1   Introduction

There is a current trend in the automotive industry that vehicle electronics moves from the support of non-critical applications, e.g. for more comfort, to controlling even safety-critical applications, such as systems for collision detection and avoidance, engine management, X-by-wire steering and braking and many others.

As the portion of electronics in vehicles grows and the usage of modern wireless communication devices increases, electromagnetic compatibility (EMC) becomes more and more important. It is a challenging task for system designers to minimize electromagnetic susceptibility (EMS) against electromagnetic emissions (EME) from other electronic devices like mobile phones or GPS systems. Every new system has to be designed in a way to minimize its own emissions in order to comply with the relevant standards. But in the harsh automotive environment we find many other sources of radio interference, caused by switching of inductive loads and large supply transients, also caused by high switching loads [6].

EMC considerations are necessary at each system level down to the level of integrated system-on-chip (SoC) devices which are in the focus of this paper. ASIC technology is generally moving toward smaller structures and very low voltage. As a result, very low charges are stored on each internal node, reducing the signal-to-noise ratio and reliable functionality dramatically. To guarantee EMC compliance for automotive applications, several standards for designing and testing of integrated circuits (ICs) have been passed: IEC 61967 (EME measurements between 150 kHz and 1 GHz), IEC 62132 (EMS measurements between 150 kHz and 1GHz) and ISO 7637 (transient disturbance by conduction and coupling).

Another important source of interference is particle radiation caused by alpha particles, neutrons or, cosmic rays at earth level. Energetic particles potentially induce charges in a semiconductor material which may lead to transients and subsequent soft errors in storage elements such as flipflops, latches, register and memory cells. As the feature size of integrated circuits has significantly decreased over the last decades, the susceptibility to particles radiation has increased by many orders of magnitude [1 - 3].

In this paper we concentrate on protection against so-called single event upset (SEU) faults which are correctable state or bit errors in ASICs caused by any of the above mentioned sources of interference [4, 5]. Already existing techniques [7 - 12] are briefly explained. Our new design approach addresses real-time requirements as well as the reduction of chip area and cost while maintaining a high protection level. Instead of the commonly used triple modular redundancy (TMR) we use modified double modular redundancy (DMR) with the same protection level.

In addition to SEU protection we present a new area-saving technique to protect ASICs against both SEU and single event latch-up (SEL) faults. Latchup generates a shortcut between power supply and ground in CMOS devices and is potentially destructive. SEL protection by technological means is generally suited to protect latchup-sensitive CMOS devices and can be extended by current-supervising circuits [14, 15]. Additional SEL protection by design techniques, as we propose, is required for highly dependable real-time capable systems, for example military vehicles in a radiation-contaminated environment.

2 Common Fault-tolerant Design Methods

The goal of fault-tolerant design is to improve the system dependability, usually by preventing signals and/or storage elements from changing to a wrong state. This can be done e.g. by special redundant circuits and related circuit dimensioning. Alternatively, it is also possible to accept an incorrect state and to use a redundant circuit to correct it immediately. The advantage of the second method is that it is independent of the type of interference and therefore eases the design procedure.

Redundancy is always required for any fault-tolerant design method. The higher the redundancy level, the higher is the protection level but the larger is also the chip area, power consumption, and cost. Therefore we have to trade off between the redundancy level and the dependability requirements in the automotive industry.

Furthermore, fault-tolerant design methods can be used at different levels in system hierarchy, as on the level of the whole system, at the level of system components (ASIC device or sub-systems within an ASIC) or even down to the gate logic level. The higher the level the easier the design process is. But the main problem for protection on higher design levels is the increased effort for synchronizing a faulty sub-system to its redundant counterpart. Typically this is the reason for loosing the real-time capability of the system because it becomes necessary to stop the running system task or system clock during the synchronisation [4, 5]. Therefore in this paper we concentrate on protection at the gate logic level.

Dependent on the application and on the available sub-systems within an ASIC, a fault-tolerant design can be limited to safety-critical parts only. This always makes sense if the cost factor is very important and redundancy for non-critical systems would unnecessarily increase the chip area. This possible aspect of cost saving is mentioned here but will not be considered further in the paper when the fault-tolerant design techniques are described.

2.1 Logic Design using Triple Modular Redundancy

Different methods to mitigate SEUs have already been published [7 - 12]. An often used example is the triple modular redundancy (TMR) design with voter. The voter is typically realized as majority voter which selects the majoritarian signal on its three inputs. Fig. 1 shows the TMR principle consisting of a combinational logic block with a number n of input signals, connected to three redundant D type flipflops and one voter. The n input signals are either device inputs or connected to other voter outputs. A complex logical system typically consists of such basic elements, with a varying content and size of the combinational logic blocks.

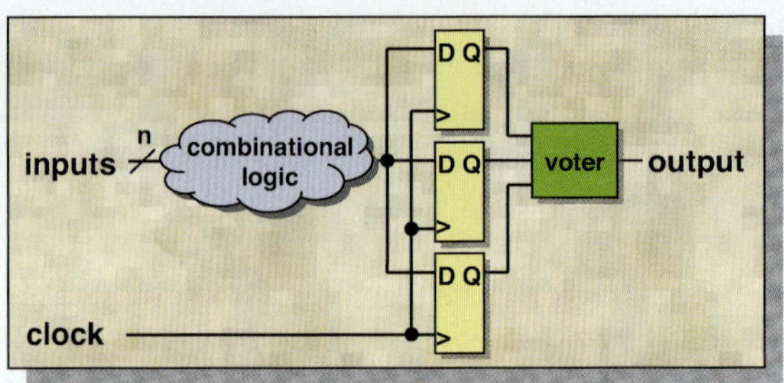

Fig. 1. Simple TMR flipflop

Fig. 2 shows a full TMR approach using additional triple modular redundancy for the combinational logic as well as for the voters to avoid non-correctable multiple bit upsets (MBUs). This happens if a transient pulse changes the input signal of all (or at least two) redundant flipflops simultaneously during an active clock edge. For similar reasons the clock tree and the reset tree (not shown here) should also either be triplicated or otherwise protected.

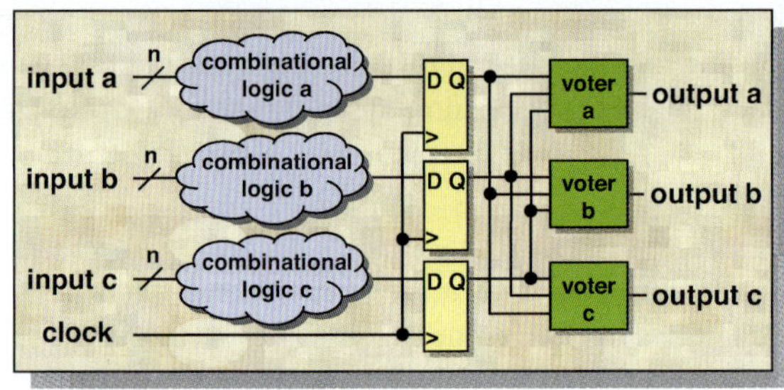

Fig. 2. Full TMR logic

Triple modular redundancy, especially full TMR, gives the logic design a high level of fault protection (even against static faults) but at the expense of increased chip area, power consumption and cost. Further advantages are the relatively simple design technique and the real-time capability because a non-static bit upset is corrected automatically in the following clock cycle.

2.2 Fault-tolerant Memory Designs

Similar TMR principles can be used for on-chip memory protection. In this case complete memory blocks are triplicated and the outputs are voted. Because the voting between redundant cells within the memory block itself is not easy, additional effort is necessary to guarantee synchronization without affecting the system real-time behaviour. Therefore and because of disadvantages mentioned above (increased area and cost) other protection techniques are better suited.

A different possibility to protect memories is to use special memory cell designs. There are many proposals available [7 -9] which will not be described here in detail. A shared problem of these techniques is that interference pulses during a memory access (for example, in the address logic or read amplifiers) are not automatically mitigated and can lead to faults. This makes additional effort necessary to reach the intended protection level. We present in this paper memory protection techniques using double modular redundancy (DMR) and explain the approach in the following Section 3.

3   SEU Protection using advanced DMR

Double modular redundancy means we have only two modules with the same functionality. If one module experiences a fault event, some additional logic is required to select the unfaulted module output for continuous functional system behaviour. We present such a logic which enhances a simple DMR design to show the same protection level as is achieved by TMR designs.

3.1   Logic Design using Double Modular Redundancy

To obtain a DMR system, we can start from the TMR case and reduce the number of flipflops and use a modified voting scheme. The resulting DMR design can replace the simple TMR logic in (Fig. 1) and is shown in (Fig. 3). The proposed DMR voting works differently compared to that of a standard TMR design. If, for example, both flipflop outputs show different values, an SEU is simply assumed and the (majority) voter output does not change the previous value. This is realized by a feedback connection to the third voter input, effectively turning the voter into a new storage node. All erroneous signal changes on one flipflop output between two active clock edges are mitigated in this way.

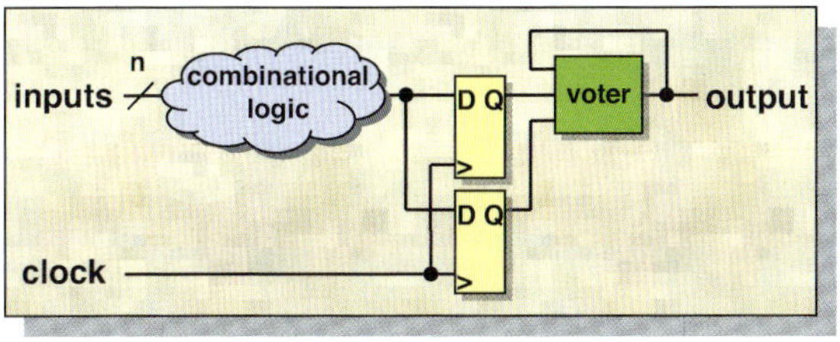

Fig. 3.   DMR flipflop Example

The redundancy is significantly reduced compared to the conventional TMR design. However, both designs cannot repair the error if a transient pulse changes the flipflops input value during an active clock edge. In this case, all flipflop outputs will switch to the wrong state. One solution is to introduce delay elements at the flipflop inputs to avoid that a transient affects the inputs simultaneously. Alternatively, a special latch design is described in [7]. In contrast to the simple TMR logic design, transients during an active clock edge, resulting in one erroneously not-switching flipflop, are not detected. In this case the second correctly switched flipflop output would be considered as a fault. Furthermore, a simple DMR flipflop as in (Fig. 3) will not mitigate static or long-time errors in one flipflop. In these cases the voter output will also become static (depending on the second flipflop, the value will sooner or later reach the erroneous state).

Another extended approach uses additional full redundancy in the combinational and voter logic (Fig. 4). Full redundancy avoids the problem of simultaneous transients on both flipflop inputs. Moreover the DMR voter is given additional multiplexer logic to avoid errors leading to a not switching flipflop output as described above.

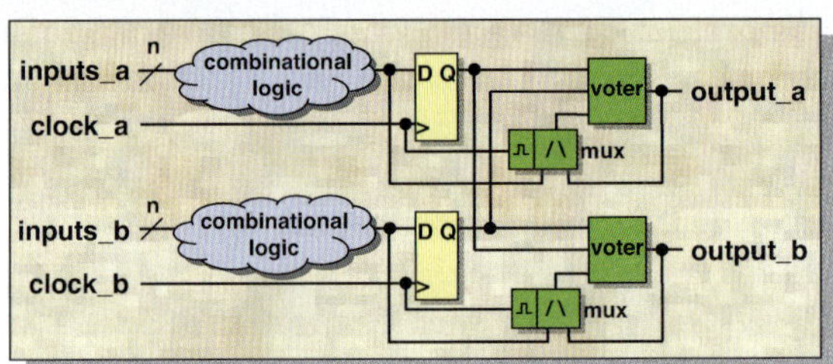

Fig. 4.   Full DMR Example

The multiplexer select input signal is derived from the clock signal and selects the D input of the flipflop for a short time after the positive (active) clock edge. Such full DMR example can tolerate long-time and even static faults of one flipflop but, of course, cannot protect the redundant non-affected flipflop against further faults during this time. The synchronization after SEUs occurred is done automatically. After long-time faults, the time needed for synchronization depends on error propagation in the affected path. Single events are synchronized immediately after the next clock cycle. This design example is interesting because it can also easily be used for SEL protection as explained later in Section 4.

3.2 Memory Design using double Modular Redundancy

Memories need a different protection strategy as was already mentioned in Section 2. A simple protection method is the use of coding techniques instead of duplicate memories. A code uses additional check bits for every memory word. Depending on the number of check bits, an error-detection-and-correction (EDAC) logic can detect and possibly correct a certain number of bit errors. To avoid that multiple bit errors (MBUs) accumulate over time, periodically read-correct-write cycles should be performed for the whole memory. This so called "scrubbing" could be done automatically by the EDAC, either during inactive memory cycles or via a second memory port to guarantee the full real-time behaviour for ordinary system memory accesses.

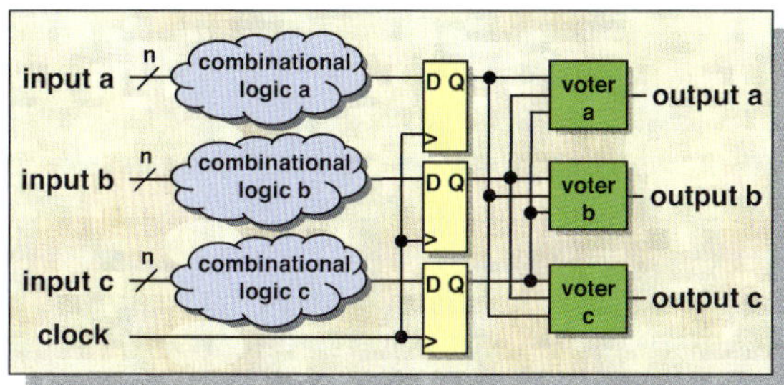

Fig. 5.   Memory SEU Protection (Data Flow)

In Fig. 5 a code-protected memory is shown using d data bits and c check bits. The coder and decoder (CODEC) performs data coding/decoding and the EDAC error detection/correction functionalities. The necessity of a second port is optional as mentioned above and depends e.g. on the expected frequency of memory accesses. As a further advantage, errors in address or control lines to the memory block are verifiable with a second port and some additional compare logic. If more bit errors are detected than the EDAC can correct, a system exception is generated. The CODEC/EDAC itself can be protected as already explained for logic DMR. Compared to unprotected systems the CODEC/EDAC logic will increase the memory access time somewhat and therefore marginally decrease the speed of memory accesses.

If the CODEC/EDAC in Fig. 5 is protected by full DMR design as shown in (Fig. 4), we have two redundant CODEC signals corresponding to a single signal to the memory. Two redundant inputs can be connected directly by wire but two redundant outputs must be combined by some logic such as an additional voter. Any interference happening after the code is generated by the CODEC logic and leading to a SEU in a memory cell will automatically be removed after the word is checked by the EDAC logic.

In a substantially different approach, two separate memories can be used to implement a full DMR design. As seen in Fig. 6 the memory is now duplicated. In contrast to a DMR protected EDAC version as shown in Fig. 5, now the EDAC logic is additionally designed to compare both sets of memory output data. Data can be corrected automatically (independent of system access) if erroneous data in one memory are detected. This helps also to reduce the number p of check bits for each memory word, for instance to a simple 1-bit parity check.

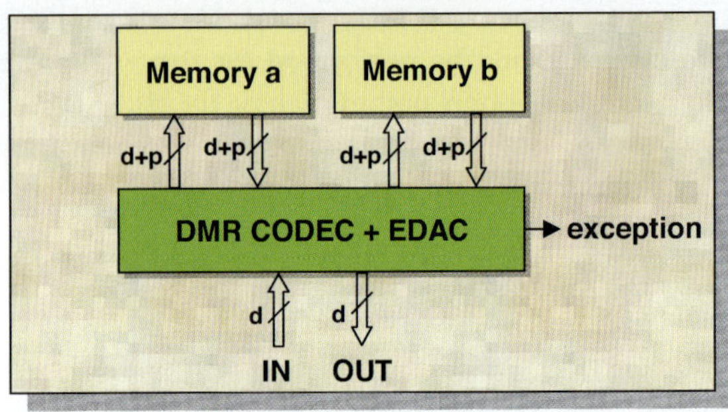

Fig. 6.   Full DMR Memory Protection (Data Flow)

More difficult to solve are problems generated by long-time or static faults before coding data with the CODEC logic. Differences in the redundant input data to the CODEC/EDAC logic lead to system exceptions because such faults are not correctable. One solution is to place a CODEC/EDAC logic unit close to the data processing unit in such a way that only coded data is available everywhere else in the ASIC. This data can be checked and corrected by a simple EDAC logic before storing it into or reading out from the memories and/or registers. It must be considered that between the location of a static fault and the next check/correction point, the logic must be protected against SEUs. But even if no further SEU protection is used, such an error is not hidden and the problem can be handled at a higher system level.

A redundant memory allows simple synchronization in spite of real-time requirements. It is also important to point out that access errors to the memory, such as faults in the memory address decoders, are easily detectable in contrast to a (single-ported) memory as shown in Fig. 5. Of course, this full DMR memory design needs more area than the example in Fig. 5, but a further advantage becomes visible if we consider additional SEL protection as described in Section 4.

3.3 ASIC Design Flow

It requires significant effort to design a fault-tolerant ASIC using standard design tools. Currently, already existing unprotected designs must be re-designed from the top level without any possibility of re-using previously generated netlists. Therefore we are working on an improved design flow to generate fault-tolerant designs from the netlist level in a semi-automatic way.

Our activities are divided into two steps. First, a newly developed tool will create a DMR netlist from an existing netlist and will prepare it for simulation using an existing testbench. Afterwards the layout can be generated and tested in the regular way, yielding a fault-tolerant design using SEU protection.

If additional SEL protection is needed, another tool at the layout level is required as will be explained later in Section 4. We first explain the technique to protect ASIC designs against SEU and SEL simultaneously.

4 Combined SEU and SEL Protection

This section gives an overview of methods for combined SEU and SEL protection, covering highly dependable as well as real-time capable system properties, even suited for special automotive applications such as in radiation-contaminated environments. Combined SEU and SEL protection is also an important issue in the aerospace/space industry, for detectors for particle accelerators and in other applications.

The main characteristic of SEL protection is to detect the high current generated by latchup (i.e., a short circuit between power supply and ground) to switch off the power of the affected subsystem and, after short recovery time, to switch on the power again. Such SEL and subsequent power cycling usually leads to loss of information or, at least, to loss of time needed for re-establishing the system state. Therefore it is difficult to meet real-time requirements.

Existing methods to handle SELs are centralized on- or off-chip current monitoring circuits to switch off the power if a current limit is exceeded [14, 15]. The disadvantage is that after a subsequent power-on it is necessary to extensively re-establish the system behaviour, hereby possibly experiencing an unavoidable data loss. Furthermore so called micro SELs (i.e. those with relatively low latch-up current) are not detectable if their additional current flow does not exceed the monitored operating current limit.

As a new approach we propose a combination of SEU and SEL mitigation on the design level. We take advantage of the redundancy that is introduced for SEU protection to simultaneously provide SEL protection. In contrast to already existing solutions, our approach handles not only SEUs but also SELs in real time without affecting other connected systems nearby. The following examples concentrate on DMR designs but can easily be modified to protect TMR designs in the same way.

4.1 Separately powered redundant Modules

A basic example showing how to provide SEL protection to a system-on-chip is shown in Fig. 7. The system necessarily must be doubled and SEL controllers and power switches are added. Both sub-systems (SYSA/B) and controllers (CTLA/B) are separately powered (VDDA/B) via power switches (SWA/B) whereby each controller monitors the current of the other side (signals DETA/B). If a latchup occurs, power is switched off and on again (signals LUPA/B to SWA/B) and all lost information of the affected subsystem is afterwards copied from the unaffected system (signals LUPA/B to SYNCA/B and all storage node I/O signals DATAA/B).

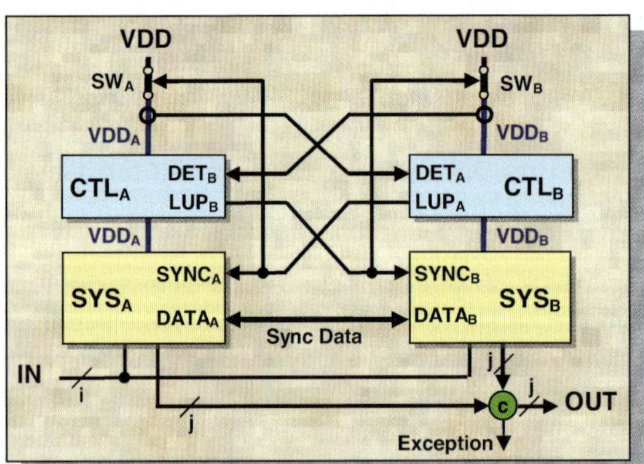

Fig. 7.   Principle of SEL Protection

This principle of SEL protection does not automatically support combined SEU and SEL protection. If the system requires additional SEU protection, both redundant systems must embed SEU protection independently or, at least, compare their outputs to detect system faults. Here synchronization without affecting the real-time behaviour is very difficult.

4.2 SEU and SEL Protection for Logic and Memories

An example using combined SEU and SEL protection for logic circuits is shown in Fig. 8. Here, the systems (SYSA/B) are not independently protected against SEUs as was the case in Fig. 7. Instead each system consists of one-half of the SEU protected system according to Fig. 4 and the system connections are between n flipflop outputs (FFA/B) and n voter inputs (VA/B) in both directions.

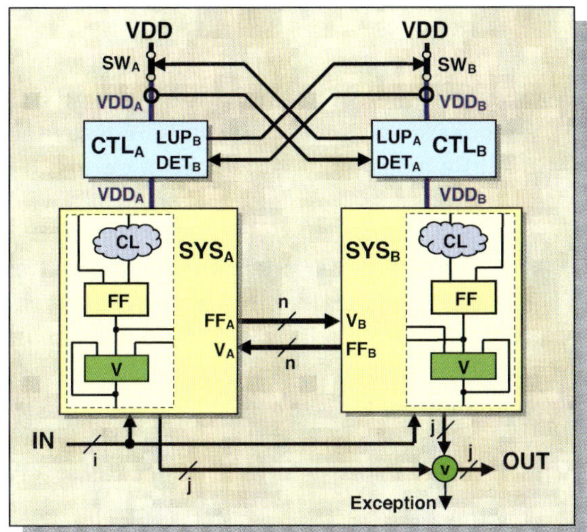

Fig. 8.   Combined SEU/SEL Logic Protection

In contrast to Fig. 7 the synchronization after SEL is done automatically as was already seen in Fig. 4 by the connections between voters and flipflops of the redundant sub-systems. Power-off time can be considered a long-time error and depending on this time, the automatic synchronization after SEL could need several cycles because of error propagation. But this effect can be reduced by additional effort in the voter selection logic or by data coding and checking as mentioned in Section 3. At the point when the redundant outputs must be combined to a single signal, a special separately powered (voter) logic is required, able to detect and ignore a faulted static input signal. Compared to the basic example in Fig. 7, here much less redundancy is needed to attain combined SEU and SEL protection. The design is fully real-time capable.

A combined SEU and SEL protection for memories is easily achievable if power switches and controller, as seen in the previous example in Fig. 8, are added to the full DMR design according to Fig. 6. To mitigate SEL related faults, the second separately powered memory block is needed in this case.

As a side-effect the power switches for SEL protection can be used for power gating to reduce power consumption by switching off temporarily inactive parts of the system. Furthermore, if the high protection level is temporarily not necessary, one complete redundant path can be switched-off. Or, as an alternative in this case, the redundant systems can be used in parallel to increase the system performance.

4.3 Design and Layout Considerations

In particularly for SEL protection, it is advantageous to split the layout into many small layout areas, each controlled by smaller power switches, in order to reduce the SEL affected areas. This significantly reduces the probability of a second SEL (or SEU) in a logic area that is redundant to a logic area where another SEL is already in process. Furthermore smaller controlled logic areas decrease the current supply per power switch and may increase the SEL controller sensitivity with respect to the previously mentioned micro latch-ups. In ASIC designs one possibility is to use cross points of vertical and horizontal power lines for placing the power switches. This reduces the additional area consumption to a minimum or even to zero because such cross points are normally not used for cell placements.

All layout modifications needed for adding SEL protection to a full DMR design will be done automatically by another tool currently in development. By making sure that the required modifications do not influence the timing parameters of the system, we ensure that no further simulations for verification will be necessary.

5 Summary

We have presented fault-tolerant techniques to protect ASIC designs in the automotive environment for safety-critical applications. It has been shown how a high protection level can be combined with reduced chip area and cost by using an improved DMR design. For special highly dependable system requirements, additional protection against latchup effects can easily be incorporated.

One of the most important features is a fault-tolerant design which does not affect the real-time behaviour of the system. We are developing design tools to support and ease the design process for such fault-tolerant designs. The final goal is to take already existing designs and make them fault-tolerant for safety-critical applications.

Our focus is to protect ASIC designs by special design techniques and not by technological means. Our proposed ideas are independent of the used technology but will provide additional safety if any other protection measures are already employed on the level of technology (e.g. silicon-on-insulator - SOI) and/or layout geometry (guard rings etc.). We do not try to prevent SEUs or SELs in ASICs but handle them in an efficient way and in real-time.

Currently we are developing a fault tolerant ASIC implementing the presented SEU and SEL protection techniques. Subsequent EMC and radiation tests will show the effectiveness of our approach.

References

[1]   Shivakumar, P. et. al, Modeling the effect of technology trends on the soft error rate of combinational logic, Proceedings International Conference on Dependable Systems and Networks (DSN '02), Pages: 389 - 398, 2002.

[2]   Baumann, R., The impact of technology scaling on soft error rate performance and limits to the efficacy of error correction, Electron Devices Meeting (IEDM '02), Digest, Pages: 329- 332, 2002.

[3]   Nelson, V.P., Fault-tolerant computing: fundamental concepts, Computer, Volume 23, Issue 7, Pages: 19 – 25, 1990.

[4]   Baleani, A. et. al, Fault-tolerant platforms for automotive safety-critical applications, Proceedings of the International Conference on Compilers, Architectures and Synthesis for Embedded Systems, Pages: 170 – 177, 2003.

[5]   Manzone, A. et. al, Fault tolerant automotive systems: an overview, On-Line Testing Workshop, Proceedings, Pages: 117 - 121 , 2001.

[6]   Ohletz, M.L., Schulze, F., Requirements for design, qualification and production of integrated sensor interface circuits for high quality automotive applications, Advances in Sensors and Interface (IWASI 2007), Pages: 1-7, 2007.

[7]   Rockett Jr., L. R., An SEU-Hardened CMOS Data Latch Design. IEEE Transactions on Nuclear Science, NS-35(6):1682-1687, 1988.

[8]   De Lima, F.G., Designing single event upset mitigation techniques for large SRAM-based FPGA devices. Universidade federal do Rio Grande do Sul, http://www.inf.ufrgs.br., 2002.

[9]   Whitaker, S. R., Single Event Upset Hardening CMOS Memory Circuit, U.S. Patent No. 5,111,429, 1992.

[10] Gambles, J. W. et. al, Apparatus for and Method of Eliminating Single Event Upsets in Combinational Logic, U.S. Patent No. 6,326,809, 2001.

[11] Gambles, J. W. et. al, Radiation Hardness of Ultra Low Power CMOS VLSI. 11th NASA Symposium on VLSI Design, 2003.

[12] Zhang, M., Shanbhag, N.R., Design of Soft Error Tolerant Logic Circuits, Workshop on System Effects of Logic Soft Errors, 2005.

[13] Schoof, G., Kraemer, R., Jagdhold, U. Wolf, C., Fault-tolerant Design for Applications Exposed to Radiation, Proceedings of the Conference DASIA 2007 - Data Systems in Aerospace, 2007.

[14] Layton, P. et. al, Single Event Latchup Protection of integrated Circuits, Proceedings Third ESA Electronics Components Conference, ESTEC, 1997.

[15] Czajkowski, D. et. al, Radiation induced single event latch-up protection and recovery of integrated circuits, U.S. Patent No. 6,064,555, 2000.

Gunter Schoof, Michael Methfessel, Rolf Kraemer
Innovations for High Performance Microelectronics
Im Technologiepark 25
15236 Frankfurt (Oder)
Germany
schoof@ihp-microelectronics.com
methfessel@ihp-microelectronics.com
kraemer@ihp-microelectronics.com

Keywords: fault-tolerant design, automotive, safety, EMC, EMS, SEU, SEL, DMR, TMR

Short PWM Code: A Step towards Smarter Automotive Sensors

L. Beaurenaut, Infineon Technologies

Abstract

Digital sensors offer the possibility to increase globally the functionality of automotive systems and at the same time to further reduce their cost. The Single Edge Nibble Transmission (SENT) digital communication protocol is a promising low-cost solution for communication between sensor satellites and a microcontroller. This paper presents the Short PWM Code (SPC) protocol aiming at extending the SENT communication link in terms of functionality, performance and cost efficiency. SPC provides application relevant functionalities which are not included in SENT while still keeping as close as possible to the original protocol. In particular: bi-directionality, synchronicity and bus capability.

1   Introduction

Over the last decades, embedded automotive systems have had their complexity continuously increasing in order to meet new requirements in terms of emissions and safety. The introduction of still more stringent emission regulations combined with the need for reducing fuel consumption and CO_2 emissions, generates new challenges in terms of engine efficiency. Besides, the introduction of "by-wire" technology and the standardization of Safety Integrity Level for automotive applications mean that new concepts must be implemented in order to have safe and dependable systems. At the same time, the recent success of low cost cars has proved that the introduction of technical innovations should also be paired with a reduction of system costs. One consequence of this is the need for smart digital sensors.

1.1   Migrating towards Digital Sensors

Most of the sensors used today in diesel and gasoline engine management solutions are analog sensors [1]. For example: temperature sensors (temperature after the air filter, after and before the Diesel particulate Filter (DPF), before the Selective Catalyst Reduction (SCR), etc.), pressure sensors (Manifold Air Pressure (MAP), In Cylinder Pressure, rail pressure, oil pressure, boost pressure, pressure differential across the DPF, barometric pressure, etc), air flow sensors, position sensors (Pedal and Throttle, etc.), piezo sensor (knock), and so on.

An analog sensor delivers a voltage (typically in the range of 0 to 5 Volts) which is sampled by an ADC embedded on the main microcontroller of the Electronic Control Unit (ECU). The number of ADC channels used depends on the complexity of the ECU, the application class addressed (low end or high end cars) as well as the control strategy chosen by the ECU supplier. Indeed, different approaches are here possible, relying either on physical measurements or making intensive use of mathematical models. Nevertheless, systems needing fifty ADC channels or even more are not rare in Engine Management. On the other side ADC modules are expensive to integrate into a microcontroller because they can not be shrinked in a similar way as digital logic.

Digital sensors offer interesting ways to reduce costs. Basically, an analog interface can only provide one data as a piece of information, i.e. the sensed physical value. Digital sensors can be clustered into a single module, which communicates with the ECU via a single bus. On the physical layer, the costs of the interconnect can significantly be reduced through the use of a single cable set. Bearing in mind that cables and interconnects may represent more than 25% of the total cost of an engine system, the potential savings are tremendous.

1.2   Single Edge Nibble Transmission

SENT is a 3-wire digital communication interface developed by the SAE SENT Task Force. SENT is an open standard [2]. It is intended as a replacement for the lower resolution method 10 bit ADC and PWM and as simpler low cost alternative to CAN or LIN.

SENT is a unidirectional point-to-point communication scheme, connecting a sensor to a controlling device (in practice a microcontroller on an ECU). After power-on, the sensor starts transmitting SENT frames asynchronously to the ECU. The signal coming from the sensor is transmitted as a series of pulses with data measured as falling to falling edge times. The 4 bit data packets are encoded into nibbles. A typical SENT frame contains 6 data nibbles (24 bits data) plus a Status nibble (whose use is optional and in most of the cases application specific). The baud rate achievable by a SENT channel is in the range of 24 kBit/s (3 µs tick time). SENT having been de-signed for low cost systems, it allows significant clock variations on the sensor's internal clock. Therefore, a

SENT frame starts with a calibration pulse, which allows the receiver to compensate the +/-20% variations of the sensor's clock and therefore accurately decode the incoming data. A typical SENT frame is depicted in Fig.1.

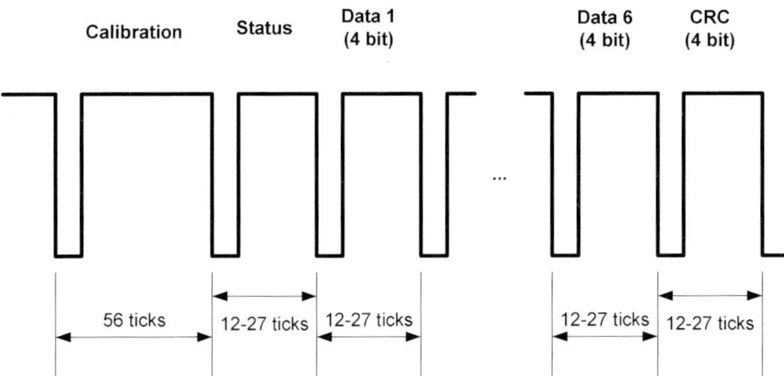

Fig. 1.   Typical SENT Frame

SENT gives the possibility to increase the functionality of the sensors. For example, in order to implement efficient On Board Diagnosis (OBD) strategies, self checks have to be run periodically by the sensor so that it can verify its own functionality. It is of course possible to notify a malfunction to an ECU using an analog way. However, the number of information that can be coded that way is in reality pretty limited due to noise. A digital communication link on the other hand gives a lot more flexibility. A large number of status information can be computed in the sensor and be read on request by the ECU microcontroller.

Another practical advantage of SENT over analog sensors is that data integrity is inherently increased by the use of digital coding. Digital communication links are less sensitive to environmental noise, and data integrity is further increased by the CRC included at the end of a SENT frame. With such features, the Safety Integrity Level (SIL) of the whole system can be increased.

Accuracy of the sensor is another aspect to be taken into account. Actually, digital logic is already massively used in today's automotive sensors. With MEMS technology, it is possible to couple a sensor cell (signal acquisition) digital signal conditioning blocks, such as filters, in a single chip. A common example is pressure sensors (MAP and BAP) where the acquired signal is processed internally in such a way that the sensor delivers a temperature independent voltage as a linear function of the pressure. So the sensor processes internally digital values and reconverts them into an analog voltage. The performance loss in terms of accuracy is obvious, compared to the fact that the same digital value could be transmitted directly to the ECU

Several communication interfaces are competing today as the solution for tomorrow's automotive sensors. SPI will probably be the preferred digital interface for on-ECU sensors, due to the wide support of this protocol in terms of hardware and software. For off-ECU sensors, SENT is a solution which is on its way of becoming well accepted by the automotive industry. SENT has however not yet reached the acceptance level nor the penetration rate as established communication standards like RS232, SPI or LIN have. However, SENT has been designed so that a simple capture/compare unit can be used to acquire the frames sent by the sensor to the ECU. Therefore, virtually every microcontroller in powertrain and safety applications can, with the adequate software driver, support SENT. It is expected that in 2010, most Engine management ECU in high end cars will support 2 to 6 SENT channels. This number could grow to 8 to 12, or even higher, as early as 2012

2   Short PWM Code

2.1   SPC Motivation

The aim of SPC is to provide application relevant functionalities which are not included in SENT, while still staying as close as possible to the original protocol and not increasing system costs. In particular:

- ▶ SPC is bidirectional. Commands can be sent from the microcontroller (SPC master) to a sensor (SPC slave).
- ▶ SPC supports synchronous transmissions. Data transmission is initiated by the master. As a result, the data rate delivered by the sensor is determined by the master. This offer multiple advantages: fixed sample rate, optimization of the CPU processing load to the application needs, possibility to synchronize the data flow of two sensors , etc.
- ▶ SPC supports multi-slave systems: several sensing elements can be connected to the ECU via the same set of cables, resulting in significant cost reduction
- ▶ SPC supports both 3.3V and 5V voltage domains (whereas standard SENT only supports 5V). This allows using SPC with practically almost any microcontroller designed for safety or powertrain applications, without additional on-board adaptation circuitry.

2.2 SPC Frame Definition.

SPC has been kept as much as possible compatible to the SENT J2716 standard. Therefore, no major effort is to be expected to design solutions which support both SENT and SPC. Fig. 2 gives an overview of an SPC frame. It can be seen that the frame definition is based on a standard SENT frame, but which is preceded by a trigger pulse and to which an end pulse has been added.

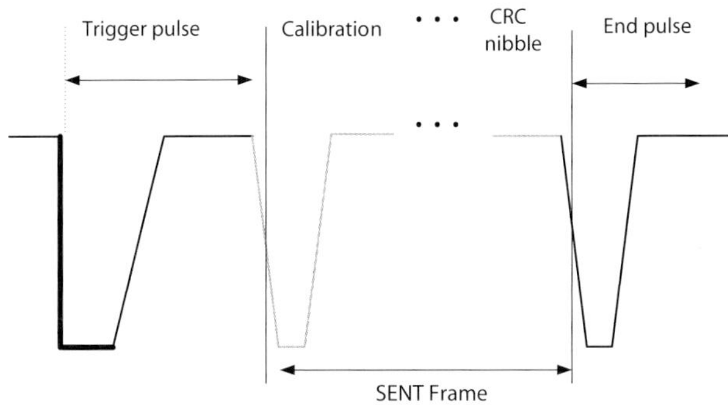

Fig. 2.   SPC Frame.

An SPC transmission is initiated by a trigger pulse. SPC is a synchronous communication link in the sense that a frame transmission is always initiated by the master. The SPC master takes the ownership of the signal line and generates the falling edge of the trigger pulse. It then holds the signal at a low level during a defined time. The master then releases the line and let the slave drive it. At the end of the trigger pulse, the slave starts sending a regular SENT frame. The trigger pulse has a defined length, which depends on the operating mode of the SPC link. In standard SENT communication, the information is coded within the time between two falling edges. The moment where a rising edge occurs within a nibble is of no significance. In SPC, the master codes information within the low time of the trigger pulse. That way, commands can be sent from the microcontroller to the sensors. The slaves can decode this information and react accordingly.

The motivation for having a dedicated trigger pulse is that SPC (as standard SENT is) is based on the time encoding of the information. Therefore, the exact timing of the nibbles is of primary importance. Consequently, special care has to be taken that there are no timing problems incurred by having different output stages and input trigger levels on the master and slave sides. Internal delays on both sides should not affect the transmission of the data. For example, if one had chosen to use the calibration pulse as a triggering pulse, the difference in the timing of the falling edges could generate an error when decoding the calibration pulse. In worst case, this could lead to the corruption of the complete data contained in the frame, due to the computation of a wrong correction factor. In order to avoid those timing issues, the triggering part and the data transmission part of the SPC protocol have entirely been decoupled. After being triggered, the slave transmits an entire SENT frame, which includes a regular calibration pulse and therefore allows the master to accurately determine the internal unit time of the slave.

Since SPC is synchronous, the time interval between two frame transmissions is not defined. In order for the CRC nibble can be properly decoded, it is necessary to add an additional end pulse at the end of the frame. This end pulse is only here to provide a falling edge so that the CRC nibble duration can be measured.

SPC is based on the fact that the signal line can be either driven by the master (during the trigger pulse) or by the slave (rest of the frame). This can be easily realized on electrical level by open-drain output structures. This also allows having several slaves sharing the same physical lines (Fig. 3). Using open drain outputs instead of push-pull outputs has the additional advantage that multi-voltage domains can easily be set up. As the protocol is only based on one single edge, only this slope must have a well defined behaviour.

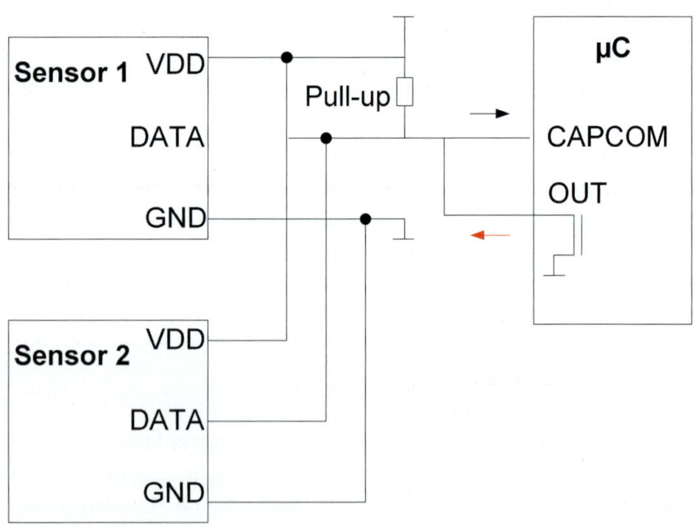

Fig. 3.   SPC Physical Layer

2.3 SPC Modes.

SPC defines four operating modes: synchronous mode, command mode, bus mode and long command mode.

In the synchronous mode, the master triggers a slave with a very short trigger pulse. This mode allows the master to request data at a specific moment (Fig. 4a). If multiple slaves are used in parallel for redundancy reasons, it is possible to acquire the data of both slaves at the same moment, significantly improving the comparability of the transmitted values.

The command mode can be used to transmit simple command information to a single slave by modulating the low time of the master trigger pulse (Fig. 4b). These command bits can be used by the slave to modify internal registers dynamically and adapt its function accordingly. Possible examples include an adaption of the content of the data frames, a change in a physical parameter in intelligent sensor slaves or to request status/diagnostics information from the slave. The amount of information that can be transmitted in this command mode depends on the oscillator variation in the slave and the acceptable trigger pulse length. Two bits of information can typically be transmitted.

The bus mode SPC uses a similar coding as the command mode, but this time the master low time is used to code the ID information. Several slaves can then be attached on the same bus line. Each slave only responds if the transmitted ID (encoded in the master low time) corresponds to their internally stored ID. Simple bus systems can therefore be built and both wiring and master computing resources can be economized. Similarly to the command mode, the number of slaves that can be uniquely identified is affected by their relative oscillator clock spread. Up to four sensors could be attached to the bus, the trigger pulse overhead of the protocol still being kept acceptably short.

The long command mode (Fig. 4c) differs from the other modes in the sense that the ECU is the transmitter, and the sensor the receiver. After addressing the sensor, the ECU interrupts the calibration pulse by pulling down the line. The ECU transmits its own nibbles on the selected sensor time base. The ECU has to adapt the sensors timing from an earlier received cycle.

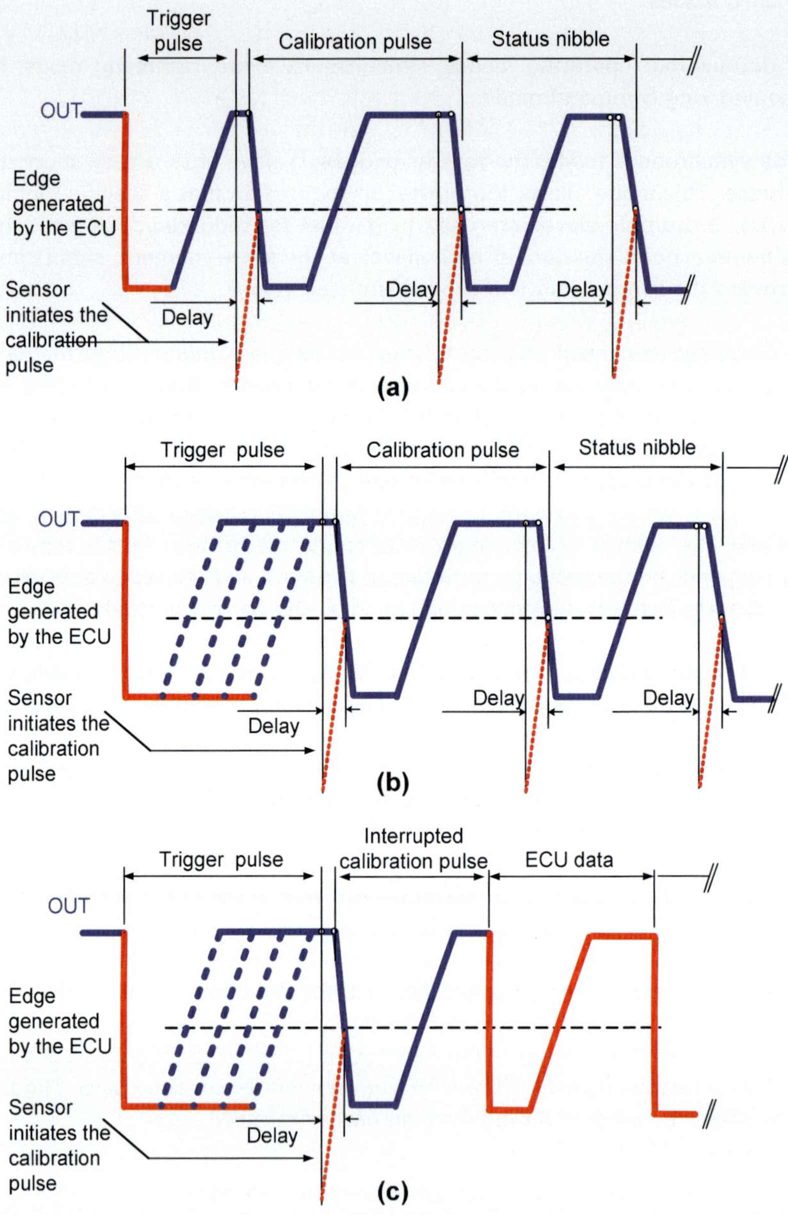

Fig. 4.   SPC Trigger mode (a), Command & Bus mode (b), Long Command mode (c)

The command code and the long command mode aim at covering different kind of functionalities. The command mode allows sending very simple commands only. For a Hall sensor, this could for example be used to change the measuring range of the sensor. For a pressure sensor, it could be to a diagnostic for the sensing element. The long command mode on the contrary allows a complex question / answer communication dialog between the ECU and the addressed sensor. One use case would be tuning protection, where the ECU could challenge regularly a sensor (e.g. MAP) to verify its authenticity.

3 Practical Implementations of SPC

3.1 From SENT to SPC

Only a few modifications are needed to adapt a SENT system to SPC. At hardware level, the electrical characteristic of the interface has to be changed from push-pull to open drain. Open drain capability is standard and supported by most, if not all, microcontrollers dedicated to powertrain and safety applications. On PCB level, this only means reducing the value of the pull-up resistance (Fig. 5) recommended by the SENT specification to a lower value, in order to ensure that signal transitions are fast enough.

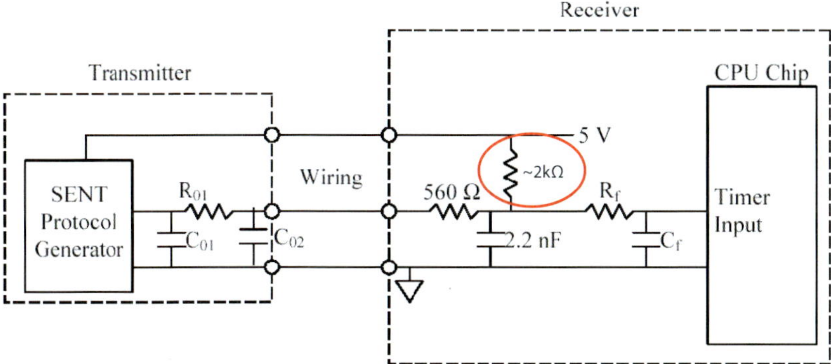

Fig. 5.   SPC Circuit Topology

The protocol decoding itself on the receiver side is almost identical to SENT. Due to the fact that the sensor data is only transmitted on request, the CPU load needed for a SPC channel will be smaller than for a regular SENT channel. All in all, a SENT system can be easily upgraded to SPC without major changes in the ECU design.

This makes the SPC approach especially attractive compared to current-based solutions, which are being discussed today as a low-cost alternative to existing voltage-based interface. A two-wire based interface may seem an attractive solution due to the cost savings on the cables. However, several issues need still to be answered in order to design a viable solution. First of all, there is no well accepted solution today that can cover both powertrain and safety requirements. PSI-5 [3], which would be the best candidate due to its wide diffusion in safety applications (e.g. side airbag), still needs significant modification in order to be powertrain suitable (for example: operation at low battery, initialization time, etc.). On the other side, the gain in terms of system cost is arguable: the savings for the cables are significantly compensated by the additional needs of an ASIC generating the wanted current levels. Besides, technological drawbacks have to be overcome, like higher heat dissipation in the sensor or the capability to offer a fast communication channel from the ECU directed to the sensor. For all those reasons, it is likely that a mature two-wire solution will not emerge in the short term.

3.2 SPC in Powertrain and Safety Application.

The SPC technology has been already successfully integrated in Infineon's TLE4998 Linear Hall Sensor, which in use in Electronic Power Steering (EPS) systems. This sensor provides a signal proportional to the measured magnetic flux. In an EPS application, two sensors are used redundantly to sense the torque information. The synchronicity provided by SPC is here necessary in order to have correlated measurements between the two sensors. Besides, the status nibble is used to transmit information about the internal status of the sensor: reset and overvoltage conditions as well as the magnetic range that is used for the data acquisition. The six data nibbles are used to transmit a magnetic field value and a temperature information. Compared to other interfaces such as analog or PWM, achieving typically below 12 bit data resolution, a higher resolution of 16 bit can be transmitted here. Additionally, the temperature information can be used for plausibility checks or additional processing in the ECU.

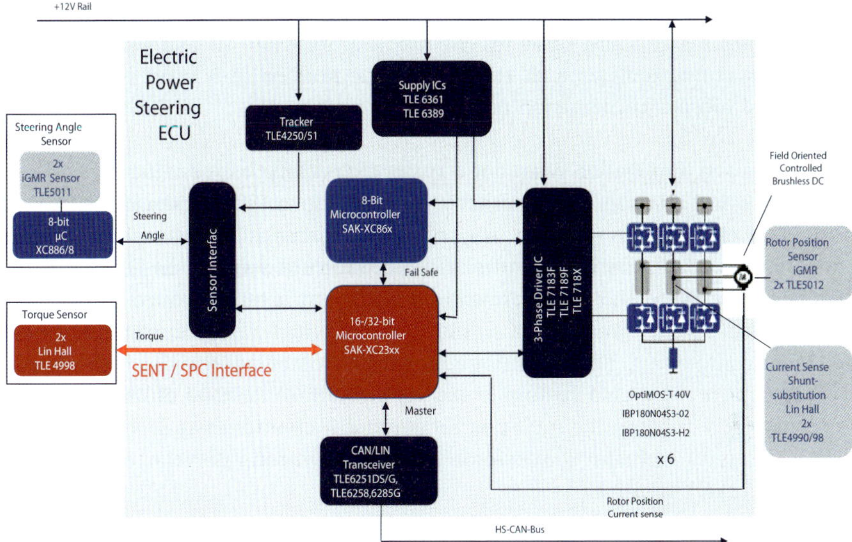

Fig. 6   Electronic Power Steering System

SPC can moreover be used wherever SENT is relevant, and therefore benefit from the wider acceptance of this protocol. In engine management applications, this concerns actually sensors located outside the main ECU and whose output is sampled at low data rate. Manifold Air Pressure (MAP) and Mass Air Flow (MAF) sensors are good candidates. Other subsystems in the engine are also seen as possible candidates for SENT: aftertreatment - the pressure sensors monitoring the load of the diesel particulate filter (DPF) or fuel delivery – the fuel rail pressure sensors.

One limitation of SPC is the effective bandwidth available on the bus. The trigger pulse and the end pulse increase protocol overhead. In the case of synchronous mode, the duration of those pulses can be kept quite small (in the range of 70 µs, assuming 3 µs tick time). One of the main ideas of the synchronous mode being anyway not to use the full bandwidth available (as a standard SENT sensor does by sending back to back frames), this constraint is probably acceptable in most of the cases. For the bus mode, this overhead can be significantly higher, due to the longer trigger pulse. In order to take into account the variation of the internal clocks of all the slaves attached to the bus, the IDs must be coded in the trigger pulse in such a way that no wrong interpretation of the timing occur. This additional timing margin will grow together with the number of sensors on the bus. In the case of four sensors, the trigger pulse may last as long as 230 µs, assuming 3µs tick time. Besides, the bus bandwidth being shared between several sensors, the bus mode is not suitable in the case where the sensors' data need to be read at a fast rate. In the cases where fast update rates are needed, appropriate measures could relax the timing constraints such as: decreasing the nominal tick time, reducing the number of nibbles transmitted per frame, etc..

As long as the application does not require extremely fast readings of the sensors, SPC could be used in the cases where, outside an ECU, several sensors are physically close to one another. In engine management for example, one could consider the MAP, the MAF and some temperature sensors. In DPF systems, this could be the differential pressure sensor and a temperature sensor. SPC sensors can be clustered into a single module, which communicates with the ECU via a single bus. The advantage of sensor clustering, besides costs, is that the integration of such a system is greatly simplified. Instead of having to deal with multiple sensors having specific electrical characteristics, an ECU communicates with a cluster via a single and standardized interface, almost like a "plug and play" solution.

The next natural evolution step to sensor integration would then be sensor fusion, where several sensing capabilities are integrated into a single monolithic device. The integration for example of temperature and pressure sensing is today technologically possible. One could even consider a fully integrated Electronic Throttle Control (ETC) module where Hall-effect based position detection is combined together with MAP and temperature sensing

4 Conclusion

Automotive systems will make a wider use of digital sensors in the future. SPC is a mature concept already implemented in Infineon's Hall sensors products. It provides an answer to some of the main challenges the automotive industry is facing: system cost reduction, tuning protection, OBD, data integrity. Rather than a new protocol, SPC has to be seen as an evolution of SENT, extending significantly its capability. The acceptance level of SPC and SENT in automotive systems will be directly correlated to the degree of interoperability between components from different suppliers. Standardization on both hardware and software levels will be a key factor of success.

References

[1]   Leteinturier, Xie, Zhou, Zhang, Tanh, Hou, Advanced gasoline engine management platform for Euro IV & CHN IV emission regulation SAE 2008-01-1704, 2008.

[2]   SAE SENT Task Force, „Single Edge Nibble Transmission for Automotive Applications", SAE J2716, 2008.

[3]   PSI-5 website, www.psi5.org.

Laurent Beaurenaut
Infineon Technologies AG
Am Campeon 1-12
85579 Neubiberg
Germany
laurent.beaurenaut@infineon.com

Keywords: SENT, powertrain, sensor, safety, SPC

Low-cost Approach for Far-Infrared Sensor Arrays for Hot-Spot Detection in Automotive Night Vision Systems

K. F. Reinhart, M. Eckardt, I. Herrmann, A. Feyh, F. Freund
Robert Bosch GmbH

Abstract

Sensor data fusion of active near infrared (NIR) and passive far infrared (FIR) for reliable detection of vulnerable road users in future warning automotive night vision systems requires for low-cost, mid-resolution FIR sensor arrays for hot spot detection. We present a new cost efficient technology for FIR arrays adopting a volume proven integrated MEMS process for the production of a suspended thermo-diode array. In contrast to established bolometer production all steps of the process developed are fully semiconductor compatible as the sensor element formation is an integral part of the read out IC processing and does not require ASIC backend processing with dedicated equipment. Vacuum wafer-level packaging compatibility further reduces cost. In a first step the proposed process has been verified with small integrated FIR arrays consisting of 42x28 pixels. The FIR array development reported is part of the EU FP7 project 'ADOSE'.

1   Introduction

Next generation automotive night vision systems for driver assistance will improve the safety of vulnerable road users with active warning signals and in future systems also automatic system action. Reliable detection with low false alarm rates is essential for such systems.

In contrast to stand alone FIR night vision with expensive high resolution bolometers, a fused NIR/FIR system combining high resolution NIR and lower resolution FIR, allows excellent image display quality using the active NIR image from affordable good resolution CMOS image sensors. In addition, this combined system facilitates reliable identification of vulnerable road users, supported by hot spot detection from an FIR-array sensitive in the 7-14 µm wavelength range.

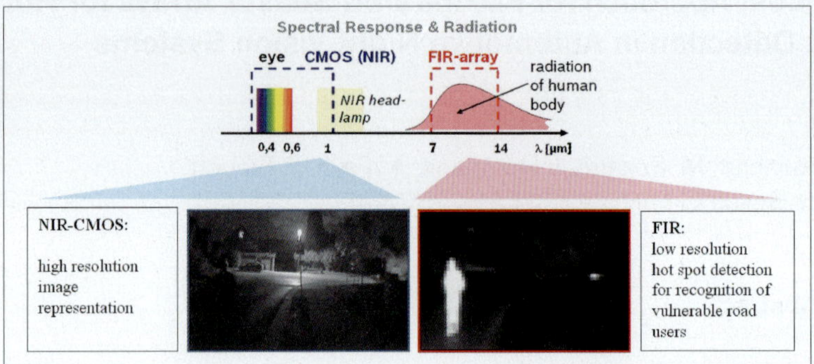

Fig. 1.   Multi-spectral approach for warning night vision with NIR / FIR data fusion (top). Identical night vision scene with different sensors (bottom); CMOS-NIR imager without active light (left) and low resolution FIR-array (right).

Various different un-cooled sensors are in the market or under development [1]. Despite somewhat reduced resolution requirements for hot spot detection compared to FIR-imaging, automotive sensor cost-demands remain a challenge for the sensor technology. Present established un-cooled bolometer technologies e.g. based on vanadium oxide VOx [2] or amorphous silicon [3] do not allow meeting the stringent cost targets for FIR add-on sensors – mainly because of the expensive manufacturing process requiring back-end processing of completely finished read-out ASIC wafers within a dedicated production environment due to incompatibility of sensor material or process-flow with standard semiconductor manufacturing. Additionally batch vacuum wafer-level packaging is not applied for these bolometer technologies. Other alternative FIR concepts proposed, like [4], apply silicon-germanium multilayer films for the sensor requiring temperature budgets above the limits to which a substrate carrying a fully processed ROIC can withstand. This means to employ sophisticated and expensive transfer processes for the sensor film onto the ASIC wafer.

In this paper we propose a fully CMOS production compatible process adopting a volume proven integrated MEMS process and vacuum wafer-level packaging. This process - described in more details in section 3 - together with the reduced resolution requirements of a non-imaging FIR add-on sensor for hot spot detection allows us to meet the aggressive automotive cost targets.

2   Requirements

Several relevant use-cases for an automotive night vision system comprising active CMOS-NIR imaging and additional FIR hot spot detection have been analyzed within the European project ADOSE. The requirements for the FIR–add-on sensors for hot-spot detection have been derived there and are listed in Tab. 1. It was derived that an array resolution of 100 horizontal and 50 vertical pixels should already be sufficient for detecting a person as a hot spot at a specified minimum viewing distance of 120 m.

FIR camera requirements		*Remark*
Horizontal Field of View (FOV):	*± 12°*	*For data fusion with NIR*
Angular Resolution:	*4,18 pixel / °*	*Defined by smallest object to be resolved @ 120m*
Object Temperature resolution:	*< 500 mK*	*for hot-spot detection; no greyscale image display* *NETD < 300mK for chip @ F#1 optics*
Frame Response:	*> 12,5 Hz*	*for 3 verifications of object in the NIR image*
Array Size:	*100 x 50 pixels*	*Defined by FOV and angular resolution*
Wavelength Range:	*7-14 µm*	*Spectral emission maximum of vulnerable road users*

Tab. 1.   Typical requirements for a FIR sensor in warning night vision based on FIR/NIR data fusion.

Concerning the sensor cost, only few tens of Euros in cost are acceptable for the whole add-on sensor comprising the FIR-array chip, optics, electronics and packaging, as the sensor is an additional part of the warning night vision system. FIR array and optics are the dominating cost drivers. Although in the ADOSE project low cost FIR optics is addressed as well, in this paper we will focus on the technical results from a new process set up for considerably reducing the array costs for small to medium resolution FIR arrays.

3 Sensor Technology

3.1 Sensor Concept

In order to allow integration of the sensor element processing into a semiconductor process flow we use mono-crystalline silicon for the sensor material together with a suspended thermo-diode detector similar to [5,6,7]. Additional advantage from using diodes instead of a resistor array is the inherent decoupling of the matrix without need for a pixel transistor and for stacking the sensor above the electronics. This allows us simultaneous production of the sensor element within the readout-ASIC manufacturing process.

The basic concept for generating a suspended semiconductor device is illustrated in Fig. 2. Latest generations of surface micro-machined pressure sensor process modules support integrated circuits in an epitaxial mono-crystalline silicon layer above a pre-structured vacuum cavity [8,9]. This process was only slightly adapted and the FIR pixel designs have been made to cope with the layers available from the ASIC process. We also found the absorption of pixel's layer stack to be sufficient for about 50% absorption which allowed us to avoid further absorption structures like the expensive thermal radiation collectors described in [6]. This design allows fully exploiting the cost savings from synergy with high volume semiconductor manufacturing.

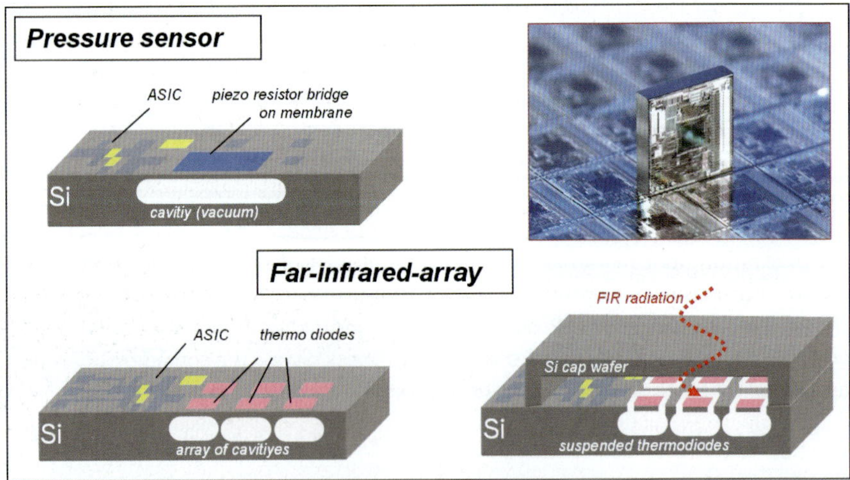

Fig. 2.   A production proven integrated MEMS process for surface micro-machined pressure sensors (top) is adapted for manufacturing suspended thermo-diodes for FIR sensor arrays with a single additional mask step and wafer-level vacuum packaging (bottom).

3.2 Process Flow

First a cavity below an epitaxial layer of mono-crystalline-quality silicon is formed by means of local porosification of silicon, followed by deposition of an epitaxial silicon layer. Cavity formation occurs due to thermal rearrangement of the porous region in a high temperature process [8]. With the substrate prepared in this way a standard ASIC process can be run in the epi-silicon layer generating the read-out ASIC and the p+/n-epi sensing thermodiode array simultaneously during a single process sequence. Finally openings of the cavity are produced by reactive etching the epitaxial layer and simultaneously removing the silicon under the suspension arms in order to generate suspended thermodiode islands only connected to the substrate by two bridges of dielectric material and contact metal. This is required for thermally and electrically decoupling the diodes from the substrate. The general processing sequence is shown in Fig. 3.

A final vacuum encapsulation on wafer level finishes the process and provides easily testable and dice-able chips suited for direct application in a chip-on-board assembly. The cap wafer is micromechanically pre-structured with a cavity in the active array area and with openings to keep the bond-pad region free. As bonding material we use standard screen printed seal glass. This established MEMS process is production proven in high volumes e.g. for accelerometers or gyros and can be used for mono-crystalline FIR-arrays without problems in contrary to other bolometer materials like VOx, which do not withstand the temperature budgets involved.

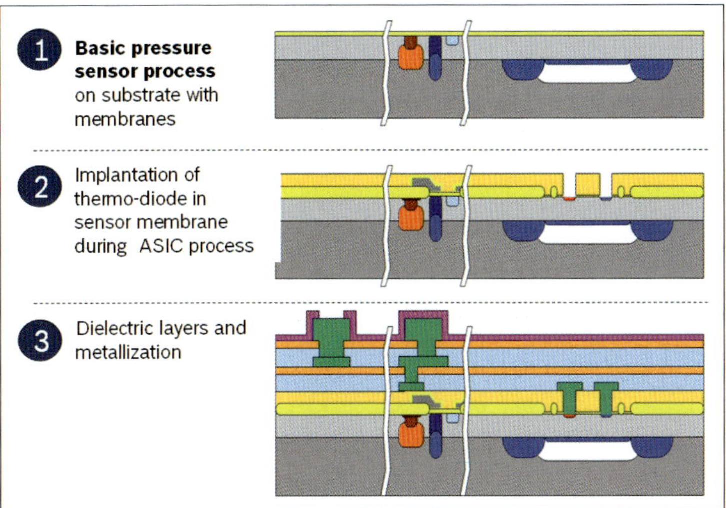

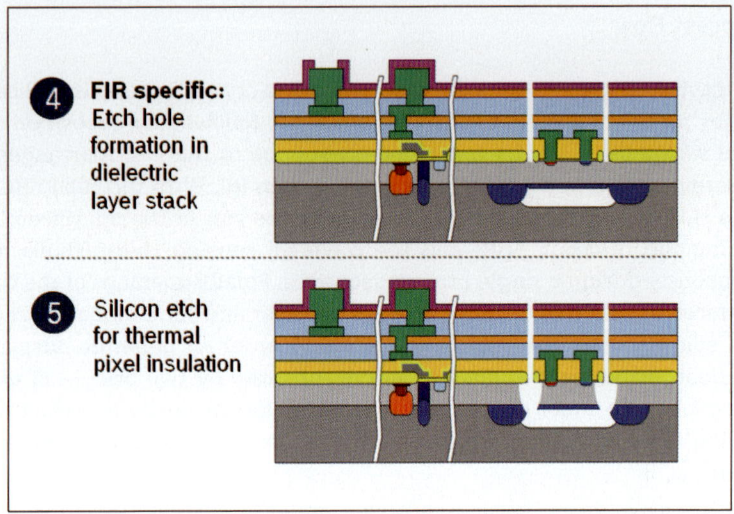

Fig. 3.   Basic process flow for the FIR sensor: Steps 1, 2 and 3 schematically represent the steps adapted from the integrated pressure sensor; 4 and 5 show the cavity opening as FIR-array specific process
step.

3.3 Sensor Test Design

Thermal insulation of the detector structure is the key parameter in any
thermal FIR-detector design in order to achieve a high temperature increase
from the small amounts of thermal irradiation on the pixel. The heat conduction through the suspension arms dominates the quality of thermal insulation
besides radiation losses and heat conduction through the residual gas. We use
L-type suspension arms (Fig. 4, left) giving the best compromise of thermal
insulation and pixel area reduction due to the in-plane construction. The suspension and pixel surface material consists of the dielectric layer stack out of
the semiconductor process and incorporates also the connection lines to the
diodes suspended below the pixel surface. The contact-lines are made in the
ASIC's metal 1 layer material and add the largest contribution to the thermal
leakage. In order to achieve the same electrical characteristics for the pixel and
the substrate temperature reference diodes, we use the same suspended diode
design for both, but have the suspension arms thermally short-circuited in the
case of the reference diode. Fig. 4 shows the simulated temperature increase
due to the same incident radiant power of 50 nW for both structures. The small
thermal response of the reference diode is acceptable and is further improved
if the diode is additionally shielded with a metal layer against thermal irradiation.

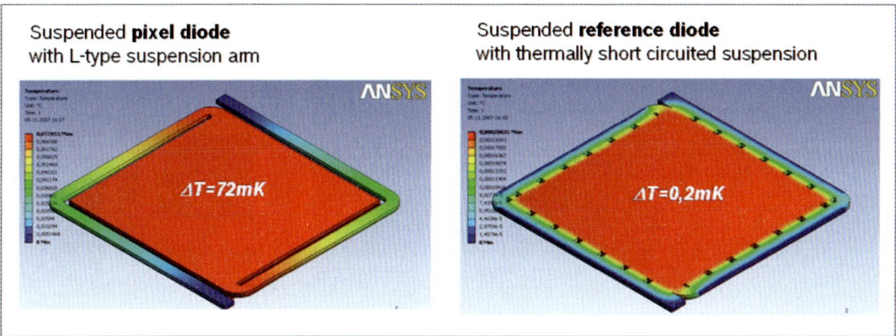

Fig. 4.   Pixel diode and reference diode with simulation of the temperature increase from an incident radiant power of 50nW.

As test device for verification of the process and concept we designed and realized small integrated arrays with 42 x 28 pixels at a pixel pitch of 225 µm with a supporting monolithic addressing and readout circuit for the diode array.

4   Results

4.1   Technology Run

Suspended diodes (see Fig. 5) produced in the test run have been characterized on wafer-level also with regard to temperature behaviour. Fig. 6 shows the characteristics of a typical diode for three temperatures measured on wafer level. The thermal sensitivity of the diodes with constant current operation at 1µA was extracted to be -2.3 mV/K at 25°C.

Responsivity was measured after wafer-level vacuum packaging with 7-14 µm thermal radiation and was found to be in the range of 150 V/W for simple single diode pixel designs without any further means like absorption adaptation, gettering or anti-reflection coatings on the silicon window.

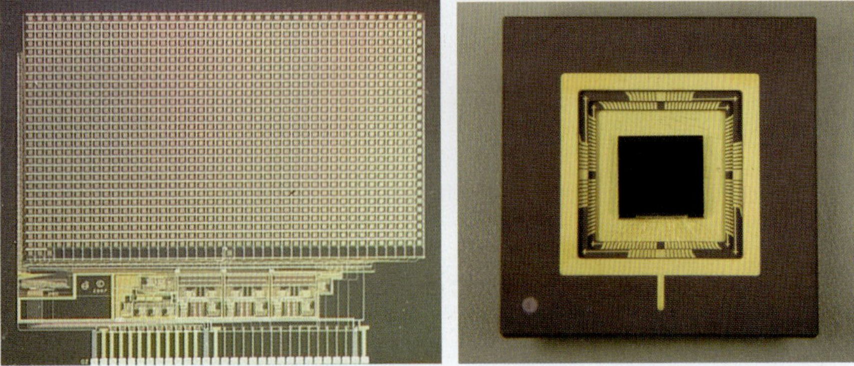

Fig. 5.   Photo of the 42 x28 FIR diode test array made with the new process and final vacuum wafer-level packaged chip bonded into a test package.

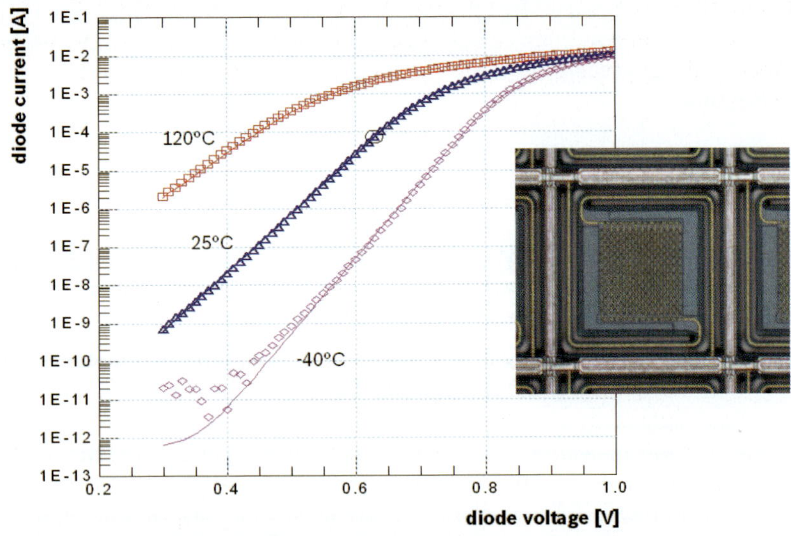

Fig. 6.   Current/voltage diagram of a FIR-Diode at different temperatures (left; points: measurement; lines: diode model); photo of a pixel with a suspended thermo-diode (right)

4.2   Results from Integrated Test Arrays

The processing and integration compatible technology has been verified with first small arrays consisting of 42 x 28 pixels and an integrated electronics enabling simple sequential readout of the individual pixels of the array. The electronics provide a column addressing de-multiplexer and multiplexed constant current sources driving the pixels from the row side. Fig. 7 shows the basic readout implemented. A low noise differential amplifier provides +40 dB pre-amplification of the signal difference between the selected pixel diode and the reference diode. All other signal processing was made in a flexible external electronics with an FPGA providing timing, addressing, first level offset correction algorithms as well as data transfer to a PC.

Fig. 7 shows a picture taken with the 42x28 array after 4 times pixel interpolation and a cosine shaped contrast increase. In the optical setup we used an Umicore GasIR® doublet lens with a focal length of 9 mm and f/1. The frame rate was restricted to 5 Hz due to the pixel sequential read-out principle. This test device with relatively poor thermal insulation due to the full dielectric stack and the metal tracks present on the suspension arms, together with the single diode pixels allowed us resolving object-temperature differences of about 1K.

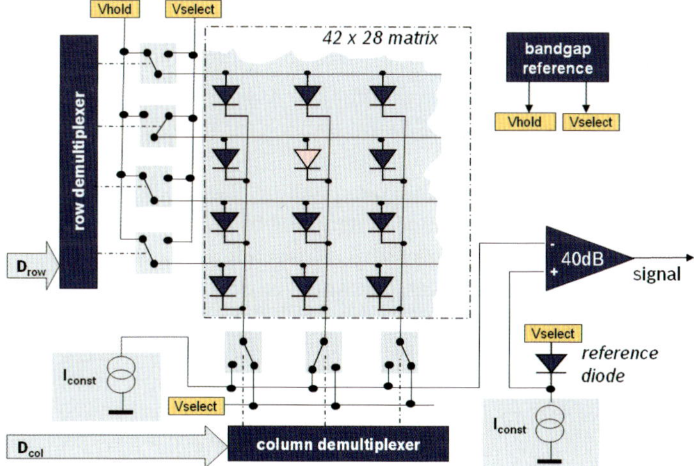

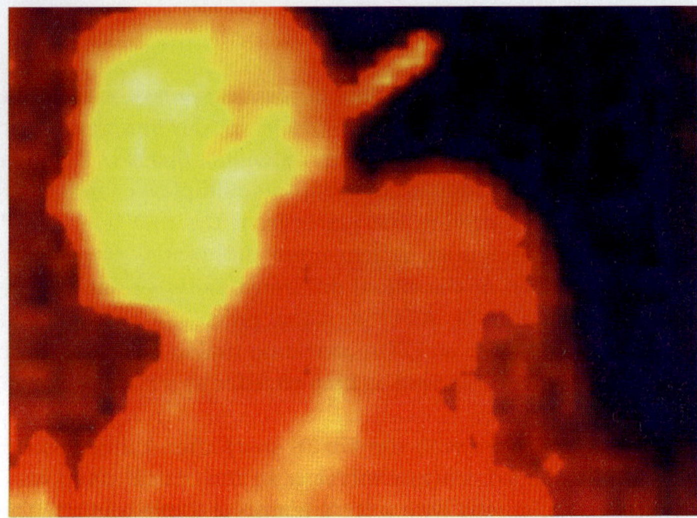

Fig. 7.   Basic integrated readout circuit with diode serial addressing for testing (top) and picture taken from a person with the chip (bottom).

5   Discussion and Outlook

In order to meet the automotive requirements discussed in chapter 2, further
improvements together with a pixel shrink are currently under development in
the EU FP7 project 'ADOSE'. The automotive design will also implement faster
column-parallel array read-out. Several options for overall sensitivity improvement exist to compensate the unwelcome effects resulting from shrinking
the pixels. As the noise of the mono-crystalline thermodiodes is less than the
amplifier input noise, multiple diodes per pixel in series can be applied to boost
responsivity. The residual gas pressure originating from seal glass out-gassing
during the encapsulation process presently is another limitation if no getter
material is used. We expect from future seal-glass free wafer-level encapsulation processes to gain at least a factor of 2 in sensitivity even without getter
material deposition. Additional improvements of the pixel's thermal insulation
are possible by reducing suspension thickness or by using poly silicon with
reduced thermal conductance as contact material instead of the metal.

6   Summary

A new cost effective and fully semiconductor compatible process has been described which is suitable to manufacture far-infrared sensor arrays for hot spot detection in automotive night vision. A first integration run with the new process was successful and yielded functional integrated 42x28 FIR-arrays. Development of a 100x50 FIR array according to automotive specification is ongoing in the ADOSE Project.

The research leading to these results has received funding from the European Community's Seventh Framework Programme under grant agreement n° 216049, relating to the project 'Reliable Application Specific Detection of Road Users with On-Board sensors (ADOSE)'.

References

[1]   Akin, T., 'CMOS-based Thermal Sensors', Advanced Micro and Nanosystems, Vol 2, CMOS-MEMS, pp 479-512, Editors: H. Baltes, et. al., WILEY-VCH, ISBN: 3-527-31080-0

[2]   Terre, W., et. al., 'Microbolometer development and production at Indigo Systems', Proc. SPIE, Vol. 5074, pp 518-526, 2003.

[3]   Tissot, J.L., et. al., 'Un-cooled microbolometer detector: recent developments at Ulis'; Opto-electronics Review 14(1), pp 25-32

[4]   Kvisteroy, T., et. al., 'Far Infrared Low-Cost Uncooled Bolometer for Automotive Use', Advanced Microsystems for Automotive Applications 2007; pp 265-278, Editors: Jürgen Valldorf and Wolfgang Gessner, VDI, Springer Berlin (2007).

[5]   Kimata, M., et. al., 'MEMS-based un-cooled infrared focal plane arrays'; Transducers & Eurosensors '07, Lyon, pp 1357-1360 (2007).

[6]   Kimata, M., et. al., 'SOI diode uncooled infrared focal plane arrays', Proceedings SPIE, Vol. 6127, pp 61270X (2006).

[7]   Eminoglu, S., et. al., 'A low cost 128x128 un-cooled infrared detector array in CMOS process'; J. Microelectromechanical Systems Vol. 17. No.1, p20-30, (2008).

[8]   Armbruster, S., et. al., 'A novel micromachining process for the fabrication of mono-crystalline Si-membranes using porous silicon', Digest Tech. Papers, Transducers 2003, June 2003, Boston, USA, pp 246-249 (2003).

[9]   Adam, B., et. al.; A New Micromechanical Pressure sensor for Automotive Airbag Applications', Advanced Microsystems for Automotive Applications 2008; pp 259-287, Editors: Jürgen Valldorf and Wolfgang Gessner, Springer Berlin (2007).

Karl Franz Reinhart, Martin Eckardt, Ingo Herrmann, Ando Feyh, Frank Freund
Robert Bosch GmbH
Corporate Sector Research and Advance Engineering
Microsystem Technologies
CR/ARY3, P.O Box 10 60 50
70049 Stuttgart
Germany
karl-franz.reinhart@de.bosch.com
martin.eckardt@de.bosch.com
ingo.herrmann@de.bosch.com
ando.feyh@de.bosch.com
frank.freund@de.bosch.com

Keywords: far-infrared, sensor array, automotive night vision, infrared sensor technology, MEMS

Smart Sensors: Enablers for New Safety and Environment Initiatives

F. Berny, A. Glascott-Jones, L. Monge, e2v

Abstract

Future automobiles will be faced with many constraints based on environmental and safety legislation. In addition further demands such as cost reduction, differing fuel types and improved driver comfort will be made. At first view many of these demands are conflicting. This paper describes the role that sensors and in particular the interface Application Specific Integrated Circuits (ASIC) or specialised Application Specific Standard Products (ASSP), play in meeting these exacting requirements. The latest advances in sensing element technology must be matched with an equivalent improvement in the performance of the readout electronics to produce the complete sensor package. Requirements of several sensing elements are reviewed and how the ASIC design responds to these challenges will be explained.Examples in the fields of power train efficiency using precise angle measurement and pressure sensors will be given. Other applications such as maintaining maximum night time driver vision and the monitoring of fuel sources for maximum efficiency will be outlined.

1   What is a Smart Sensor?

A smart sensor can be defined as a measurement component that contains all the intelligence required to function with minimal external interface. This obviously must include a sensing element and communications interface; it will often require an internal microprocessor and internal compensation/linearisation algorithms and along with this comes the ability to provide internal diagnostics, to be able to reconfige. the device or to have the device totally autonomous until an event occurs, such as an over threshold value and the sensor will communicate this value.

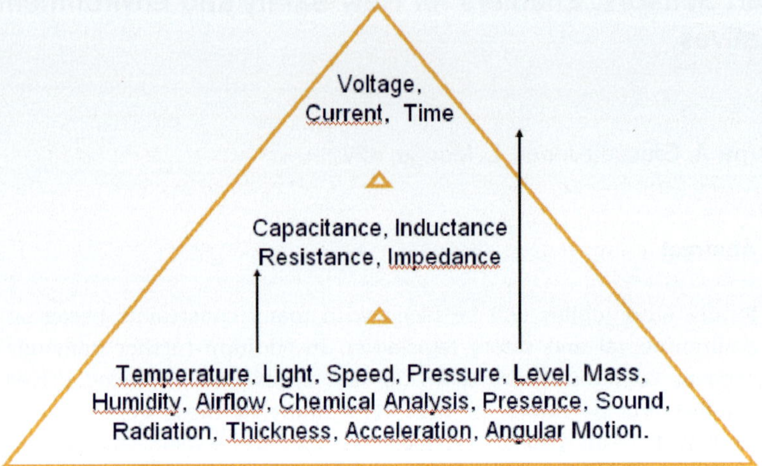

Fig. 1.   The Sensing Pyramid

Obviously the sensing element itself is vital, within the readout electronics the only parameters available are voltage, current and time. However as illustrated in Fig. 1, the more useful parameters to be measured are outside this. Some sensing elements convert the parameter directly to voltage or current, such as a photodiode. However, most parameters require an intermediate step using a transition via a capacitive, resistive or inductive sensing element. To convert changes in these values requires an analogue processing stage.

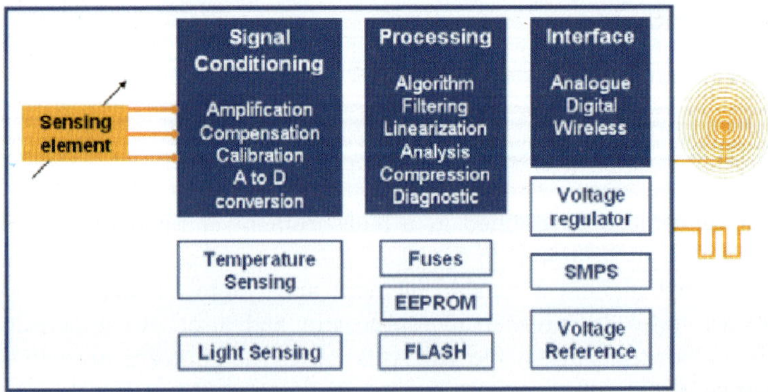

Fig. 2.   A smart Sensor

Of course once a microprocessor is used within a smart sensor for the processing of data, then the signal derived from the sensing element must be converted into digital form. This then necessitates analogue to digital conversion.

The major functions of the electronics required within smart sensors are illustrated in Fig. 2.

Many applications and sensor types require precision results and the requirements for the analogue processing elements become demanding – examples of this can be found in capacitive MEMS sensors where the capacitance change is very small. Other applications for resistive sensors used for measurement of small pressure changes for fluid level and flow applications see a typical voltage change of less than 50 uV.

The addition of microprocessor intelligence within the SMART sensor allows many functions:
- ▶ Linearisation of sensing element response;
- ▶ Background monitoring – for warranty validation and for lifetime operating conditions information;
- ▶ Self monitoring to detect fault conditions and smart calibration of offsets and gain, during the product's lifetime.

Of course for cost reduction within systems that already contain one or more microprocessors there would be no need for an internal microprocessor. In this case the electronics within the sensor could contain the analogue front end and simply a digital interface, calibration coefficients and other parameters could be stored in internal EEPROM.

Another important reason to use more processing power within the ASIC/ASSP is that this may allow a lower quality (and cheaper) sensing element to be used. This can be seen in force feedback MEMS sensors where the feedback enhances the linearity of the system.

Finally the interface function is highly important and could range from a simple PWM output up to a more complex LIN node.

2   Safety and Environmental Issues

There is increased requirement for safety within the automotive environment. A goal of the European community is to reduce road fatalities by 50% before the end of 2010 in addition for the European consumer; safety is one of the most important characteristics when buying an automobile Ref [1]. Technology can provide systems to avoid roll over, to detect and avoid pedestrians and to pre-warn the driver of an imminent accident.

That automobiles are adequately maintained is another area where the smart sensor can help, the USA TREAD Act requires that all tyre pressures are automatically monitored. This provides a dual safety and environmental role since it is reported that an under pressure of 50 kPa can result in a 2% increase in fuel consumption. Ref [2]. Along with this reliance on technology comes the requirement that the electronic systems are reliable and trustworthy, this results in the need for standards such as IEC 61508. In conjunction with these safety concerns come environmental issues. The drive to reduce emissions runs on three tracks; increase engine efficiency, increase vehicle efficiency and clean up emissions. Parts of the solution to increasing engine efficiency requires the accurate measurement of engine parameters, ranging from angle of crankshaft and angle of valves used within the engine system – such as the exhaust gas re-circulation valve. Accurate measurement of pressure, for example inlet air pressure, fuel pressure or exhaust pressure, will also provide increases in efficiency. The search for fewer emissions has resulted in the use of different fuels, the characterisation of which is important for the smooth running of the engine; timing parameters will be changed dependant on the fuel in the tank. The goal of reducing emissions leads to the use of electric vehicles, powered entirely from batteries. These systems require extensive monitoring of the state of health, battery monitoring systems are becoming common on standard vehicles to aid efficiency.

Examples of the electronics interface behind all these systems will be given later in the paper. The next section discusses the tools available within an ASIC design to provide these functions.

3   The ASIC Design Toolbox

One of the most important design considerations when undertaking a sensor interface design is knowledge of the sensor type. Many measurements can be made using either resistive type sensing elements or capacitive and obviously the sensor interface circuit topology will change considerably. Equally within the capacitive range of sensing elements knowledge of the interaction between sensing element and readout circuit is essential to attain the required performance. As such, having a top down design approach (from system spec to ASIC building blocks spec) is essential to ensure the overall system consistency and competitiveness.

3.1 Specifying the ASIC/ASSP

As a first step of the top down approach, one must ensure the ASIC correct specification. This is particularly true when a complex sensing element is providing the input signal. Indeed sensing elements such as MEMS could have a different response depending on the readout I.C. which would result in non-linearities in response. In order to anticipate such effects there is a need to build electromechanical models to anticipate and compensate for these mechanisms. Based on these high level models, one can then more easily specify the signal conditioning front end to ensure the best system performance. The best tools to perform this job are the ones that enable mathematical modelling of the system elements. They provide fast simulation time, and thus ease system parameters optimisation by successive iterations.

3.2 ASIC Technology

As a second step of the top down approach comes the ASIC implementation based on a silicon technology process.The first consideration for technology choice is whether the process is automotive qualified. AEC-Q100 is the standard set for the stress test required for automotive qualification. The choice of ASIC technology is affected by the requirements of the device. The Fig. below shows some of the tradeoffs. Hence components requiring more digital content would be directed towards a 0.18u process. 0.35u would be used for components requiring more analogue or high voltage tolerance.

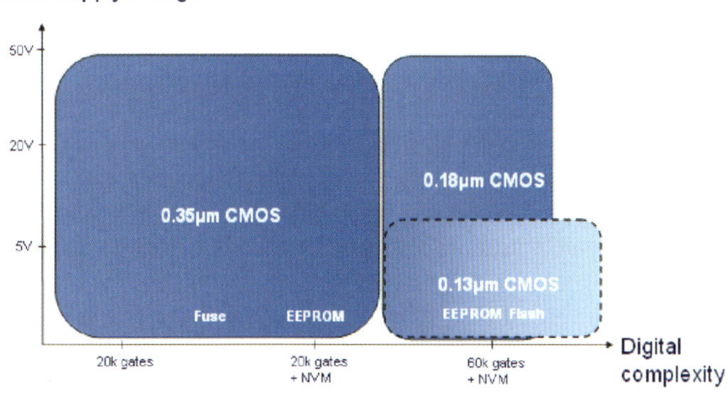

Fig. 3   Automotive ASIC Technology

The more mature 0.18u and 0.35u processes are commonly chosen because of their automotive qualified status, high voltage capability and because of their good cost/performance ratio.

Within some automotive environments operating temperatures of +150°C and greater can be encountered, it has been found that careful design using standard processes can result in a fully performing product. However using a Silicon on Insulator (SOI) process is always an option especially for even higher temperatures.

3.3 Cost of ASIC

The ASIC cost is directly related to the size of the die, the cost of a wafer is fixed for a particular volume and technology and so the bare die cost is proportional to the number of good die that can be produced from a wafer. The yield will also play a part in the expected cost, other factors include test time and number of masks.

3.4 I.P. Requirements

To be able to interface with the range of different sensing elements that can make up a SMART sensor requires a large portfolio of silicon Interlectual Property (I.P.). If this I.P. is already proven within silicon then so much the better in order to reduce the ASIC or specialised ASSP development duration and reduce development costs by increasing reuse. As mentioned before, MEMS capacitive sensing elements require extremely low noise techniques and advanced use of sigma delta modulators. Resistive sensors will often necessitate low offset and low drift techniques which can include chopper stabilisation and auto-zero circuits. Often in the automotive environment, sensors will not operate at very high data rates so in many cases the sigma delta converter offers the best performance.

From the digital perspective the ability to offer a range of solutions from the digital sparse ASIC which may only use fuses for programmability up to advanced 32 bit processors. The ability to offer non –volatile memory for program code storage and calibration parameter storage is essential. Indeed given the high performance of the 32 bit processor along with the NV Memory and peripherals that can be embedded within the smart sensor it becomes possible for the smart sensor to undertake some of the tasks normally performed by the application level microprocessor. Interface requirements can range from simple analogue output to complex buses such as LIN or CAN, however in general within an automotive context the sensor will not contain the CAN interface. LIN being more prevalent along with newer standards such as SENT and PSI5. Wireless standards such as Zigbee can also be embedded in the ASIC or specialised ASSP.

Amplifier Reference	Noise	Offset	Current consumption	Signal Bandwidth (Hz)	Extra feature
Standard CMOS	5 µV/√Hz@10 Hz 30 nV/√Hz@10 kHz	<10 mV	50 µA	10-10 k	
E2v medium offset	250 nV/√Hz 1 Hz to 10kHz	200 µV	150 µA	1-8 k	Auto zero for offset cancellation
E2v ultra low offset and noise	6 nV/√Hz 1 Hz to 10 kHz	1 µV	1 mA	1-1 k	Chopper stabilized
E2v ultra low noise	7 nV/√Hz 1 Hz to 200 Hz	100 µV	2 mA	3-200	Chopper stabilized

Tab. 1.   Selection of e2v Amplifier I.P.

Application	Number of effective bits	Order	Sampling frequenz	Output rate
Voltage acquisition	16	2	2 MHz	8 kHz
	20	4	256 kHz	1 kHz
Capacitive interface	19	2	8 kHz	16 Hz
	12 to 20	2	128 kHz	64 Hz
Digital radio	16	3	140 MHz	500 kHz

Tab. 2.   Typical e2v ADC I.P.

3.5 The Design Process

A typical automotive ASIC design process will take up to 24 months from sign- off of specification to production ramp up at the customers, however first silicon will be available in under a year – much less if standard off the shelf I.P. blocks can be used.

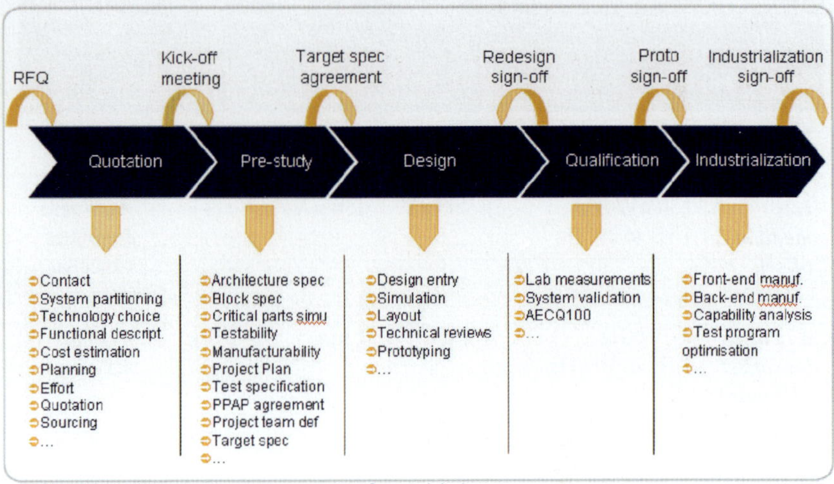

Fig. 4 The ASIC Design Process

4 Examples of Smart Sensors

4.1 Angle Measurement

Requirements to comply with new laws aimed at reducing pollution will demand improvements in engine control which will mean that all flaps or valves which are currently ON/OFF will need to be measured to high accuracy Ref [3]. One precision method of angle measurement uses magneto-resistive elements in a bridge formation and a rotating permanent magnet mounted just above the sensor. The sensing elements change resistance as a function of the angle of the magnetic field and produce two output signals in Sine and Cosine form; the ratio of the two signals can de used to provide the angle of the permanent magnet.

The ASIC is an example of a full smart sensor since it uses high performance analogue processing and a dedicated digital processor to perform the arc tangent function using a CORDIC algorithm. The output interface can be either analogue or one of a number of digital serial busses. The smart capabilities of this sensor lie in the continual detection of sensor state. The ASIC detects and signals error conditions such as sensing element disconnection, over voltages and loss of magnetic field. In addition an auto calibration feature enables compensation of the offset and gain of each channel, the calibration parameters being stored in EEPROM. Other features include a facility for storage of calibration parameters for separate sections over the angular range of the sensor enabling sensing element and even mechanical non-linearities to be compensated

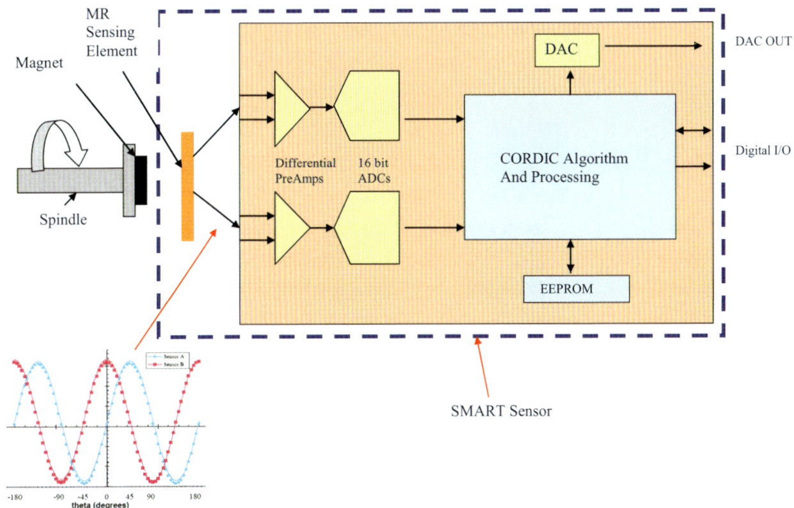

Fig. 5.   A SMART Angle Measurement Sensor

4.2 Pressure Measurement Using MEMS

MEMS capacitive sensors are well adapted to measurement of pressure, the change in pressure causing a movement of a diaphragm which results in a change in capacitance. MEMS manufacturing techniques bring advantages such as reduced sensor size and reduced cost to fabricate.

However the small size of the sensor means that the corresponding change in capacitance is small (the order of 100 fF). Precise capacitance measurement requires excellent ASIC noise performance as described in the next section, the most appropriate technique considering the small value of capacitance change is a sigma delta modulator.

More precise measurements of pressure within the engine system will mean that engine control becomes more efficient which will improve engine efficiency. However, the variation of capacitance is non-linear with pressure and also with temperature. Hence, this sensing element is fully reliant on the smart features within the readout ASIC, these provide a programmable third order compensation algorithm which maps between capacitance and pressure followed by a third order compensation for temperature. To ease component integration the ASIC embeds a temperature sensor accurate to 0.1°C (after calibration).

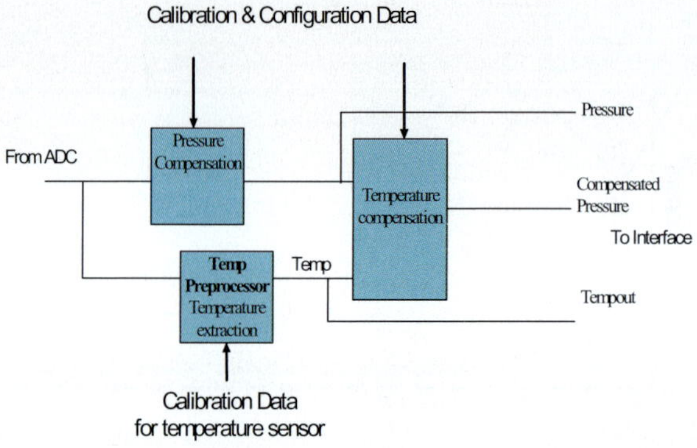

Fig. 6.   Capacitance to Pressure Compensation

This system can also be used in precision altimeters and GPS application. Its sensitivity is such that pressure differences due a change in altitude of 10cm can be measured. These kinds of compensation techniques are also applicable to resistive sensors.

4.3   Tilt Measurement

MEMS accelerometers can be used to measure the orientation of a vehicle to provide information to an electronic stability system. One other application is tilt angle measurements which can be used to adjust headlight aim. If we consider the requirements of such a system:- The figure for the change in angle of a headlight beam with two passengers seated within the car is 0.7 deg. Ref[4] So, we take as our goal the ability to detect changes of angle of the order of 0.5 deg. The sensitivity of such a system varies with the angle. Displacements about the horizontal will produce a larger component of g in the plane of the sensor than those about the vertical. For the present example we take an example of a nominal inclination of 5° and consider changes about this.

Hence the nominal projection of g onto the axis of the sensor is

$$g \times \sin(5) = g \times 0.0871 \tag{1}$$

$$g \times \sin(5.5) = g \times 0.0958 \tag{2}$$

A change of 0.5 degree gives

And so an acceleration change of 8.7 mg

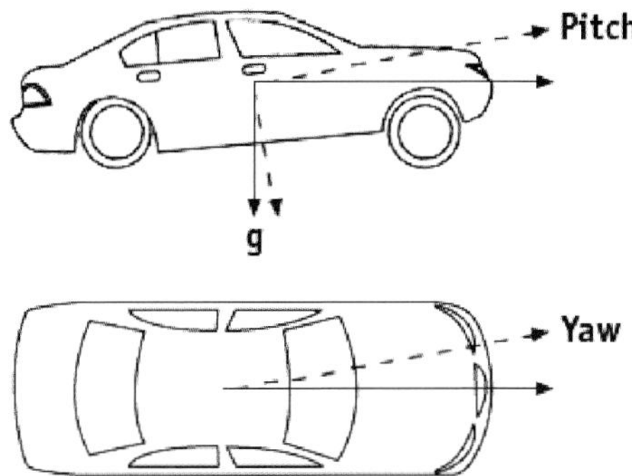

Fig. 7.   Illustration of pitch angle and yaw angle

A typical accelerometer will show a capacitance change of 0.2pF /g. Hence the total change in capacitance for this inclination is 1.74 fF. The best capacitive readout devices using sigma delta techniques can attain a noise floor of $3aF/\sqrt{Hz}$. If we take a total bandwidth of 100Hz this gives 30aF root mean square (rms) in total. Taking the peak-peak noise value as 6.6 x the rms value, this gives a total peak-peak noise of 0.2fF. Hence using these advanced readout devices the angle resolution is easily obtained; in fact a resolution of less than 0.1 degree is possible.

(Note: $1\,pF = 1 \times 10^{-12}\,F$ $1\,fF = 1 \times 10^{-15}\,F$ $1aF = 1 \times 10^{-18}\,F$

4.4 Battery Management

The knowledge of the state of health of a battery is a useful parameter to:
- Protect the cells or the battery from damage;
- Prolong the life of the battery;
- Maintain the battery in a state in which it can fulfil the functional requirements of the application for which it was specified.

In themselves battery management systems provide an important service to the reliability of an automobile. Battery management systems help to reduce energy consumption and saves approximately 2.4 grams of CO_2 per mile. Ref. [5]. In addition such systems become almost essential when used in vehicles using idle-stop. Idle–stop technology gives an immediate reduction in emissions by turning off the car's engine when the vehicle comes to a stop and restarting when it is time to move off. Obviously the engine management system would need to be confident that the battery contains enough charge to start the engine. Battery management systems have two drivers for ASIC usage; one is size constraint – the sensing system should fit on top of the resistive shunt used to measure the current and the other is cost.

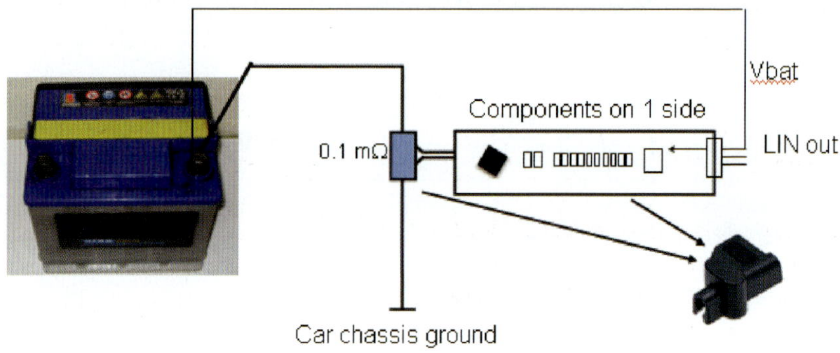

Fig 8.   Battery Management System Physical Location

The smart battery management system will require a precision analogue front end since a wide range of parameters need to be measured the current can range from quiescent values of as low as 10 mA through normal operation of perhaps 10 A and as high as 100 A for the start current. It will require significant data processing capabilities to be able to model the battery's behaviour and also significant amounts of non-volatile memory to store calibration parameters for the battery. The LIN interface is standard for this type of application. Similar measurements are required for Hybrid vehicles using Li Ion battery packs. In this case the ability to stack each measurement ASIC at the progressively higher voltage of each cell of the battery pack becomes important.

5 Future Developments

Further developments in smart sensors could take the form of Driver Assistance Systems for collision avoidance using optical techniques. Significant digital processing would be required for this application.

Fuel quality measurement will provide increases in efficiency; quality can be measured using I.R. spectroscopy, density or dielectric constant measurement.

In addition the future will see samrt sensors that are self powered using energy scavenging techniques which extract energy from thermal or motion sources. So if we consider that the interface could be R.F. , some samrt sensors will become totally wireless.

References

[1]   CVIS July RACC Automobile Club, 2007.

[2]   European Tyre and Rubber Manufacturers June, 2007.

[3]   Granig,W; Hartmann,S SAE -01-0397, 2007.

[4]   Jianzhong Jiao NAL Koito NHTSA Workshop , 2004.

[5]   MarketWatch Oct Hella Energy Management Solutions, 2008.

Franck Berny, Andrew Glascott-Jones, Laurent Monge
e2v
Mixed Signal ASICs Business Unit
Ave. de Rochepleine BP 123
Saint Egrève Cedex.
France
franck.berny@e2v.com
andrew.glascott-jones@e2v.com
laurent.monge@e2v.com

Keywords: ASIC Design, ASSP Design, Sensor interface, SMART Sensor, angle measurement, pressure measurement

Airbag Electronics: from Single Building Blocks to Integrated Solutions

H. Geitner, M. Ferraresi, STMicroelectronics

Abstract

Airbag electronic systems measure various sensor signals and control several different actuators in an effort to prevent passenger injuries in case of accidents. This paper describes the electronics needed in such an application (acceleration sensors, bus interfaces, uC, power supply, squib drivers etc.) and the development of airbag systems as well as their increase of complexity over time. Acceleration information is delivered via PAS, PSI5 or DSI busses by high g acceleration sensors installed on vehicle corners and pressure sensors installed inside the doors. This information is validated by the microcontroller vs. centrally located inertial sensors. Precisely sampled acceleration values are used to identify a crash situation, in which case the firing circuit is enabled and a firing signal is sent to deploy the airbags. Building blocks like voltage regulator/power supplies and squib drivers have to provide the energy at the right time to ignite the airbag balloons mounted in the front and sides of the vehicle and tighten seat belts. Over time, new features have been added to the system - including rollover protection with airbag curtains, rear crash protection (whiplash prevention), and superior advanced crash detection algorithms to fire multi-stage airbags. All this functionality needs to be addressed by effective, safe, reliable, cost-effective electronics. Integration of the functions in a few application specific ICs and the fact that all function blocks and necessary semiconductor processes are available in one company makes the creation of synergies and the implementation of cost-effective silicon strategies possible.

1   Introduction

The basic rationale of airbag systems is to reduce the risks of injuries or death for the passengers of a vehicle during a crash. The idea was born a long time ago – in the 1960's, researchers were looking for an enhancement of the seat belt, which had been introduced in the 1950's by Ford [2] and by 1959 was available as a more optimized three-point front shoulder belt by Volvo. To further improve the safety of passengers, a new system had to be introduced to optimize the life saving effect of the seat belt – the airbag system. Wearing seat belts was (and in some countries, still is) a burden that many people did not want to bear. The airbag system therefore had to be burdenless, invisible, and effective.

The airbag system was originally an inflatable bag for the driver only, and in the case of an accident, the airbag electronics initiated the inflation of the airbag. Airbag electronics contained a mechanical sensor and many discrete components, filling a 19" rack in the early days of design. In the late 1980´s, the electronics became more and more integrated, and tier one suppliers started to work on application-specific integrated components (ASIC). With the integration of the functionality into ASICs, the system cost could be significantly reduced, and a wider distribution of airbag systems was possible. Not to forget that in many countries the benefit of airbag systems lead to a mandate for airbag systems in cars (and trucks later on) . In the US, the airbag for both front seats became mandatory for cars starting in the model year 1995. The positive effect of airbag systems, especially in combination with seatbelts, also helped to increase the awareness of the benefit of an airbag equipped car.

2.   Airbag System

A standard airbag system today consists of
- ▶ sensors that measure the reduction of velocity during a crash (acceleration sensors);
- ▶ the cable/communication between the sensors and the electronic control unit (ECU),
- ▶ the ECU itself,
- ▶ the pyrotechnical inflators, and
- ▶ the airbag.

The number of bags per vehicle has increased significantly, and today, cars with more than 10 airbags can be found. With the increasing number of airbags, the number of sensors also increased. Today it is common to have at least 3 mono axial acceleration sensors in the periphery of the car body and one bi-axial acceleration sensor in the ECU. To improve the performance and reaction time of the system, pressure sensors are also used to sense side impacts. The interpretation of a pressure sensor signal is quicker than the signal of an accelerometer, but the mounting location of the pressure sensor is inside the door. Side impacts that don´t deform the door need acceleration sensors to be detected.

2.1 Electronic Control Unit (ECU)

The ECU is located in the middle of the car and contains the majority of the electronic components of the system. The function of the ECU is to quickly read the data from the sensors (satellites) and process the data in the airbag algorithm. The algorithm analyses the signals and does some validation by comparing the incoming signals with each other. In the case that the algorithm detects a strong deceleration, it determines the angle of impact and calculates the severity of the crash. In modern multistage airbag systems, an input will give the algorithm more details about the passengers – today this input mainly comes from capacitive seat sensors. Their tasks are to register if there are persons sitting (passenger seat) and whether this is a full size person or a small child. CMOS cameras in the future that analyse drowsiness/sleepiness of the driver may provide this additional information to the restraint system. Depending on the weight/position of the persons, the algorithm has to calculate how forcefully the bag needs to deploy in order to protect the persons in the car.

2.2 ASICs and ASSPs

The hardware functions performed by the ECU (3.3 V...5 V) include handling/reading the data from the satellite sensors (accelerometers and pressure sensors) via a serial 2-wire current bus (PASx, PSI5, DSI etc.), apply the algorithm in a reliable micro controller, and verify the functionality of the main microcontroller with a second microcontroller. Additionally, it drives the actuators via bus or directly by immediately providing a high current (1,2/1,75 A, 40 V, 2 ms) to ignite the numerous pyrotechnical inflators of the airbags and the seatbelt pre-tensioners. These tasks require different semiconductor technologies: microcontrollers need memory and are low voltage devices, the communication with the various busses needs higher voltages (5 V...12 V). Data from the sensors and the status of the airbag system has to be forwarded to other systems in the car (i.e. emergency call) via a CAN or LIN bus. And all systems have to be safe and work reliably over the life of the car. It became clear that integration in dedicated ASICs on suitable processes is the solution. Also, with the commoditization, there is a chance to benefit from application specific standard products (ASSPs), where a higher total volume leads to lower cost.

2.3 Satellites

With the introduction of peripheral sensors (satellites), the problem of "how to transmit the signal reliably and quickly to the host?" was evident, and soon a Manchester encoded transmission was considered ideal for robustness, efficiency, speed, and EMR/EMI. The sensors inside the satellite and the signal conditioning need power - because of the distance between the ECU and satellites, a minimum of wires was requested. Today´s busses are based on two wires only, and the power is also transmitted to the satellite on the same wires as the communication. Over time, there where many protocols and physical layers defined – PASx , BST, BOTE, DSI, and recently PSI5. All of them are similar, but not completely compatible. The market requirements and the commoditization, however, ask for standards, and now we see two major busses that are left: PSI5 and DSI.

2.4   Application Challenges

The main application challenges in the airbag system are measuring quick events, slow communication, fast reaction times, and sensible actuation. Of course the increased complexity (10 sensors and 20 actuators can be found in many cars today) and the failsafe concepts are some of the reasons why special IC solutions are required.

One of the most critical scenarios is a side crash in which the large thorax/side bags have to be inflated within 10 ms and the decision to inflate has to be done within 3...5 ms [3].

The typical communication speed on the busses from the satellites is about 100 kbit/s – a satellite with 16 bit data output can send a theoretical maximum of 18 samples in 3 ms - but in reality it is less than that. The system has to decide, if it is an abuse case or a crash, and then immediately open the side airbags based on a few measured samples. To make it as safe as possible, various cross checks are needed in such a situation. Not only does the system software need to be optimized, but also the hardware is dedicated to this application. Let´s have a look at the signal chain and its challenges from the sensor to the actuator.

3.0   MEMS Satellite Sensors

MEMS-based sensors have been widely used in automotive applications, and accelerometers in airbag applications have been available since 1994 (Volvo). Accelerometers, gyroscopes, pressure and flow sensors in the electronics for power train, body, and chassis today still represent the largest market for MEMS sensors in terms of revenue. One of the major market opportunities in automotive for accelerometers is to sense crashes and deploy airbags. In particular, the increasing number of airbags that is required to boost passenger safety is accompanied by an increased number of satellites.

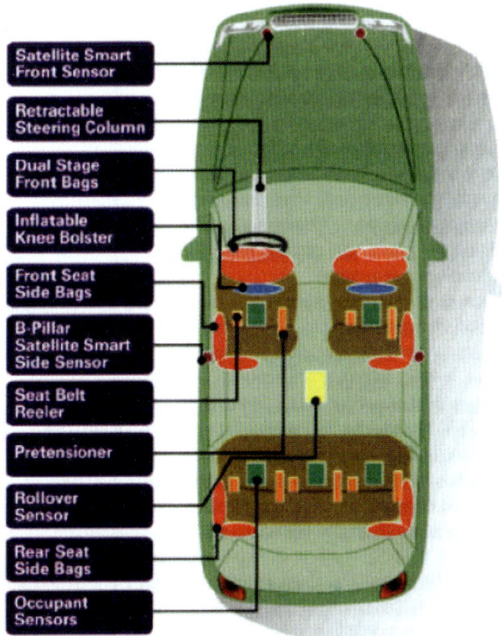

Fig.1.   Satellite Sensors are mounted in the front, side and sometimes side/rear and/or rear. Number and mounting location of the satellites depend on the size and construction/stiffness of the car body and available mounting space.

According to ST's internal analysis, the market evolution of satellite accelerometers in the next 5 years is expanding with a CAGR of 13% in volume, which, due to price erosion, corresponds to a CAGR in value of around 6%:

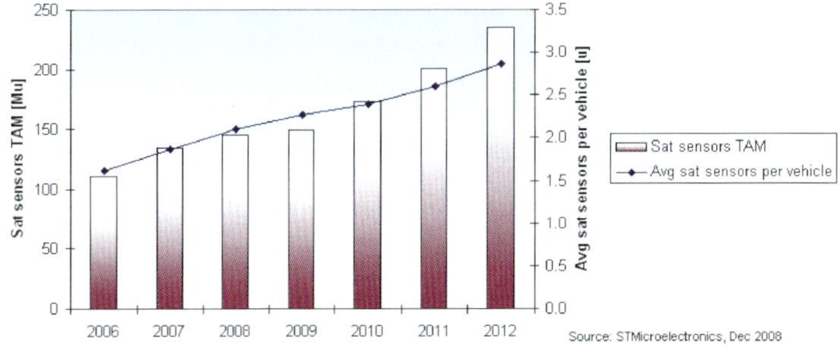

Fig. 2.   Market trend of satellite sensors

It is therefore a high-volume market that is driven quickly towards a standardization imposed by the two existing consortia (DSI and PSI5), where also production capability and efficiency will play a decisive role on the road to success. STMicroelectronics, already a major semiconductor supplier to the automotive industry, leverages a consolidated and efficient 8″ based MEMS manufacturing machine, and is directing efforts and resources to rapidly become a reference in the market.

3.1 Brief Technology Overview on MEMS @ STM

The micromechanical acceleration sensor is designed with a silicon technology called ThELMA, which stands for Thick Exitaxial Layer for Microactuators and Accelerometers.

ThELMA™ is a surface micromachining process based on a 0.8 μm technology. Surface micromachining, as opposed to bulk micromachining, is an additive process requiring the building up of various layers of materials that are selectively left behind or removed by subsequent processing [1]. Hereafter is a quick comparison among several technologies used in the planar silicon process:

Category	CMOS	Bulk	Surface	ThELMA
Min Litho [μm]	< 0.1	3	1	0.8
Structural Material	N/A	MonoSi	CVD PolySi	Si Epipoly
Sacrificial Material	N/A	MonoSi	Si Di-Oxide	Si Di-Oxide
Cleanroom class	1-10	10-100	10-100	10-100
Wafer Diameter	8″ – 12″	4″ – 6″	6″ – 8″	6″ – 8″
Mask Levels	>20	8-10	>25	8-10

Tab. 1.   Quick comparison among different Si technologies

The overall sensor is made of two wafers, bonded together: one with the sensing structure, the second with the main function of protecting the tiny mechanical formation during the injection moulding process. While historically the bonding between cap and sensor was made with a glue made of glass-based material, STMicroelectronics is pioneering metallic bonding, which is the use of an intermediate metallic layer as the bonding layer between the two silicon wafers.

Metallic Bonding on ThELMA allows

- ▶ reduction of the MEMS die size
- ▶ replacement of the glass frit with a safer material, completely environmental friendly
- ▶ an electrical interconnect of the sensor and the cap wafer
- ▶ improvement of the vacuum control inside the seal cavity (hermeticity)
- ▶ use of standard semiconductor processes to pattern the seal frame and therefore improve manufacturability

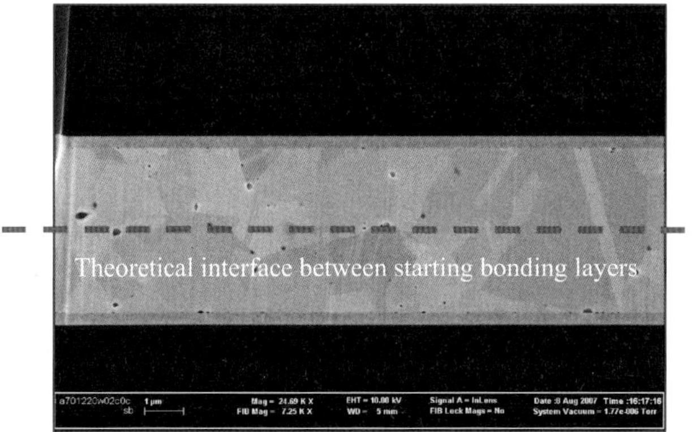

Fig. 3.   Focussed Ion Beam (FIB) Cross Section of a Au-Au thermo-compression joint

3.2 MEMS Sensors in Satellite Accelerometers

In airbags systems, the sensors located in the car might be subjected to shocks that range from 25 g (g = 9.81 m/s^2) up to 500 g. Covering this extremely wide range with a single structure poses some technical issues related to the rigidity of the structure versus its sensitivity. ST MEMS are of capacitive type: they convert an acceleration signal into a capacitive variation, which is processed by an Interface IC to give useful output voltage information proportional to the acceleration seen at the input.

We have therefore defined two different single-axis sensors to cover the entire required range:

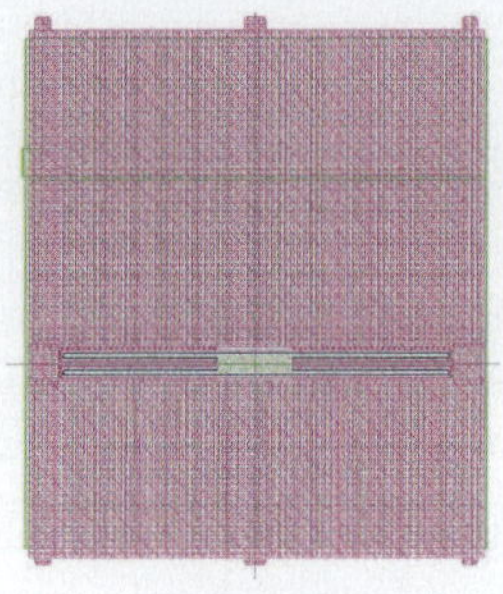

Parameter	Typ Value
C_0 *[pF]*	0.4
Sensitivity [fF/g]	0.4
F_{res} *[KHz]*	10
Q factor	1.5
g range [g]	25

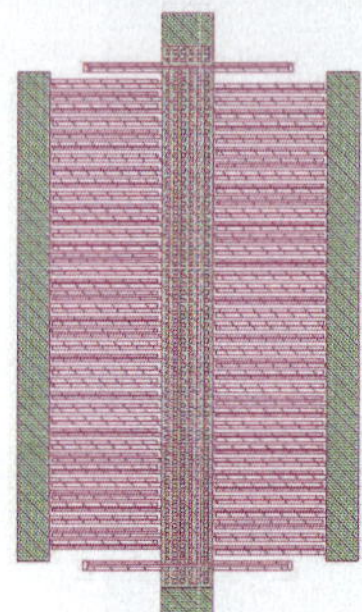

Parameter	Typ Value
C_0 *[pF]*	0.4
Sensitivity [fF/g]	0.15
F_{res} *[KHz]*	20
Q factor	3
g range [g]	$120 \sim 500$

Fig. 4.   Layout and parameters of the structures, sensitive respectively along the z (orthogonal) axis and along the in-plane (x or y) axes:

Note that, for all type of sensors belonging to this family, the value of C_0, the sensing element rest capacitance, is chosen to be the same. This makes it possible, for any g-range requested, to adopt the same signal processing IC. This allows quick adaptations to changing requirements, respectively reducing risks in case modifications to the sensor system are needed. The block diagrams of the PSI5 ASIC and of the DSI ASIC (Fig. 5) under development in STMicroelectronics using high-voltage silicon technologies are hereafter indicated:

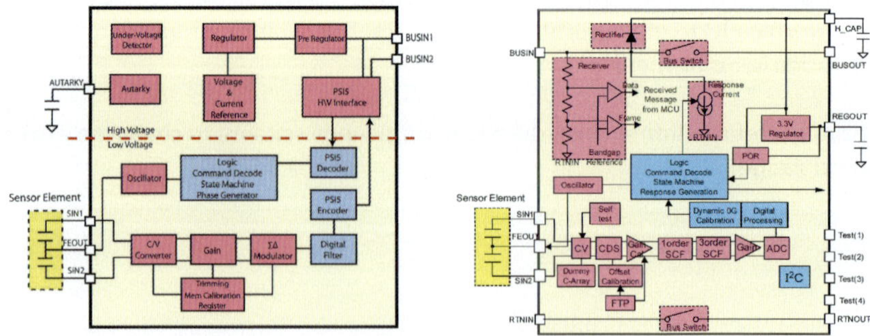

Fig. 5.   Block Diagrams PSI5 ASIC, and DSI ASIC

In both ASICs, we can identify 4 common blocks:
- ▶ the analog front end to the sensor
- ▶ the analog to digital converter
- ▶ the memory
- ▶ the sub-blocks implementing the communication protocols

Since both PSI5 and DSI require sustaining voltage levels in the range of 16 V and 40 V respectively, the challenge for silicon manufacturers is the integration of low voltage processing, memory, and the high-voltage portion needed for the communication interface.

Packages selected for these two ICs, where sensor die and interface die will be placed side-by-side with reverse frame forming for EMI reduction, are standard SOIC full-moulded packages, with 14 or 16 leads.

4   ECU Electronics

Depending on the complexity of the safety system, different functions are embedded in the ECU. The common blocks are sensor interface, power management, microcontroller and squib drivers.

STMicroelectronics has been supplying semiconductors since the early days of airbag systems and began integrating function blocks for customer specific ASIC projects. Half a decade ago, the standardization reached a level where ASSPs could be offered to the open market, and some early adopters counted on "not invented here" technologies. Despite commoditization, still today the ASIC share in this segment is larger than that of ASSPs. A very clear trend towards further integration exists with the target of a 2 chip system solution (μC + U-chip) or even a fully integrated System on Chip (SoC).

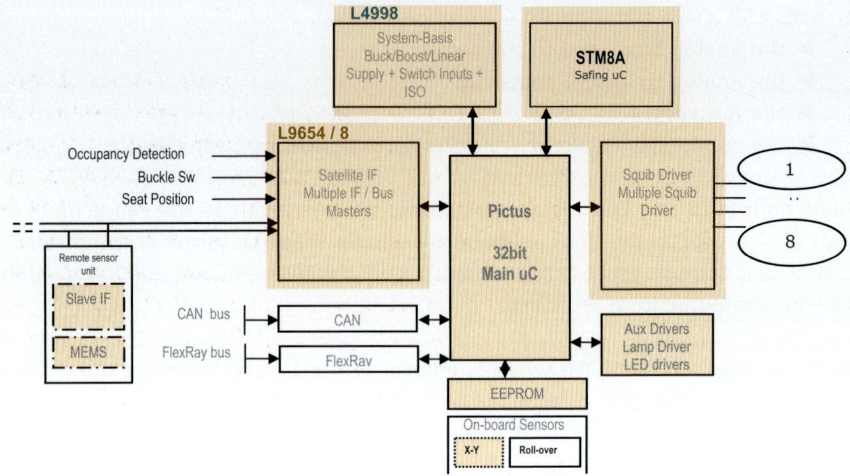

Fig. 6.   Several building blocks can be integrated into ASSPs / ASICs if the appropriate semiconductor processes are available. Processes like BCD (Bipolar-CMOS-DMOS) allow for the integration of power and analog as well as digital circuitry effectively on a single chip.

4.1   ECU Mounting Location

The preferred mounting location for an airbag ECU is in the middle of the car, where the stiffness in lateral and longitudinal direction is high. This gives the safety to the sensor, which is typically an ECU based, dual axis accelerometer, large, fast and reliable signals. In cars with rollover protection, there is also a roll rate sensor included. This gyroscope measures the rotational speed around the longitudinal axis of the car and measures the motion in the case of a rollover accident. This signal can also be used in the ESC/brake system for rollover prevention.

The central mounting location is also ideal for the ESC system which uses a yaw gyroscope as main sensor. A very good position for this gyroscope is close to the center of the car. Both systems could share the same space in the vehicle (-40C to +85C) for a more cost-efficient solution.

5 Actuators

An airbag system has to deal with several actuators. The major actuators are the inflators, which are triggered with a remarkable energy pulse (1,7A, >20 V,<2 ms) from the squib driver. As there are up to 8 squib drivers in one IC, it is not only important to provide this energy at several channels consecutively, moreover there must not be any cross talk between the channels. The seat belt pre-tensioner also uses pyrotechnical actuators, and in some cases, it may use electric motors as well, increasing the complexity of the IC even more. In the future, we may also see actuators lifting the hood if an impact of a pedestrian is detected.

6 Processes

To optimize the result of integration, it is essential that the semiconductor supplier has the necessary experience and processes available. STMicroelectronics started its BCD processes in 1984 with a 4 µm BCD1 process, and over the years has developed a very wide range of processes - BCD2, 3, 4, 5, 6 and is now in development with BCD9 − 0.13 µm

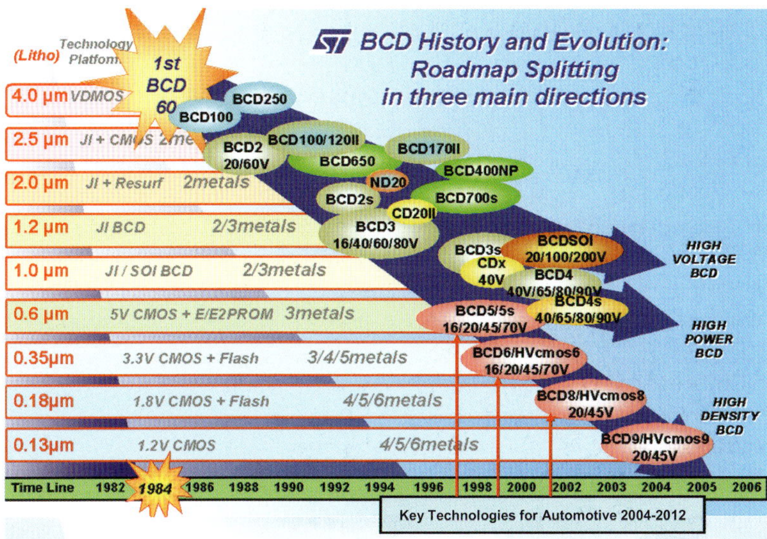

Fig. 7. Continuous improvement in process technology allows cost optimized solutions for ASICs and ASSPs

7   Outlook

Airbag systems will further benefit from integration – acceleration sensors and bus interfaces will merge; in the ECU, building blocks like bus interface and squib drivers will merge. Even power management functions can go into such a bus/squib/power IC. This reduced number of ICs in the ECU improves space, energy consumption, cost and last but not least, the quality of the system.

As airbags are mandatory systems, and ESC will become mandatory soon, it can be assumed that the industry will merge airbag and ESC controls into one ECU. The complexity of such an ECU will increase; hence there will be a significant potential for savings if integration continues. Furthermore, we will see advanced driver assistant systems (ADAS) like adaptive cruise control (ACC) and CMOS camera clusters interacting with the airbag/ESC system to go one further step towards accident free driving.

References

[1]   B.Vigna, More than Moore: micro-machined products enable new applications and open new markets, IEEE 2005.

[2]   Ching-Yao Chan, Fundamentals of Crash Sensing in Automotive Air Bag Systems, SAE, 2000.

[3]   Horst Bauer, Bosch, Kraftfahrtechnisches Taschenbuch, 1999.

Hubert Geitner
ST Microelectronics
Werner von Siemens Ring 5-7
85630 Grasbrunn
Germany
hubert.geitner@st.com

Marco Ferraresi
ST Microelectronics
Via Tolomeo, 1
20010 Cornaredo (MI)
Italy
marco.ferraresi@st.com

Keywords: airbag, ASIC, ASSP, MEMS, satellite, metallic bonding, DSI, PSI5, squib, safety

New Functions, New Sensors, New Architectures – How to Cope with the Real-Time Requirements

T. Kramer, R. Münzenberger, INCHRON GmbH

Abstract

The detection of the omni-present event chains in embedded applications goes far beyond functional modelling and static analysis. Once identified the analysis of their dynamics reveals a lot of data about the system like stability, critical paths or load reserves for future extensions. By using task-models and a real-time simulation tool the detection and analysis of event chains is very easy. Especially in distributed and collaborative development environments this is very helpful in reaching not only functional perfect systems but also delivering a high level of real-time quality.

1   High Quality in Spite of Growing Complexity

The majority of the upcoming smart system technologies and Information and Communication Technologies (ICTs) in the automotive industry combine existing and new sensors, build new distributed functions and require networked architectures. They span multiple domains and break up the existing one-vendor-one-box-principle. But how can such a complex system be architected and built to reach high quality levels in short cycles while combining several technologies and contributing parties?

The aspired high quality covers different aspects. The dominant aspect is functional quality. This is well served by established methods and tools like model driven development. Often the process starts with requirements gathering, model creation, implementation in code. It can end at different levels of integration. Parallel to this a variety of quality tests are specified and executed to assure that the system matches the requirements at a high functional quality. But quality also demands reliability and robustness. The system shall generate the correct functional results at the desired point in time – reliably, repeatable and under all circumstances. These timing and robustness aspects do not share the same level of tool support so far. Nevertheless they do have a tremendous impact on the system's quality and on the development time.

This paper will describe how distributed embedded systems' soft- and hardware can be modelled to describe their dynamic real-time behaviour. These models can be executed and visualized on the real-time simulator chronSim or analyzed by the real-time validator chronVal. The execution timing, events and sequence of functions and communication is calculated and displayed in various diagrams. The worst case and distribution of execution times can be determined for single functions or complete systems.

In an example from the automotive industry a system will be discussed, where a function is implemented on multiple control units using FlexRay for inter processor communication. The application tasks generating and receiving data run asynchronous to the FlexRay bus on separate control units due to application specific cycles.

2   Event Chains and their Timing Aspects

Automotive electronic systems are built from various sensors, actuators, communication busses and controllers. They are highly integrated and distributed systems with real-time requirements performing discrete and continuous control tasks. To work properly the systems have to execute the functions in a logcally correct order with the specified reaction to the stimulation. But not only the reaction has to be correct, it also has to occur at the right point in time.

A proper method to describe and test the system's reaction to stimuli is to follow the signals and functions along the various paths through the systems. The stimulus can be a pressed button or a sensor signal. But it can also be a group of signals or a defined system state that will lead to the reaction. And the reaction can be dependent on the history of these subsequently calculated system states. As a result there are many different paths through the system.

The system depicted in Fig. 1 highlights a path from the sensor 1, through ECU1, the bus, into ECU2 to the applications B and C. Finally the signal is sent over a communication interface to the actuator. But there are many more paths in that system. A stimulus from the sensor group will trigger the preprocessing that sends the combined signal data to ECU2. From here multiple reactions can be initiated and could stimulate ECU4, thus describing more paths.

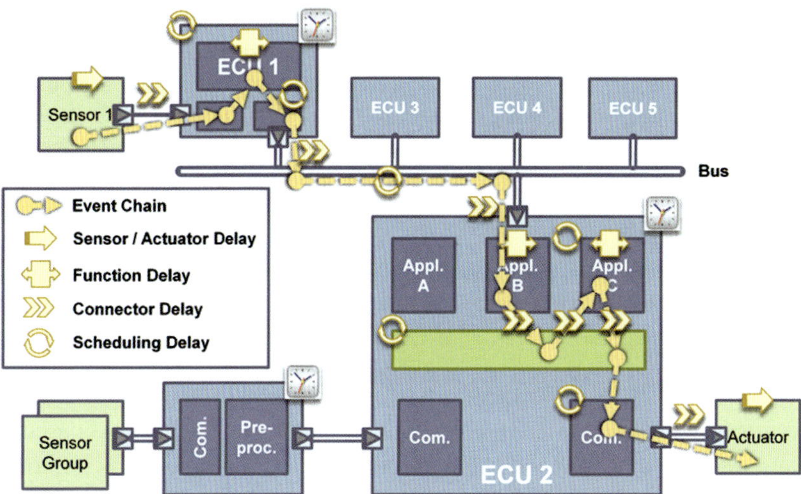

Fig. 1.   Block diagram of a distributed system highlighting the delays along the complex event chain.

2.1   Event Chains

Every step along the path through the system will create an event. If these events are ordered in causally relation one will get a chain of events. Event chains (aka effect chains; in German: ‚Wirkketten') are a commonly used term to describe the sequence of steps along the critical path performed to fulfill a certain functionality. The event chain has a timing requirement defining the end-to-end time from stimulus to reaction for the function to be correct. Having a common understanding what the event chains are in a system, all developers and architects can design, implement and test the functions properly.

Event chains are system immanent, but not necessarily well known and documented. In simple systems they are easier to identify and understand. In more complex and distributed systems the event chains often are unknown or only determined for fractions of the complete chain. That is because complex systems are developed by multiple groups and even companies. [2] The necessary information is as distributed as the system. The architects and developers often use function models and event chains only with the scope and detail level that is relevant for their own work. The outcome is that the complete information about the path is subdivided over all involved development groups.

Another cause for imprecisely identified event chains is the difficulty to determine the critical path in a system. The example in Fig. 1 has two sensor inputs, one with ECU1 in between sending the sensor data to ECU2 over a bus, the second sensor is a group of sensors with an additional preprocessing unit. Will the preprocessing or the bus communication add critical delays and jitter to the system timing, thus causing the path to be real-time critical? Especially when signals from different sensors are merged, when parallel processes are providing data for a joint calculation, it is not simple to determine that chain of events causing the critical timing.

Static methods and tools cannot answer these questions. With function models the consecutive steps along a path can be analyzed according their functionality and logic. It can be proved that the functions are called in correct order and a stimulus can produce the correct system reaction. But this implies that all signals are sent and received only once and that the functions could be executed immediately. In contrast, when running the software on a dynamic system environment, the sensor could be triggered multiple times before the preprocessing has finished. The application can be blocked by higher priority tasks, never finishing the control loop calculations before the next cycle starts. With static analysis it is possible to see what happens to the data, but not when. That makes it very difficult to determine what dynamic timing behavior will cause one or another path to be real-time critical.

2.2 Time Durations of Event Chain Steps

Every component in a system has an execution time tied to it. Each sensor needs time from sensing the physical value to transforming it into an electrical signal. Every command will be executed on the microcontroller in a certain amount of time. So the execution time of every event chain (end-to-end timing) is the sum of all components' delays.

Very often the time slices comprising the end-to-end timing vary because of the physical effects in the sensor or actuator. Other delays are dependent on the then active system state. If for example a communication bus is blocked by higher priority messages, the message carrying the sensor signal will be delayed until its message wins the arbitration. Furthermore the messages currently on the bus are the result of the states of all involved functions communicating over the bus. Thus the message delay is a very dynamic factor.

There are many possible delays in an event chain:

Sensor or Actuator Delay – Depending on the physical principle and the sampling method the reaction time will be constant (e.g. opening an electro-mechanical valve) or vary over a cycle time. For example when using image acquisition with a camera running at 50 frames per second, the acquisition can be just finished or may just have started. So the image can be accessed immediately or at worst after 20 milliseconds.

Function Delay – The software function commands executed on a microcontroller can be a constant set of cycles or may vary depending on the algorithm and values (e.g. division by 1 or by 0.9).

Connector Delay – The software functions or modules communicate over interfaces. These interfaces may interconnect directly with global variables or even use multiple abstraction layers to access different communication methods including in-vehicle busses or internet protocol. The connector delay can therefore depend on the scheduling of the communication layer, the bus scheduling and the current bus load.

Scheduling Delay – Sharing resources (µC, busses, etc.) requires scheduling to grant or block their access. Requests a step in the event chain a shared resource, the delay depends on multiple factors like priority, time phase compared to other functions' cycles, mapping order within the tasks, and many more. And while a function is executed, the scheduler might suspended it before completion due to higher priority tasks or interrupts.

2.3 Impact of Multiple Clocks

On top of the above mentioned variable delays the fact that the system has multiple clocks adds another dynamic timing factor to the critical path. Clocks and periodic events play an important role in the timing of event chains. The components involved to perform the function often have their own cycles and time bases. That might be the revolutions of the crank shaft or wheels, the cycle times of the real-time bus FlexRay and not to forget the CPU clocks in the ECUs. The applications on the CPUs are often triggered periodically by the scheduler deriving its time base from the individual CPU clock. Every clock and cycle has a start time, cycle time and phase. All three parameters can vary showing drift, jitter and sometimes may drop out.

The examples in Fig. 2 and Fig. 6 show the effects, that drifting clocks produce. Event chain latency will vary significantly and even 'jump' from the best to the worst case in one cycle. Having passed the data to the next step right before the time slot, in the next cycle it can miss the time slot. The data transfer will have to wait for a complete cycle of the receiver to take place.

To reduce the variation of all cycles and the resulting large jitter in the end-to-end timing some of these clocks can be synchronized to a master clock. Or, what has been the choice in the past, the application is designed to run on one CPU only and is using an event-based communication. With the growing number of functions being distributed over a networked system, this design choice has become rare.

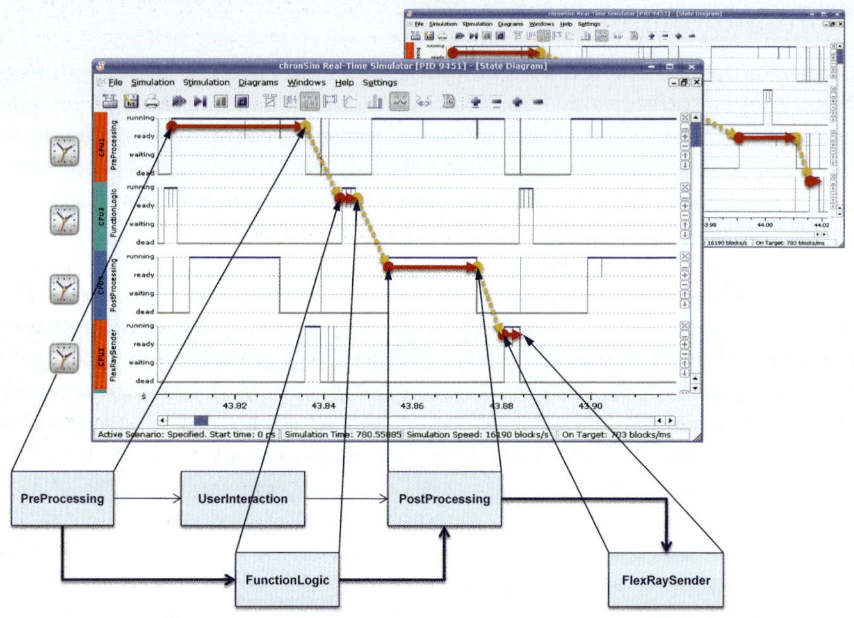

Fig. 2.   Task state diagram with short and long event chain

Fig. 2 depicts an example how drifting clocks can lead to very short and much longer event chain latencies. The screen shots from chronSim show the states of the tasks executing the functions referenced in the block diagram underneath. Since CPU1 (red; 1st and 4th plot), CPU2 (green; 2nd plot) and CPU3 (blue; 3rd plot) have own clock sources they may drift against each other. So the cycle phases of the tasks can be ideal for the event chain, the functions can be executed 'back-to-back' and the latency will be small. But as shown in the small screen shot, the adverse cycle phases will lead to large latencies. None of the situations represents a malfunction according to the designed behavior – but the event chain's latency might violate a timing requirement.

3. Creating a Real-Time Simulation Model

One approach to develop an embedded system is to define a soft- and hardware architecture, write all code, build the hardware, integrate the system and test if it fulfills the requirements. If it does not pass the tests look for faults, correct them and start all over again. Obviously this trial and error method cannot be the best way to reach the goals. And for the functional requirements this has not been the chosen method. Instead functions are modeled, the models tested and refined until they could be used to generate the function code. The software then is tested once again using SiL (Software in the Loop) before it is deployed onto the target hardware.

But for the timing requirements the trial and error method sometimes must be chosen since the timing behavior cannot be measured until the code is executed on the target hardware. This leads to a hen and egg dilemma. The architect has to design a system with timing characteristics he only knows, when the system is built and the timing can be measured. So real-time optimized design processes will iterate over the complete development cycle.

The proposed method to overcome this dilemma is to use real-time modeling to describe the systems and timings. The models can be simulated and analyzed focusing the timing aspects without ignoring the functional dependencies of the event chains. The real-time simulator executes the real-time models and enables the architect to experience the timing of his designed system before it has been implemented. By applying additional stimuli and load to the model the designer can use the simulator to perform sensitivity analysis and test the robustness of his design.

The real-time requirements of the design can be tested with the same models using a real-time validator. The analysis of the system with the validator will show the critical situations leading to the requirement violations. Otherwise it reports the remaining system performance that is available for further expansions.

3.1 Task Model

Real-time properties and requirements covering all system levels have so far not been standardized or made widely available. For a complete system description there was no versatile tool or data format in use until now. Task models can close that gap and enable embedded system developers to cover also the timing aspects when specifying their system using modeling. [1]

A part of the task model consists of model code in standard C or C++. Here the execution times of tasks, functions and software modules are entered. It can be derived by abstracting the target code and annotating it with the target execution time on the level of ISRs, tasks, functions or other software modules. Or it can be the starting point of a project having a skeleton code describing the software structure. Throughout the development it will become more detailed and finally may even be replaced by the target code itself.

```
ISR(ISR_Rotation)
{
    DELAY(30,unit_us);
    ActivateTask(Task_Rot);
}
```

```
TASK(TASK_Rot)
{
    DELAY(150,unit_us);
    TerminateTask();
}
```

Example of the most simple implementation of a *Task-Model* for *OSEK*.

The `ISR_Rotation` is executed in 30µs and `Task_Rot` in 150µs.

```
TASK(TASK_TT_5ms){
    DELAY(300, unit_us);         ①
    Schedule();                  ②
    while(!finished()) {
       DELAY(100, unit_us);}     ③
    exectime = 10*data_size;
    DELAY(exectime,unit_us);     ④
    DELAY(gaussian(500, 10),unit_us);  ⑤
    TerminateTask();
}
```

Example of a more complex *Task-Model*

① : constant execution time
② : scheduling point
③ : control flow dependent execution time
④ : data dependent execution time
⑤ : probability distribution of execution time

Fig. 3.   Simple and versatile task model

Fig. 3 lists on the left side a simple task model for an OSEK OS. The macro DELAY() replaces the function code with its time budget. So the ISR will run for 30 microseconds and then use an OS function call to activate a task. This task will run for 150 microseconds as long as it is not suspended by a higher priority task. If the timing behavior of the modeled system has to be defined more in detail, there are all possibilities available that C offers. The right example lists several methods to vary the execution time depending on system states or other stimuli. This enables dynamic reactions of the simulation model showing the same timing as the real system.

Event chains are specified directly in the task model. By using the EVENTCHAIN() macro the designer marks each step in a chain of causal dependent events. The current instance of the event chain execution as well as the current step within a specific instance can be specified by free variables. By passing their values between different nodes via a communication bus (e.g. within a CAN message) even causal dependencies across ECUs are possible.

```
ISR(ISR_Rotation)
{
    DELAY(30,unit_us);
    ActivateTask(Task_Rot);
}
```

```
TASK(TASK_Rot)
{
    DELAY(150,unit_us);
    TerminateTask();
}
```

Example of the most simple implementation of a *Task-Model* for *OSEK*.

The `ISR_Rotation` is executed in 30µs and `Task_Rot` in 150µs.

```
TASK(TASK_TT_5ms){
    DELAY(300, unit_us);      ①
    Schedule();               ②
    while(!finished()) {
        DELAY(100, unit_us);}  ③
    exectime = 10*data_size;
    DELAY(exectime,unit_us);   ④
    DELAY(gaussian(500, 10),unit_us); ⑤
    TerminateTask();
}
```

Example of a more complex *Task-Model*

① : constant execution time
② : scheduling point
③ : control flow dependent execution time
④ : data dependent execution time
⑤ : probability distribution of execution time

Fig. 4.   Sample code of the event chain in Fig. 2

Fig. 4 lists an example task model that generates the output shown in Fig. 2. The event chain starts with the task PreProcessing. The creation of a new message leads to the manual increase of an instance counter. The instance counter is used to mark step 0 of the event chain. The instance counter wrapped into the message is transferred to the second task FunctionLogic, then to the task PostProcessing and finally to the task FlexRaySender. In every task the value of the instance counter, the specific step and the actual point in time is recorded. The result is shown in the task state diagrams for the three different microcontrollers CPU1, CPU2 and CPU3. Only this or another analysis of the simulation results reveals the causal and temporal order of the event chain through the system.

To complete the system description the hardware, peripherals and stimuli are described using graphical user interfaces and libraries provided by the simulator tool. The scheduling method and software architecture with the definition of processes, priorities and e.g. stack sizes can be entered as well or be imported from standard OS configuration files like OIL (OSEK Implementation Language).

task models can be adapted to the abstraction level, the development phase, the focused timing detail of all involved development partners. A network designer can model the communication across multiple nodes without having to have detailed models of the nodes' internals. Accordingly the developer of an application part within one node can model his parts very detailed. But he reduces the communication details and peripheral timing behavior models to simple stimuli sources as far as they influence his applications timing.

3.2 Gathering the Timing Information

For the correct real-time simulation and validation the net execution times of the software modules are required. Depending on the abstraction level these can be the execution times of the tasks and ISRs as fixed values. At a higher detail level the times can be split down to each function or even command. Having mentioned the hen and the egg dilemma, this timing data is not available until the system is developed and integrated.

When designing a system from the scratch the architect has to set certain cornerstones defining the general functionality and timing of the system. At this point he chooses what will be executed in parallel or in sequence. For every function or task the architect sets timing quotas. Throughout the implementation more details become available and can be added to refine the real-time model.

If prototypes or legacy systems are available the timing data can be derived from measurements [3]. By adding instrumentation code at critical points (e.g. start and end of software modules) the system behavior can be traced with timestamps to identify what system state occurred when. The drawback of this monitoring technique is twofold: – the instrumentation commands add code to be executed and therefore change the real-time behavior of the system. – and the trace will cover only short time periods since the memory is limited showing only single system situations. Nevertheless the trace will provide realistic gross execution times of software modules in different system states. The net execution times can be derived utilizing additional scheduling information.

The real-time simulator also has an optional estimator module with processor models to estimate the execution times of target C code. When the development has progressed so far that target code and compilers are available, the estimator module will use the C code, the processor model and the compiler output to determine static execution time budgets. During the simulation the dynamic effects of pipelines and caches are applied to the static time budgets. This option has its beauty, but requires the code to be available and adds a detail level that can distract from the core problems and questions regarding the real-time behavior.

4   Event Chain Simulation and Analysis

Describing the dynamic behavior of the system using task models all fractioned information can be combined in a single real-time simulation model. The event chain segments from different development teams can be completed and highlighted in the model to be analyzed regarding their timing as a whole. Simulating and analyzing the model will help to understand the timing of each chain and show what the critical path and real-time relevant situations are.

The simulator executes the model and records all system events. In different graphs the task states and process communication are shown (see Fig. 2). The event chains can be plotted over time to see when and where large delays occur or where oversampling might cause data integrity problems.

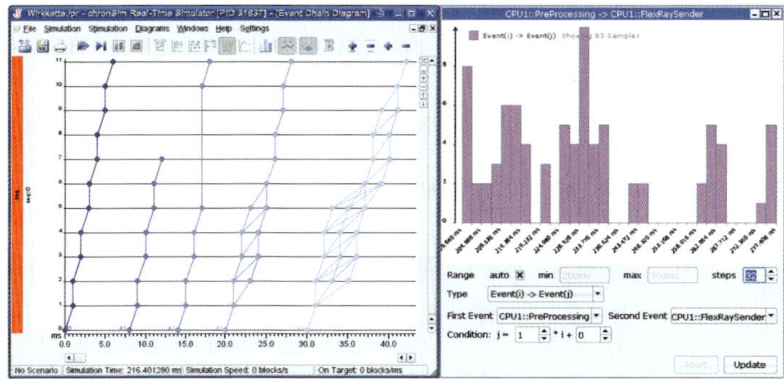

Fig. 5.   Event chain diagram and a histogram of the time between two chain events

On the Y-axis of the event chain diagram in Fig. 5 the steps along the path are listed (parameter "ChainPosition" in Fig. 4). The five plotted event chain instances show some possible timing effects. The second iteration ends uncompleted. In the third iteration some steps are never passed. The fourth and fifth iteration show multiple activations beginning with step 2. Here the event chain may continue using older or younger data since the according functions processing the data are called multiple times with the same data from step 1.

The histogram in Fig. 5 depicts the distribution of time passed between the PreProcessing and the FlexRaySender event.

5   Example Application with FlexRay

In this example two ECUs communicate over a bus to perform a control loop exchanging set values and actual values. That could be e.g. a brake control unit sensing the slippage of the wheels and requesting lower torque from the power train. The power train controller will reduce the fuel injection and reply with the actual torque value. This combination of brake and power train controllers is more efficient than just applying the brakes to prevent the car from sliding.

The event chain in this distributed control loop is spanning two controllers and a bus with an own scheduling and time base. The architects and function developers have to handle this very complex system, where parts of the event chain are defined by engineers from a different development team. Several parameters shall be identified. How long it takes from sending the torque request to the engine until the engine reports back the updated torque value. How much jitter is added through the bus communication. When one ECU fails to send the messages, how fast the system will detect the fault and switch to fail safe mode. How old the data is when it is read by the consumer application. How often messages will be lost without having a system failure.

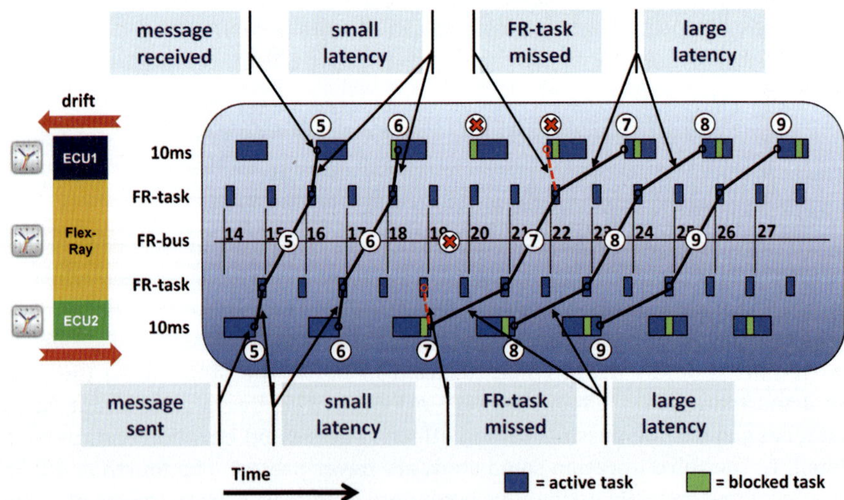

Fig. 6.   Two ECUs with drifting clocks show large message latency

Each ECU has a time triggered task architecture that runs the base software, the application and it's I/O. From these tasks only the 10 ms-tasks are shown in the diagram in Fig. 6. The deterministic FlexRay has its own time base running unsynchronized to the ECU's clocks. But the FlexRay communication software is running synchronized to the bus. Therefore the application tasks can drift against the FR-task in each ECU. In this case the ECU2 clock runs slower than the FlexRay time base. The high priority FR-tasks on the ECU start to block the lower priority 10 ms-tasks (green blocks; starting with FlexRay cycle 19). For the receiving ECU1 the drift has the opposite direction. The ECU clock runs faster than the FlexRay clock and the blocking starts already at cycle 18.

The event chains in Fig. 6 are graphed as black lines with small circles. Instances 5 and 6 of the event chain have a small latency since the producing tasks (10ms-task from ECU2) and consuming tasks (10 ms task from ECU1) send and receive shortly before and after their FR-tasks. Beginning with instance 7 the producer ECU2 cannot send the data to the FR-task in time and misses the sending in cycle 19. Instead the data is transferred delayed in FlexRay cycle 21. 'Unfortunately' at the same time the receiving FR-task on ECU1 misses the 10 ms-task because of the drift. As a result the event chain (instance 7) has a large latency and the consumer receives the data missing two iterations of the 10ms-task.

The effect of one 10ms-task missing its FR-task will occur frequently due to the drift. But having it happen on the sending and receiving side in the same event chain instance is rare and very hard to reproduce on the test bench or test drive. Nevertheless it can occur and has to be considered e.g. when a message counter would detect this as an error subsequently turning off the system.

In the under laying customer project the ECUs came from different suppliers. Each supplier could not test against this effect since he never had all components of the system. Only when both ECUs were integrated into a test system at the OEM, the complete event chain could be executed and tested – quite late in the development process. Using the task models both suppliers and the OEM could discuss the effects based on a common understanding. The improved collaboration led to discussions about ways to synchronize parts of the application to the FlexRay. Different ideas could be modeled and evaluated using the simulation much faster with fewer resources.

6   Conclusion

When complex embedded systems are developed by multiple parties, defining and fulfilling the real-time requirements is one of the most challenging tasks. task models for real-time simulation and analysis can provide the versatile basis to identify, describe and test the real-time critical paths throughout the systems. By marking the event chains in the simulation model their dynamic behavior can be observed and optimized very efficiently. The quality of real-time embedded systems regarding reliability and robustness will increase significantly due to timing optimized architectures and thorough real-time testing.

The collaboration support that task models offer becomes mandatory when AUTOSAR comes true. Software will be developed by various development teams but integrated by others. Having a model based solution to describe, architect and integrate real-time critical software from multiple parties will complete the currently available toolsets for AUTOSAR mainly focusing the functional aspects only.

References

[1]   Komarek T., et al, Developing Real-Time Constrained Embedded Software Using Task Models, in proceedings of the Advanced Automotive Electronics (AAE 2007) Gaydon www.aae-show.co.uk, January 2007.

[2]   Münzenberger R., Kramer T., Collaboration Support to Master Real-Time Challenges, in proceedings to the embedded world conference 2009, March 2009.

[3]   Kramer T., Münzenberger R., With Measurements to Real-Time Simulation, in proceedings to the embedded world conference 2009, March 2009.

Tapio Kramer
INCHRON GmbH
Lichtenbergstr. 8
85748 Garching b. M.
Germany
kramer@inchron.com

Ralf Münzenberger
INCHRON GmbH
August-Bebel-Str. 88
14482 Potsdam
Germany
muenzenberger@inchron.com

Keywords: real-time, simulation, validation, embedded systems, event chain, collaboration, requirements

Modular Inertial Safety Sensing Concepts for Functional and Environmental Fit

J. Thurau, J. Pekkola, VTI Technologies Oy

Abstract

After supplying high-performance low-g accelerometers to automotive systems for 10 years, it was time to broaden the product scope in order to utilize VTIs 3D MEMS platform for more products. Based on a combination of MEMS-tools that is effectively used in low-g high-volume production, a gyroscope and accelerometer toolbox was set-up in order to address both worlds, consumer and automotive-safety markets. The requirements of the consumer and automotive worlds are often contradictory, and synergies difficult to achieve. It can be agreed for the functional design that it needs to follow the applicational requirements – e.g. low power for consumer vs. strong and robust signal for automotive. It will be demonstrated that there are also commonalities. Regarding automotive system integration, the controversial discussion continues since functional and environmental aspects dictated by the application or by sensor fusion often lead to complex requirements. The art of bundling is to draw a line between combinable sensor classes and classes where a derivate of the sensor will fulfill deviating needs, which could be as simple as sensing orientation, temperature and vibration environmental conditions as well as signal quality requirements.

1   Inertial Sensor System Integration – 4 Architectures – 1 Target

"Form follows functions" is a famous proverb from Bauhaus architecture. For automotive inertial sensors it could be translated as "Sensor properties follow system integration concept". Historically the inertial sensors were used for Vehicle/Electronic Stability Control (VSC/ESC) only, an add-on function for the brake system. Therefore a sensor box called inertial sensor cluster with proprietary CAN bus was installed as a stand-alone ECU in the center of gravity in the car and communicated exclusively with the brake electronics. This concept was a perfect solution for the time when ESC was an option and an add-on system.

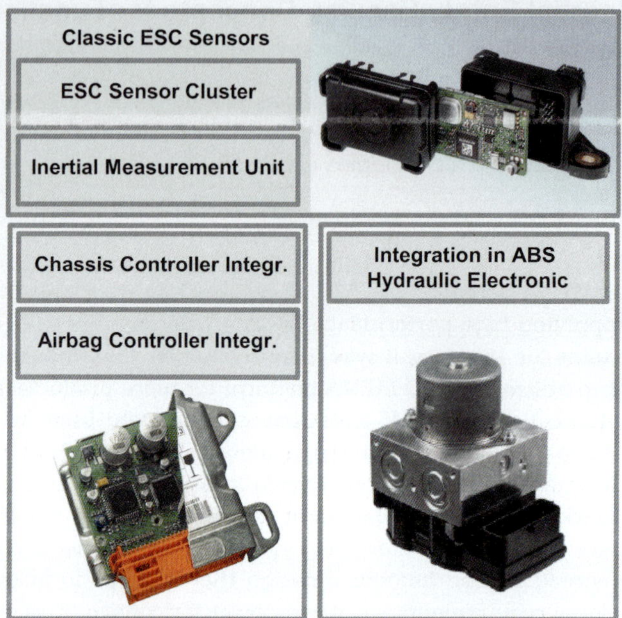

Fig. 1.   Inertial Safety Sensor Integration Concepts

Meanwhile ESC is on its way of being commoditized so that every car will soon include ESC. This is leading to the demand for better system integration due to the following reasons:

- ▶ Space allocation:
  The sensor cluster is located in a position where customers demand cup holders and other precious electronics.
- ▶ Sensor fusion:
  The inertial sensing signals are used for much more than just ESC – e.g. vehicle dynamics, navigation, adaptive cruise control
- ▶ Costs:
  Integration reduces the amount of required hardware

Fig. 1 lists the most popular ways of future integration of inertial sensing in automotive safety systems.

1.1   Classic ESC Sensors

In today's ESC systems the sensor cluster is the dominating solution with a market coverage of more than 90%. This historically grown business has momentarily the advantage of high level of commoditization so that the system cost level is acceptable and the cost of changing those systems is still higher than keeping the solution in current car platforms. The stand-alone box of a sensor cluster is a well defined system that offers the opportunity of clear interfaces with allocated responsibilities. The long track record and stable high volume production leads to significantly low PPM figures avoiding quality risks – in other words "never change a running system".

Technically the advantage is an exclusive CAN interface directly connecting the sensor cluster to the brake system so that interference from other systems can be excluded. Secondary functions in terms of sensor fusion need to be passed through the brake system towards other systems.

All the given advantages support the decision on keeping the sensor cluster as a state of the art solution. The only challenger is the offset costs driven by the extra electronics box as well as the extra power supply and communication interface that is offsetting the Bill of Material in disadvantage for the sensor cluster solution. Therefore the following solutions are winning more attractiveness.

1.2 Airbag Controller Integration – The Commodity Class

Today's airbag controllers are located at the center of gravity – most often at the tunnel of the car body. This position is the ideal position for the ESC inertial sensors also – offering a moderate vibration and temperature environment. Therefore one integration approach for the sensor cluster is the airbag controller. Beside the given advantages there are also some topics to mention which might be critical. First of all the additional components would require room so that the airbag unit would need to grow making it more difficult to integrate. Secondly, there are different life cycles of active and passive safety and while forcing those to synchronize benefits for separate optimisation are significantly reduced and later changes may require full re-qualification including crash tests. Thirdly, active and passive safety systems are typically handled by different departments at the OEM and Tier1s so that there is an additional political barrier to overcome. Another technical argument that may restrict the move of additional sensors into the airbag unit is the power consumption. The airbag unit power supply needs to have a certain buffer in order to provide the airbag unit with power after battery drop due to an accident in order to get the airbags still fired. While increasing the power budget of the ECU, additional buffer capacitors are required which increase the physical size of the airbag unit even more. The last reason that needs to be considered is that typically active and passive safety are sourced at different tier1s so that an additional interface would need to be managed by the OEM.

Overall this solution would be ideal in case that modularity would be reduced and commoditization for airbags and ESC would offer benefits to overcome the obstacles mentioned above.

1.3 Brake Electronic Integration – The Functional Combination

In a situation where active and passive safety integration is not the best option, functional integration of inertial sensors for the ESC brake function into the brake electronics becomes the best choice. Here the same economy of scale effects can be seen as in airbag integration. The system does not need to have an own mechanical box and the communication in between sensors and brake electronics is in the same unit avoiding any chance of disturbance within the wire harness. There are even more advantages in case that the car platform does provide ESC as an option only or additional optional features can be chosen in a way that the brake controller will be configured as a modular concept that does not affect other electronic units. The system is complete on its own and is therefore also qualified as one system with just one system responsibility.

The above advantages are offset by some disadvantages that can be managed. At first, the engine compartment exposure increases the environmental temperature from -40°C to +85°C for passenger compartment mounting towards +125°C. Secondly, MEMS motion sensors are supposed to measure motion so that vibration due to the hydraulic block movement and the hydraulic valve excitation tend to be measured as well. Both factors require a thorough design of the MEMS cells and signal conditioning in a way that vibration is either suppressed, mechanically compensated or handled by signal conditioning so that the conditioned signal is not showing any influence from the harsh environment. Fig. 2 shows the structure of VTIs 3D-MEMS acceleration sensors. The inertial mass is damped by squeeze film in a way that the inert gas damping is avoiding uncontrolled mass oscillation and therefore mechanical over-range conditions.

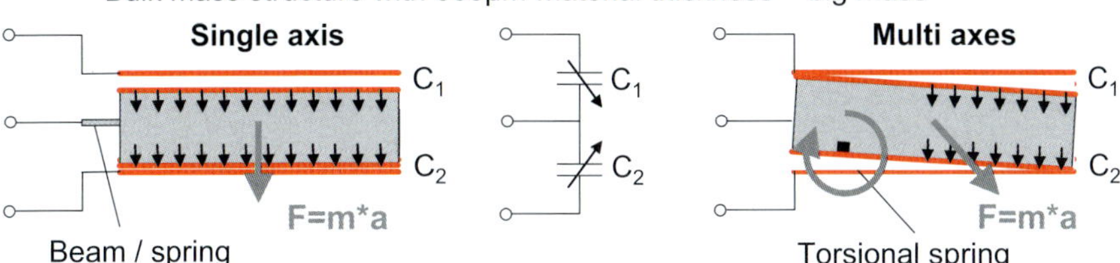

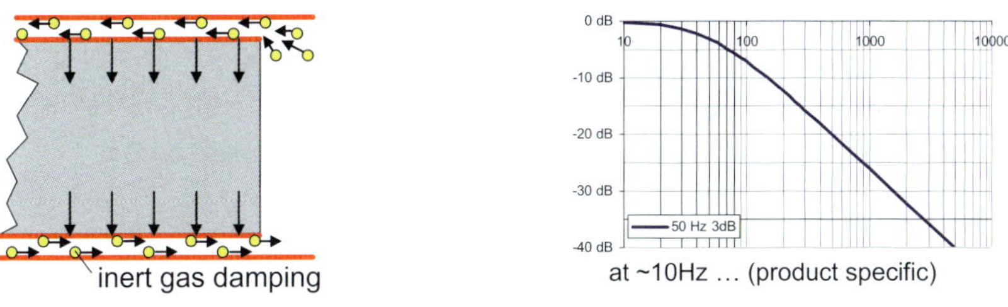

Fig. 2.   VTIs 3D-MEMS accelerometer structures and advantages for harsh vibration environment

1.4 Global Chassis Controller – The Premium Class

Beside airbags as standard and ESC as becoming standard in most of the cars in the future, there is a need for even more sensor signals in premium cars, driven by advanced vehicle dynamics systems. In these cases the OEMs are collecting the inertial sensors into one box and combining it with a powerful controller for signal conditioning, dynamical car status modelling and electrical driving of actuators. These units are called Global Chassis Controllers, Integrated Chassis Management Controller or similar. Their function is more than that of an Inertial Measurement Unit due to their computing power and driving capabilities. Typically the following functions are or can be collected:

▶ ESC (Yaw Gyro / Lateral Low-g) and Active Steering (Redundancy of Yaw Gyro / Lateral Low-g)
▶ Hill Start Assist, Electrical Parking Brake and ATM Hill Optimization (Longitudinal Low-g)
▶ Airbag optionally (Longitudinal and Lateral High-g)
▶ Rollover Protection and Roll-Compensation (Roll Gyro / Vertical Low-g)
▶ Vehicle Dynamics (up to 3 Gyro / 3 Low-g) including
▶ Active Suspension Control
▶ Pitch Control and Headlight Compensation

Neglecting the cost for such a unit it would be interesting to utilize such a powerful electronic box in all cars, hence a powerful controller, 4 Gyros, 4 Low-gs and 2 High-gs provide a Bill of Material that is justified for a small number of cars only. Therefore the above integration options or at best cascaded combinations thereof provide a better fit for all potential varieties of cars.

The advantages that can be seen are a very good chance of sensor fusion combined with the best opportunity to realize plausibility controls and therefore improved failsafe. This is an aspect of automotive sensors which is constantly gaining importance, as will be discussed in a later chapter.

2   From System to Component – Translation of Requirements

During the development of sensing element, ASIC and housing it became obvious, that the automotive requirements of safety, temperature, vibration, EMC and others are calling for dedicated measurements in order to achieve a robust product at the end that can be produced in high volume with acceptable yield. In principle it would have been possible to start with multi-axis product and with a monolithic integration of gyroscope and accelerometer within one piece of silicon. The review of the environmental factors has changed the scope towards single axis gyro and multi-axis accelerometer, leading to an optimised product for the automotive mass market. The product consists of single axis gyroscope sensing element, multi-axis accelerometer sensing element, combined ASIC and pre-molded plastic housing.

Especially the vibration requirements in the automotive environment have led to separate gyroscope and accelerometer sensing element. The gyro portion is requiring a very low internal pressure so that the inertial mass is excited with a good Q-factor in order to minimize the air friction losses. The excitation frequency needs to be higher than 10 kHz in order to avoid disturbances from the car body vibration. On the other hand the accelerometer inertial mass is damped by very effective squeeze film damping in order to provide natural vibration suppression.

Depending on the application board orientation within the car different sensing directions are required in order to address the system functionality. For the accelerometers the existing solution of the three axis sensing element with four proof masses was carried over since it addresses the orientation complexity best. Another advantage is the continuous self-test that is providing within the entire operational measurement range a real-time plausibility check due to the over-defined system of four measurement results but only three resultant acceleration outputs. Fig. 3 displays the vector addition of the four acceleration masses leading to a resulting 0-vector in case that the accelerometer is intact. This continuous self test is carried out in parallel to each signal conditioning cycle and it delivers a true real-time result for all possible static and dynamic acceleration combinations within the operational range of the sensor.

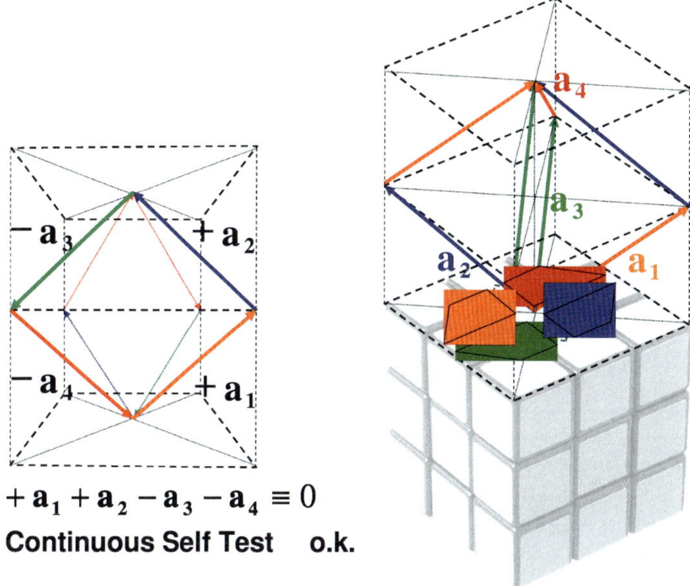

$$+\,\mathbf{a}_1 + \mathbf{a}_2 - \mathbf{a}_3 - \mathbf{a}_4 \equiv 0$$

Continuous Self Test   o.k.

Fig. 3.   Principle of continuous self test of VTI 3-axis automotive accelerometer

Additional critical features will be highlighted within the next chapter that displays that synergies and deviating requirements are both seen while considering a product range from consumer to automotive safety.

3   Synergies – Consumer vs. Automotive

3.1   3D-MEMS sensing Elements

VTI is using the inertial tool box (shown below) in order to realize pressure sensors, accelerometers as well as gyroscope sensing elements. The strategy is always a design of two functional chips – one 3D-MEMS for the physical measurement and one signal conditioning ASIC in order to optimise both sections best.

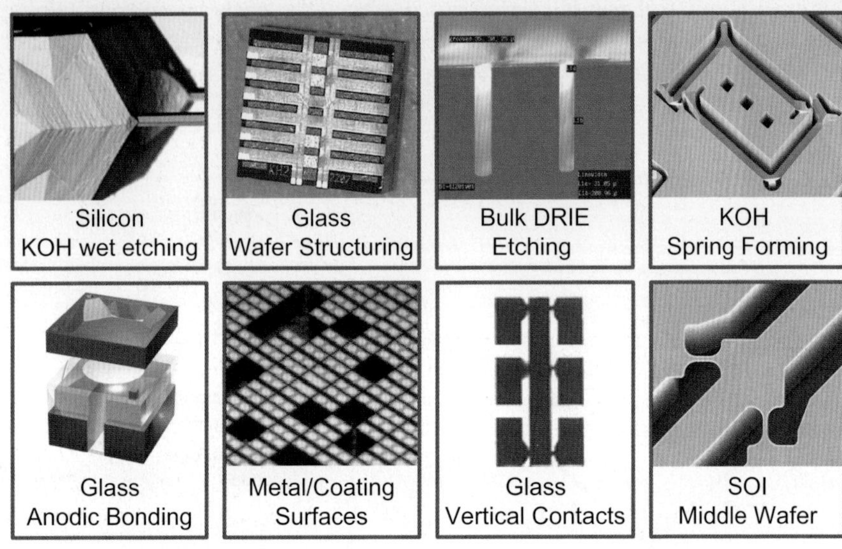

Fig. 4.   3D-MEMS toolbox

A combination of KOH etching and DRIE etching within thick functional wafers creates the typical bulk silicon micromachining structure that is legendary for the strong performance of those products. The advantage is particularly for the spring areas where a smooth transition from one material strength to the other leads to excellent shock behaviour. The top wafer structuring is making the connection via wire bonds for many contacts easy while keeping the MEMS wafer utilization at an optimum. In order to get the contacts to the top, a dedicated glass vertical contact method is used that connects the functional wafer towards the outside world through the glass of the anodic bonding layer. This technology is used for accelerometers and gyroscope sensing elements. Fig. 5 displays an explosion view of a dual axis gyroscope for consumer applications. This basic design is fulfilling the technical requirements for consumer applications.

In a second step VTI will utilize the basic structure of consumer multi-axis gyroscope MEMS structures and optimise those for automotive requirements. The main changes are related to vibration robustness and signal strength, what makes the sensing element design more complex and the sensing element surface bigger compared to consumer, where a loss of signal due to environmental factors is not considered safety critical.

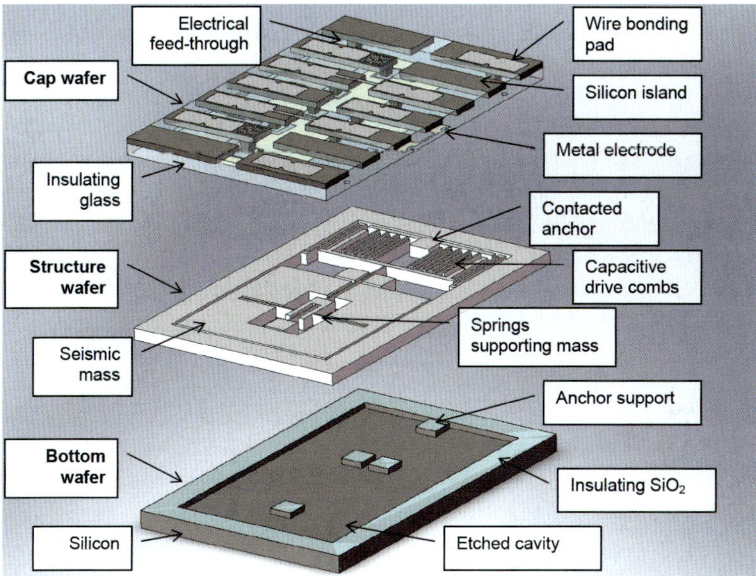

Fig. 5.   3D-MEMS toolbox – gyroscope example

The best way of achieving a robust inertial MEMS design is a symmetrical structure in order to eliminate common mode effects. This could imply in worst case a doubling of the structural size compared to consumer products.

Beside this dual axis design, many automotive applications are still requiring only a robust single axis gyroscope – or for Inertial Measurement Units also 3-axis gyroscopes. Single and three axis gyroscopes are most interesting for automotive whereas consumer applications tend to use dual and three axis gyroscopes. This leads to a situation where – especially for three-axis case – a common design approach was taken. For the automotive product the structures where made symmetrical whereas the consumer product was kept simple in order to optimise the cost for the product requirements best.

On top of the MEMS design the same processes are used to producing consumer and automotive products. The element probing is also done with the same equipment providing the same mechanical input to both product classes.

3.2   ASIC Signal Conditioning

The signal conditioning ASIC has as basic function the drive of the inertial mass for the gyroscope element and the read-out of capacities for all sensors. The principle of this function is the same no matter what the application is.

For consumer applications – and here in particular handheld devices with battery – are optimised towards low-power consumption especially in stand-by. Dedicated wake-up circuitry is added for some devices in order to realize anti-theft functionality – this function may be interesting for automotive burglar alarm as well for accelerometers. Since the signal conditioning is performed with low power in consumer applications, sample frequency as well as read-out currents are minimized. This leads to an increased noise level as well as more vulnerability towards EMC events – both are highly unappreciated in automotive. Consequently, automotive circuitries will consume more power in order to achieve a more robust and reliable output signal.

In addition to the simple functional layer requirements, automotive application are requesting dedicated failure detection mechanisms in order to address the system fail-safe standard ISO26262 and others. Those additional blocks are adding complexity and sometimes even full redundancy to the ASIC what increases the surface of the ASIC and therefore the costs.

3.3 Packaging

As size and cost of consumer devices are becoming excessively critical, very small packaging solutions are being created for consumer applications. VTI has addressed this field with its "Chip On MEMS" concept which is a wafer level packaging that will be directly soldered to PCB with solder balls. In contrast to other solutions, VTI has made the MEMS chip the connecting portion to the PCB and not the ASIC what keeps the ASIC process as standard process. The ASIC is mounted underneath the sensing element and connected via re-distribution layer towards the sensing element and outside world. Fig. 6 illustrates the construction principle. As a result, VTI was able to realize the smallest commercially available 3-axis accelerometer CMA3000 with the dimension of L = 2.0 mm × W = 2.0 mm × H = 0.95 mm.

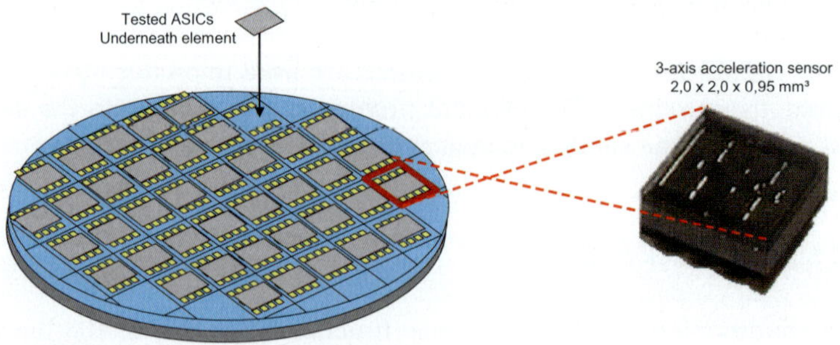

Fig. 6.   Chip on MEMS principle

Chip on MEMS is primarily considered as a consumer solution. In principle this technology withstands automotive reliability requirements as well so that after positive experience a roll-out into automotive applications is possible. The design principle requires always a smaller footprint of the ASIC than for the MEMS-cell. This would limit the functionality of the ASIC or a shrink towards smaller structures would be required in order to get the required failure detection and signal quality functionality into the same ASIC area. This factor needs to be considered as well before starting the automotive roll-out.

Automotive package is so far more conservative so that standard package solutions are typically used, like overmolded SOICs or even premolded housings. The advantage of a premolded housing is that the stress induced by the packaging material is buffered due to soft silicon surrounding the MEMS cell so that the offset performance is at the required level for automotive performance applications. Moreover, such packages offer certain advantages in terms of EMI and EMC sustainability, which are relevant in automotive applications. Several attempts to change towards other housings have ended-up in a return to premolded housings.

4   Product Roadmap - Outlook

Based on the given experience, VTI has made a decision to work with the same MEMS and ASIC processes for consumer and automotive products but with optimised chip-sets in order to meet the application need. The housing technology and assembly process is totally different today – with a future chance that automotive may follow the consumer technologies after maturity shows that the PPM levels can be reached for innovative housing solutions. Calibration and testing follows the same principle and is using some common equipment. The test coverage of automotive products is totally different compared to consumer products. At the end of the production line there are two products using as much as possible common production equipments.

	Yaw = ESC		Roll = Rollover		Pitch
	Gyro	Low-g	Gyro	Low-g	Gyro
Cluster vertical	X	YZ	Y(Z)	X	-
Hydraulic ECU	X	YZ	Y(Z)	X	-
Cluster horizontal	Z	XY	X	Z	-
Airbag ECU	Z	XY	X	Z	-
Global Chassis Controller	Z / Z	XY / XY	X	Z	Y

Fig. 7.   Application needs

Reflecting the application needs, VTI has defined the following products in order to meet the automotive safety applications (Fig. 8).

	Yaw = ESC		Roll = Rollover		Pitch
	Gyro	Low-g	Gyro	Low-g	Gyro
Cluster vertical	Gyro X / Acc YZ	or 2-axis Gyro XY / XZ	Gyro X / Acc YZ Gyro Z / Acc XY		-
Hydraulic ECU	Gyro X / Acc YZ		Gyro X / Acc YZ Gyro Z / Acc XY		-
Cluster horizontal	Gyro Z / Acc XY		Gyro X / Acc YZ		-
Airbag ECU	Gyro Z / Acc XY		Gyro X / Acc YZ		-
Global Chassis Controller	3-axis Gyro XYZ / Acc XYZ				
	Gyro Z / Acc XY(Z)		or 2-axis Gyro XY		

Fig. 8.   Roadmap for Gyro/Accelerometer combinations

The combination of a single axis gyroscope with dual axis accelerometer is the best solution for the biggest portion of the available application volume. Due to the orientation of the PCBs in the car there is a need for two different combinations of sensor orientations. The trend for future Global Chassis Controller and Inertial Measurement Units will be covered by six-degrees-of-freedom devices with their monolithic integration of three gyroscope axes and three accelerometer axes.

References

[1]   Thurau, J.; Motion sensors for vehicle dynamics – safety and comfort due to enhanced functionality, Presentation on Vehicle Dynamics Expo, Stuttgart, 2008.

[2]   Thurau, J.; Inertial sensors for vehicle dynamics control, IQPC Conference, Frankfurt, 2008.

[3]   Vilenius, T.; High-end inclinometers / Evolution of digital platform towards performance, safety and sensor fusion, AMAA Berlin, 2007.

Jens Thurau, Jan Pekkola
VTI Technologies Oy
Rennbahnstrasse 72-74
D-60528 Frankfurt
Germany
jens.thurau@vti.fi
jan.pekkola@vti.fi

Keywords: gyroscopes, accelerometers, sensor system integration, system requirements, harsh environments, sensor fusion

System Architecture for Adaptive Radar Sensors

M. Steuerer, A. Hoess, University of Applied Sciences Amberg-Weiden

Abstract

Despite the fact of permanently increasing road traffic, the number and the severity of accidents have been reduced thanks to passive and active safety systems. These systems protect road users before and during traffic accidents. Driver assistance systems inform the driver about potentially dangerous situations and intend to bring him back into the loop. The key to reliable warning and intervention lies in the performance of the sensor system.

Reliable object information requires the development of intelligent sensor systems. This article describes the architecture for an adaptive radar sensor subsystem. The architecture allows the configuration of the radar sensors dependent on the traffic situation. For this purpose, the bandwidth of the radar sensors is adjusted by means of ego vehicle speed and information about tracked objects (position and speed). Furthermore the approach improves the radar signal processing by searching objects in the radar spectrum.

1   Introduction

In modern automotive comfort and active safety applications several sensors for environment perception are integrated. These applications support the driver in a large field of driving situations, from normal driving up to support in dangerous situations. In order to reduce accidents in a permanently increasing traffic, driver support systems become more and more important. Examples for such systems are: Adaptive Cruise Control, Blind Spot Detection or Collision Mitigation.

One field of research is the enhancement of road users' safety by improving driver support systems. So called sensor management systems may help supporting this. Their task is the adjustment of sensor parameters dependent on the dynamic environment. This leads to an advanced detection probability.

A commonly used sensor type in car applications is the radar sensor. The PreCrash system presented in this article, combines three 24 GHz FMCW (frequency modulated continues wave) radar sensors to a front radar network. On the left and right side the medium range radars (range up to 40 m) are mounted, whereas in the centre line of the vehicle the long range radar (range up to 150 m) is installed. Parameterization of all sensors (e.g. bandwidth) can be changed during operation.

The three radar sensors are connected to an automotive ECU (Electronic Control Unit), which is responsible for all processing steps, like signal processing and tracking. For visualisation purposes, the results can be sent to a PC or to a display in the test vehicle. Intermediate processing results can be sent to a PC for detailed analysis.

Sensor resolution and maximum range are functions of their bandwidth: Increasing bandwidth leads to higher resolution but at the same time sensor range decreases. In consequence, close objects can be detected with higher resolution. In contrast, lower bandwidth allows the earlier detection of objects at higher distances due to higher range.

In the radar signal processing the adaptive threshold can be adjusted dependent on a given situation. It is feasible to raise or lower the threshold in certain parts of the FFT spectra. The advantage of this procedure is that lost objects can be retrieved. Thus, it is possible to search for already detected objects within the radar spectrum. With enhanced data, the fusion and tracking process can be improved as well.

This article focuses on an architecture which provides the possibility to adapt the radar sensors, their signal processing and the data fusion process.

2 Configuration Architecture

Several sensor data fusion systems have been discussed in the literature before. An overview of existing architectures is given in [1]. Not all fusion architectures are designed for sensor management purposes. A few of them include a feedback loop for sensor management, and others contain a separate function block for sensor management. In [2] multi-sensor management is formally described as: *"a system or process that seeks to manage or coordinate the usage of a suite of sensors or measurement devices in a dynamic, uncertain environment, to improve the performance of data fusion and ultimately that of perception".*

2.1 Task

Before developing the architecture it is necessary to define the goal of the system. The main tasks of the configuration system are to adjust the radar sensors' bandwidth, the radar signal processing and the data fusion process. This is summarized in Tab. 1. The table is split in two columns: The first column is labeled as "to configure" and lists the elements which will be configured. The second one called "influenced by" shows the necessary input parameters for the configuration.

To configure		Influenced by	
Level	*Description*	*Description*	*Level*
0	Rader sensor bandwidth	Car data	0
		Tracking results	2
1	Adaptive threshold	Up-Down Merge	1
		Tracking results	2
2	Sensor data fusion	Radar sensor bandwidth	0

Tab. 1.   Configuration Overview

The advantage of the configuration of sensors' bandwidth is that the range and resolution can be adapted to the traffic situation. The bandwidth customization is influenced by two parameters. The first one is given by the ego vehicle data. For example, the faster the vehicle moves the larger the sensor range needs to be. Second, the system is influenced by the results of the tracking, more precisely the closest object in front of the ego vehicle. Necessary object information is position and velocity. For example, in case of a close object in front of the ego vehicle, the bandwidth can be increased. This leads to a shorter range and to detection with a higher resolution.

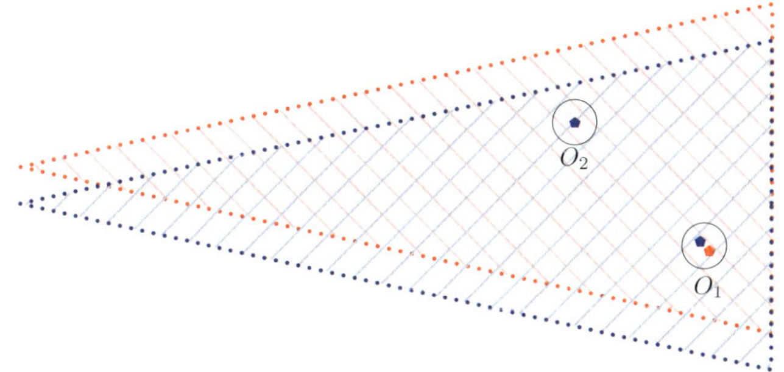

Fig. 1.   Measurement Example: Object 1 has been detected by both sensors, while Object 2 has been detected by just one sensor

For the adjustment of the adaptive threshold two possibilities are considered.

▶ The first possibility uses objects from the tracking to find objects in the radar spectrum. When objects have been predicted by the tracking filter, these objects will be searched in the radar spectrum of the next measurement cycle. If required, the threshold can be lowered at the predicted segment.

▶ The other possibility to find objects in the spectrum is to make use of the overlapping sensor detection areas. If an object has been detected by a radar sensor while another sensor with the same observation area could not find that object (e.g. due to weak amplitude caused by fading effects), see Fig. 1, the problem may be solved by searching the object in the radar spectrum using the results from the Up-Down Merge.

Finally, the sensor fusion unit is realized by means of a grid. The expansion of the cells is proportional to the radar resolution. Higher sensor resolution leads to smaller grid cells. The size of the grid cells is automatically adapted to the radar configuration.

2.1 Architecture

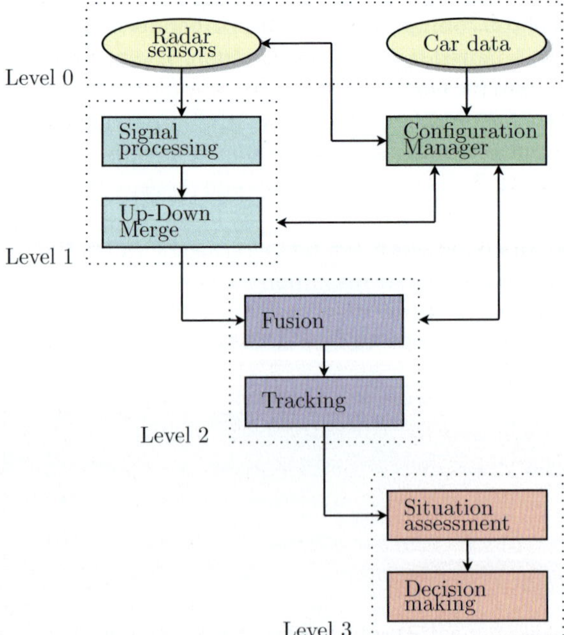

Fig. 2.   System Architecture Overview

The architecture of the entire configuration system is illustrated in Fig. 2. Similar to architectures already described in the literature (e.g. [1], [2] or [4]), the proposed architecture incorporates different levels of data processing. The four levels can be summarised: Input level (Level 0), individual data processing level (Level 1), combined processing level (Level 2) and decision level (Level 3).

The data process flow is partly bidirectional. Some information is fed back to previous levels through the "Configuration Manager". The list below describes the four different levels of data processing.

- ▶ Level 0 is the input level. All sensors and input entities are collected there. Input entities are the ego vehicle data and the three radar sensors for environment perception.
- ▶ Level 1 contains two function blocks. The first block is the signal processing unit for individual sensor processing purposes. It includes the Fast Fourier Transform (FFT), the adaptive threshold (ATH) and the peak detection (PD). The second function block is called Up-Down Merge. It is responsible for combining the up and down sweeps of a radar measurement. The Up-Down Merge results are collected separately for each sensor in so called Intermediate Lists. The Intermediate List may contain some ghost objects (objects that result from range/Doppler ambiguities of the radar FMCW measurements).
- ▶ Level 2 contains the function blocks for combined data processing. These are radar data fusion and tracking tasks. Input parameters for the fusion block are the Intermediate Lists from each sensor. The fused data will be tracked by means of a Particle Filter.
- ▶ Level 3 is the situation assessment and decision making level. Its operations are on a higher abstraction level. In this level, a conclusion about the current situation is made and the driver is informed.

The main part for the entire configuration architecture is the Configuration Manager. It controls the first three processing levels by means of feedback information. The Configuration Manager collects information from Level 1, Level 2 and from the vehicle. On basis of this information the new configuration parameters are calculated. Subsequently, the new parameters are sent to the radar sensors, the signal processing and to the data fusion function block.

2.1 Configuration Manager

Only the first three levels are relevant for the configuration process. Fig. 3 illustrates this part of the architecture from a different point of view. The architecture has been split up and the relevant function blocks are arranged around the Configuration Manager. The input elements are on the left hand side, the output elements are situated on the right hand side.

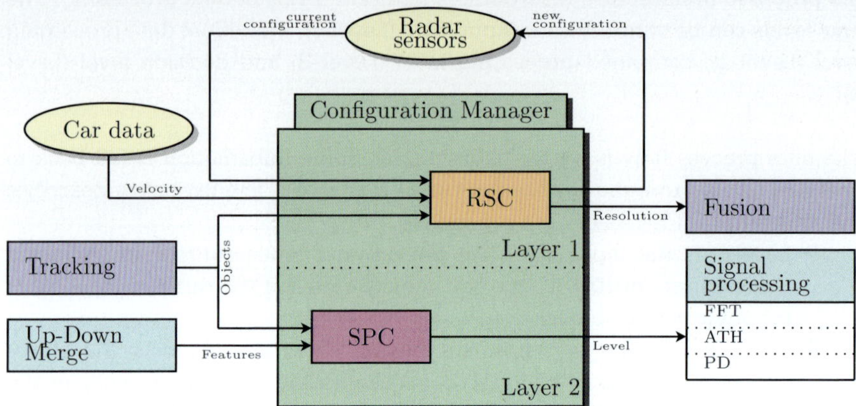

Fig. 3.　Configuration Manager

The configuration manager is divided in two layers. The upper layer (Layer 1) is responsible for the configuration of the radar sensors. In addition, it adapts the data fusion which is strongly dependent on the sensors' configuration. Thus, these both tasks have been combined into one layer. The "Radar Sensor Controller" (RSC) encapsulates the functionality for these tasks. Layer 2 affects the signal processing function block, more precisely the adaptive threshold. A so called *"Signal Processing Controller"* (SPC) performs this task. These two layers are described in more detail subsequently.

When that the system needs more configurable items the Configuration Manager can be enlarged quite easily just by adding another layer to the existing Configuration Manager. This modular structure allows keeping the entire system flexible.

Layer 1

The Radar Sensor Controller (RSC) is responsible for the estimation of the optimal sensor configuration; it is depicted in Fig. 4. The calculation happens in a so called *"Transition Controlled State Model"* (TCSM). The Transition Controlled State Model [3] is a special probabilistic Finite State Machine. As input parameter it requires the ego vehicle data (speed and acceleration) as well as the position and velocity of the relevant object provided by the tracking. A relevant object is an object which is close to the ego vehicle. On basis of these parameters the bandwidth change of the radar sensors is calculated.

It is mandatory to keep the configuration time for the sensors as short as possible, because the sensors are not able to send valid data during the configuration process. Each sensor bandwidth change takes about 60 ms. Thus, the bandwidth will be switched in three previously defined ranges. These are for the long range radar: 80 meters, 96 meters and 154 meters.

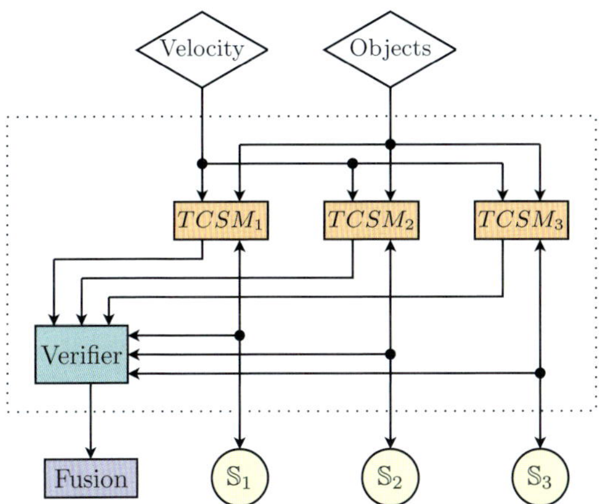

Fig. 4.   Radar Sensor Controller

The calculation of the new bandwidth is based on parameters of the ego car and if necessary on the relevant object. If no relevant object is in front of the ego vehicle, that will be executed just on basis of the ego vehicle speed. The faster the vehicle moves the larger is the range of the radar sensors to allow early object detection. The lower the ego vehicle moves, the lower is the range and the higher is the resolution to achieve detection with higher accuracy.

Another possibility is to use in addition the information about the relevant object (position and velocity) in front of the ego vehicle to parameterize sensor parameters, i.e. estimation of sensor bandwidth. In our case, a weighted value of both possibilities is calculated in the TCSM.

In addition to the Transition Controlled State Model the Radar Sensor Controller contains a Verifier. It compares the current radar sensor configuration with the results from the TCSM. This process is applied when a new configuration has been sent to the radar sensors. The connection to the radar sensors is realized with the CAN bus. When a sensor receives a new set of configuration parameters, it applies this configuration and acknowledges. The Verifier evaluates this response. In case of an error, the system either tries to configure the sensors again or sets an error state.

Layer 2

In Layer 2 the Signal Processing Controller (SPC) is situated, see Fig. 5. It processes the data for the adjustment of the adaptive threshold. Input parameters are the Intermediate List from the Up-Down Merge and the object list from the tracking unit. Items in the Intermediate List are also called features (F).

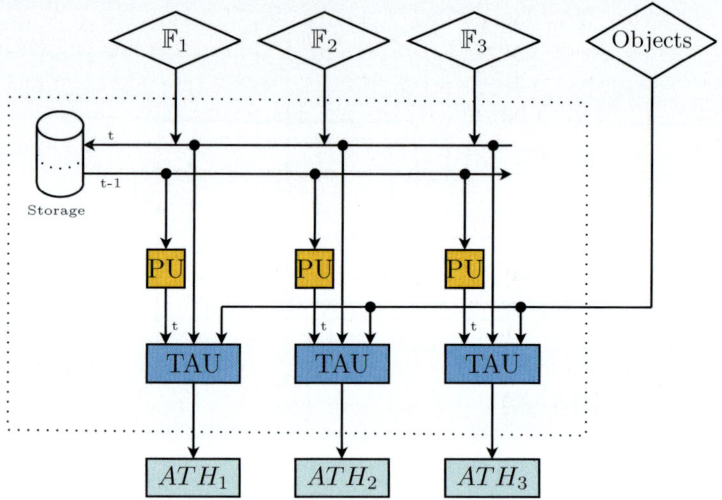

Fig. 5.   Signal Processing Controller

The improvement of the radar signal processing can be applied in three different ways. All these processing possibilities can be used either separately or together. The basic idea is to use already detected objects. A short explanation of these three processing steps is given below.

The first possibility is to use detected objects from one sensor to search these objects in the spectrum of another radar sensor with the same observation area. This is necessary to find objects that have not been detected by both sensors (cf. Fig. 1). To influence the radar signal processing of the desired sensor on an early stage, the calculated features from another sensor have to be used.

The second way is to use the predicted objects from the tracking unit. After the tracking, most of the ghost objects have been removed and the probability to find real objects increases. The positions of these objects are applied to search peaks in the radar spectrum of all sensors.

The last possibility is to predict the features in the Intermediate List within one sensor to the next time step. The position estimation of the features is calculated in the "Prediction Unit" (PU). For example an alpha-beta tracker can be used for this purpose. When a predicted object is not detected in the spectrum, the threshold can be lowered to find this previously detected object with weak amplitude.

All three steps have in common, that they lower the threshold at predicted positions. In the "Threshold Adjustment Unit" (TAU) a decision is made at which regions it is necessary to lower the threshold. The advantage of this process is that objects with low amplitude can be found again, assumed these objects or features have been detected once by a sensor.

Progress

An important point is the interaction of the layers and the function blocks. Fig. 6 illustrates this process. The measurement of the radar sensors takes place at the same time. Thus, no latency time between the sensors has to be considered. Every sensor transfers its data to the signal processing unit which processes the data consecutively.

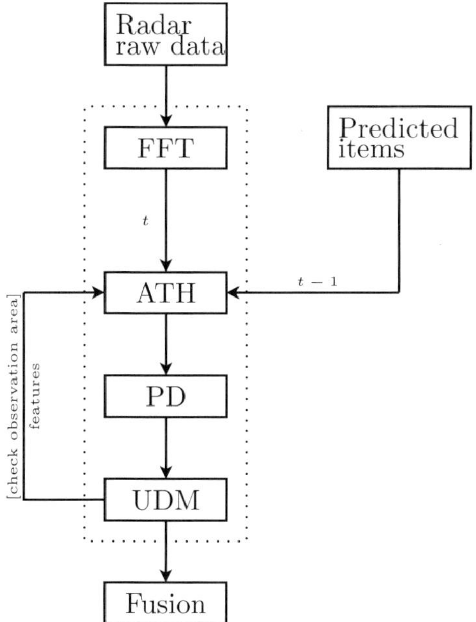

Fig. 6.   Iterative Process Flow within the Radar Signal Processing Task with Input and Output Function Blocks

The signal processing calculates the threshold, performs peak detection and executes the Up-Down Merge (UDM) in two iteration steps. The Fast Fourier Transformation is done once per process flow. In the first step the entire signal processing is executed for all sensors. Predicted features from the last measurement cycle as well as objects from the tracking are regarded immediately in the first iteration. That means the threshold is adjusted directly. Finally, the Intermediate List is generated and the first iteration is finished.

In the second iteration the Intermediate List from the first cycle is applied on all sensors with the same detection area. In case objects have not been detected in one of these sensors with overlapping observation area, the threshold is adjusted and the Intermediate List is calculated again.

After the second iteration, the Intermediate List is passed to the fusion unit. Finally, the combined potential objects will be tracked. Based on this information the Radar Configuration Controller is able to estimate the new configuration for the radar sensors. For this it uses the object list and the car data. Should it be necessary to reconfigure the sensors, the system initiates this.

3   Result

Two examples show the functionality of the system. Both examples have been recorded with the experimental vehicle in real traffic situations. The first example shows the bandwidth configuration (Layer 1) and the second the threshold adjustment (Layer 2).

Bandwidth Configuration

The result of the sensor configuration is depicted in Fig. 7. The upper two diagrams show the input parameters recorded with the experimental car. The third diagram shows the acceleration gradient. The fourth diagram shows the result of the Transition Controlled State Model in the Sensor Configuration Controller. This diagram indicates the bandwidth adjustment of the long range radar. The red graph incorporates the short state (80 m) of the long radar range, the green one the medium state (96 m) and the blue one long state (154 m).

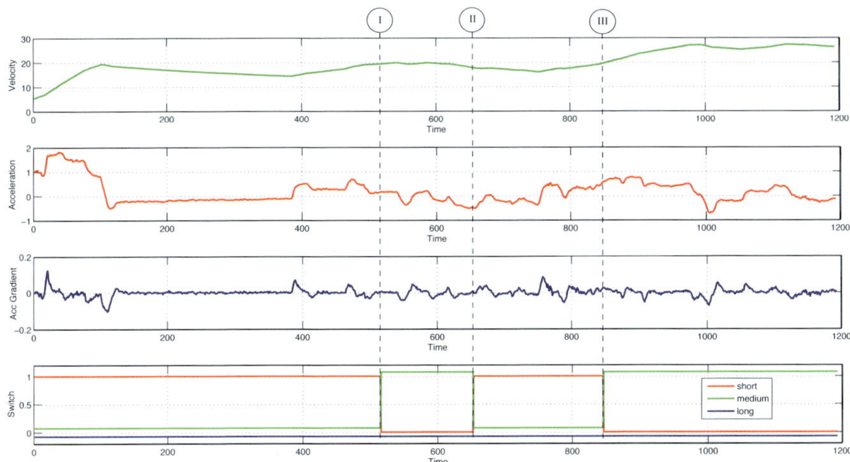

Fig. 7.   Result: Sensor Bandwidth Configuration

The system is initialised with the short state of the long range radar. At the three points (I – III) marked with vertical lines, the system changes the bandwidth of the radar sensor. First the bandwidth is changed from the short state (red) to the medium state (green). After a short while it changes back to the short state, indicated as red line which is high while the others are low. The reason for that are the decreasing velocity and the negative acceleration. The last change is to the medium state. Due to the velocity of the ego vehicle (> 80 km/h), the system keeps this state.

Threshold Adjustment

In Fig. 8, the threshold adjustment is presented. The two diagrams show the spectra of a radar FMCW measurement. The upper one is the up sweep and the lower one is the down sweep. The red graph shows the original unadjusted threshold.

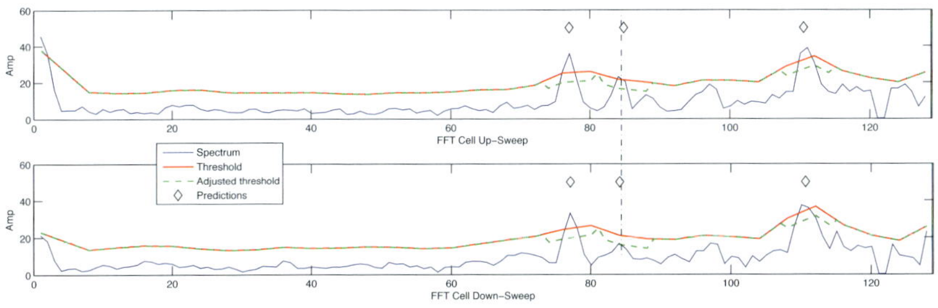

Fig. 8.   Result: Adjustment of the Adaptive Threshold

The green graph shows the adjusted threshold which has been calculated in Signal Processing Controller. Diamonds indicate the predicted feature positions from the last measurement cycle. At position 85, marked with the black line, an already detected object with weak amplitude in the down sweep is visible, while the amplitude for the same object in the up sweep is higher. In the previous step the amplitude was higher as well. Without a corrected threshold, this object would be lost. With the adaptive threshold it is detected again.

4   Conclusion

In this article the architecture for an adaptive radar sensor system has been introduced. Radar sensor bandwidth, signal processing and the data fusion process are adapted dependent on the traffic situation. Due to the requirements of modern safety applications, the architecture has a modular design which allows keeping the system flexible and extendable, e.g. to be able to integrate other sensors later on into the system.

The better the sensor data quality, the better are the results of the safety applications. Configuration systems can help to improve the measurement quality. With the possibility to influence the sensors as well as their signal processing, the data generation process can be improved significantly, with positive effects on the fusion and tracking results as well as on the decision making layer.

For adjusting the radar sensor bandwidth, information from the ego vehicle (speed and acceleration) as well as the position and velocity of the relevant object are necessary. The adaptive threshold can be regulated in certain parts by using already detected objects and features.

The benefit of varying bandwidth is that close objects can be detected with higher resolution, while objects at larger distances can be detected earlier. The advantage of adjusting the threshold is that objects with weak amplitude can be recovered.

References

[1]   Gad, A., Farooq, M., Data Fusion Architecture for Maritime Surveillance, International Conference on Information Fusion, 1, 448-455, 2002.

[2]   Xiong, N., Svensson, P., Multi-sensor management for information fusion: issues and approaches, Information Fusion, 3, 163-186, 2002.

[3]   Steuerer, M., Höß, A., Adapting Radar Sensor Resolution by Switching Bandwidth on Basis of Ego Vehicle Speed, International Workshop on Intelligent Transportation, 2009.

[4]   Esteban, J. et. al, A Review of Data Fusion Models and Architectures: Towards Engineering Guidelines, University of Manchester, University of Warwick, 2004.

[5]   Blackman, S., Popoli, R., Design and Analysis of Modern Tracking Systems, Artech House, 1999.

Marc Steuerer, Alfred Hoess
Hochschule Amberg-Weiden
Kaiser-Wilhelm-Ring 23
92224 Amberg
Germany
m.steuerer@haw-aw.de
alfred.hoess@haw-aw.de

Keywords: sensor management, radar sensor, system architecture

Appendix A

List of Contributors

List of Contributors

Appendix B

List of Keywords

List of Keywords

Printing: Mercedes-Druck, Berlin
Binding: Stein+Lehmann, Berlin